# OPTICAL CDMA NETWORKS

# OPTICAL CDMA NETWORKS

## PRINCIPLES, ANALYSIS AND APPLICATIONS

**Hooshang Ghafouri-Shiraz, D.Eng.**
*University of Birmingham, UK*

**M. Massoud Karbassian, Ph.D.**
*University of Arizona, USA*

A John Wiley & Sons, Ltd., Publication

This edition first published 2012
© 2012 John Wiley & Sons Ltd.

*Registered office*

John Wiley & Sons Ltd., The Atrium, Southern Gate, Chichester, West Sussex, PO19 8SQ, United Kingdom

For details of our global editorial offices, for customer services and for information about how to apply for permission to reuse the copyright material in this book please see our website at www.wiley.com.

*Library of Congress Cataloging-in-Publication Data*

Ghafouri-Shiraz, H.
  Optical CDMA networks: principles, analysis and applications / Hooshang Ghafouri-Shiraz, M. Massoud Karbassian.
    p. cm.
  Includes bibliographical references and index.
  ISBN 978-0-470-66517-6 (cloth)
1. Code division multiple access. 2. Optical fibre communication. I. Karbassian, M. Massoud. II. Title.
  TK5103.452.G49 2012
  621.385'1 – dc23

                                                              2011046434

A catalogue record for this book is available from the British Library.

ISBN: 978-0-470-66517-6

Typeset in 10/12pt Times by Laserwords Private Limited, Chennai, India
Printed and bound by CPI Group (UK) Ltd, Croydon, CR0 4YY

**Dr. Hooshang Ghafouri-Shiraz dedicates this book:**

To my late parents for their uncompromising principles that guided their life and for leading their children into intellectual pursuits. To my wife for her magnificent devotion to her family. To my children for making everything worthwhile. To my late supervisor Prof. T. Okoshi who made me aware of the immense potential of optical fibre communications. And finally to my research students for their excellent and fruitful research on optical CDMA which motivated us to write this book.

**Dr. M. Massoud Karbassian dedicates this book:**

To my beloved parents, brother and sister, for their support, kindness and patience at every stage of my life, since they always help me with their warm and lucent support that has encouraged me to overcome all obstacles throughout my life.

# Contents

# List of Figures

# List of Tables

# Preface

It was in the late 1980s that the concept of code-division multiple-access (CDMA) techniques was proposed in optical communications. Since then we have witnessed rapid and dramatic advances in optical fibre technologies. To provide a comprehensive and up-to-date progress report of optical spreading codes, optical modulations and transceiver architectures, we have decided to publish this new book which in fact is a broad extension to the current state of optical CDMA (OCDMA). The main objective is that this new book should serve both as a textbook and a reference monograph. As a result, each chapter is designed to cover both physical understanding and engineering aspects of coding, modulations and system structure.

With the rapid growth and sophistication of digital technology and computers, communication systems have become more versatile and powerful. This has given a modern communication engineer two key problems to solve: (i) how to handle the ever-increasing demand for capacity and speed in communication systems and (ii) how to tackle the need to integrate a wide range of computers and data sources, so as to form highly integrated heterogeneous communication networks with a global coverage.

The foundations of communication theory show that by increasing the frequency of the carrier used in the system, both the speed and the capacity of the system can be enhanced. This is especially true for modern digital communication systems. As the speed of computers has increased dramatically over recent years, digital communication systems operating at a speed which can match these computers have become increasingly important. From the electronic side, it is now apparent that the upper bound on the speed of a communication system is set by the transmission medium. An example which illustrates the rapid development in recent communication is that Today's PC generally uses PCI bus as the electrical interconnect, which can provide a peak data rate of 2133 Mbps. However, the speed of the modem normally connected to such a PC has just recently reached 24 Mbps over copper lines using ADSL2+ technology in commercial broadband access networks, about 90 times slower than the current electrical interconnects in the PC. One of the reasons for this mismatch is that modems use coax cables and these cannot operate at high gigahertz frequencies. To improve the speed and hence the capacity of the system, one does not only need to switch to a carrier with a higher frequency, but also to switch to an alternative transmission medium.

The Internet of the future requires higher bit rate and ultra-fast services such as streaming over the Internet protocol (IP), for example, video-on-demand (VoD) and TV-over-IP (IPTV). Due to its tremendous bandwidth resource and extremely low loss, fibre-optic can be the best physical transmission medium for telecommunications and computer networks. Among optical access technologies, we need to pick up the most suitable one to make full use of the large bandwidth in the fibre-optic. Meanwhile, it has to have the potential to support a random access protocol, various services with different data rates (i.e. differentiated services) and bursty traffic (i.e. IP). At the same time, there has been accelerating interest in optical transport due to the recently standardized Ethernet in the first-mile (IEEE 802.3ah) and also the establishment of the Ethernet passive optical network (EPON) as a pragmatic solution for the fibre-to-the-home technology.

A single strand of fibre offers a total bandwidth of 25 THz. More importantly, optical networks lend themselves well to offloading electronic equipment by means of optical bypassing as well as reducing their complexity, footprint and power consumption significantly while providing optical transparency against modulation format, bit rate and protocol. Fibre-to-the-home/curb/building, i.e. FTTX networks are set to become the next major successful architectures for optical communications. Not only must future FTTX access networks unleash the economic potential and societal benefit by opening up the first/last-mile bandwidth bottleneck between bandwidth-hungry end users and high-speed backbone networks, but they must also enable the support of a wide range of new and emerging services and applications, such as triple-play, VoD, point-to-point (P2P) audio/video file sharing and streaming, multichannel HDTV, multimedia/multiparty online gaming and teleconferencing.

Due to their longevity, low attenuation and huge bandwidth, passive optical networks (PON) are widely deployed to realize cost-effective FTTX access networks. Currently, PON is achieved using time-division multiplexing (TDM) single-channel systems, where the fibre infrastructure carries a single upstream wavelength channel (from subscribers to a central office) and a single downstream wavelength channel (from central office to subscribers).

IEEE 802.3ah Ethernet PON (EPON) with a symmetric line rate of 1.25 Gbps, and International Telecommunication Union – Telecommunication (ITU-T) Standardization Sector G.984 gigabit PON (GPON) with an upstream line rate of 1.244 Gbps and a downstream line rate of 2.488 Gbps represent the current state-of-the-art, commercially available and widely deployed, TDM-PON access networks. However, standardization efforts have already been initiated in the IEEE 802.3av Task Force to specify 10 Gbps EPON. Adding the wavelength dimension to conventional TDM-PON leads us to wavelength-division multiplexing (WDM) PON that has several advantages. Among others, the wavelength dimension may be exploited to (i) increase network capacity, (ii) improve network scalability by accommodating more end users and (iii) separate services.

Most of the reported studies on advanced PON architectures have considered standalone PON access networks, with a particular focus on the design of dynamic bandwidth allocation (DBA) algorithms for quality of service (QoS) support and QoS protection by means of admission control. From a networking perspective, the combination of the following three potential advantages makes OCDMA most attractive. First, OCDMA offers a larger channel count than the spectral division of WDM. Second, asynchronous transmission simplifies medium access control (MAC) as compared to TDMA. Finally, multiclass multirate services can be implemented by variable code-lengths and variable code-weight signature codes simultaneously. The motivation for OCDMA on local area network (LAN) is reinforced by the expectancy that LAN traffic patterns are characterized by burst: where a large number of users may gain access, provided that statistically fewer are active at the same time.

Alternatively, OCDMA is viewed as a candidate access protocol for passive optical network. An OCDMA-PON uses a tree topology with passive power splitters. Each optical network unit (ONU) contains an encoder-decoder pair with a unique fixed code, while the optical line terminal (OLT) may contain all encoder-decoder codes required for communication with ONUs. Unlike LAN, OCDMA-PON is not a fully broadcast system, because the signal transmitted by an ONU never reaches other ONUs directly. In this case, to avoid co-channel contention, OCDMA should be considered as an access network protocol in PON. This indicates that the existing interference can be controlled in such a way that up and down streams are established by the aid of synchronous and asynchronous modes at the same time. This technique helps the synchronous OCDMA to work in the same way as the asynchronous mode. Moreover, for a synchronous scheme, the dominating user interference can even be cancelled out by utilizing the properties of some specific codes and/or cancellation techniques. Above all, serious propositions must include the gradual migration paths from WDMA to OCDMA, which was the case recently from TDMA to WDMA.

It is believed that students, researchers and practicing engineers should be well equipped with the necessary theoretical foundation for this technology, as well as acquiring the necessary skills in applying the basic theory to a wide range of applications in optical communications and networking. There are, of course, many good books about optical components, communications and networking, while there is still a room for improvement on emerging multiple-access techniques in the optical domain like OCDMA. We are attempting to fill this gap with this book. We will be concentrating on the optical spreading codes suitable for spectral-amplitude coding and time-spreading schemes, advanced optical incoherent, coherent and hybrid modulation, transceiver architectures and analysing OCDMA potential in PON and compatibilities with current IP core networks.

Throughout this book, it is intended that the reader gains both a basic understanding of optical spreading code constructions and a comprehensive analysis of the overall system performance based on various transceiver architectures. We hope that this book will be beneficial to students aiming to study optical networks and in particular, optical CDMA, and to the active researchers at the cutting edge of this technology.

The book is organized as follows:

Chapter 1 explores common multiple access techniques in the optical domain and also the essential spread spectrum communication methods, including direct sequence and frequency hopping. The current state of the access technologies and optical networking are reviewed briefly and their existing challenges are also presented.

The exciting topic of optical encoding which covers source and channel coding in the optical domain is introduced in Chapter 2. The various code families used as spreading address sequences such as $m$-sequences, optical orthogonal codes, prime codes and congruence codes are gathered into one place in this chapter where their fundamental construction, properties and their multidimensional versions are discussed extensively. Since forward error correction (FEC) techniques and codes are considered for optical communications as well, the two common convolution and turbo code families are also explored here.

Chapter 3 overviews the up-to-date OCDMA technologies introducing different schemes of synchronous versus asynchronous, coherent versus incoherent, spectral encoding versus time spreading, wired versus wireless optical CDMA and their great merits in the access networks.

One of the most common schemes for OCDMA is spectral encoding which consists of spectral amplitude coding (SAC) and spectral phase coding (SPC), and these are discussed in Chapter 4. Various methods of encoding such as arrayed waveguide grating (AWG) structures, acoustically tuned optical filters, phase and/or amplitude masks and fibre Bragg grating (FBG) are demonstrated. Transmitters and receivers structures with comprehensive noise sources in both schemes are analysed and discussed extensively in this chapter.

As an incoherent time-spreading scheme, pulse-position modulation (PPM) with detailed signalling format and architecture is analysed in Chapter 5. In the analysis, Manchester encoding has also been analysed as a channel coding to enhance the multiple access interference (MAI) cancellation. We have also analysed the overlapping PPM (OPPM) architectures in this chapter. Additionally, the network throughput as an important characteristic of a communications network is discussed. Here also is a comprehensive analysis of a co-channel interference (i.e. MAI) cancellation technique which significantly improves the overall performance of the transceivers in terms of bit-error rate (BER) in the OCDMA networks.

Chapter 6 is dedicated to the analysis of coherent OCDMA techniques and examines the overall system performance in terms of signal-to-noise ratio (SNR) penalty as a function of simultaneous users accommodated to maintain an appropriate value of the BER for both homodyne and heterodyne detection. As a coherent modulation, binary phase-shift keying (BPSK) format is introduced and various transceiver architectures of coherent OCDMA are presented and analysed. In homodyne detection we have discussed two different phase modulations including an external phase-modulator (i.e. Mach–Zehnder) and injection-locking method with distributed feedback (DFB) laser diode. The phase limitation and the performance for two methods plus MAI and receiver noise are described in detail. The BER analysis of the optical heterodyne system using external phase modulator is also presented.

Since both coherent and incoherent configurations have advantages over each other in terms of simple architectures of incoherent and enhanced performance of coherent schemes, the hybrid coherent modulation and incoherent demodulation are analysed in Chapter 7. An MAI cancellation technique taking advantages of this hybrid method and a coding scheme are introduced in which the receiver configurations in synchronous frequency-shift keying (FSK) OCDMA network are simplified. In the theoretical analysis, the system bit-error rate is derived taking into account the Poisson effect on the I/O characteristics of the photodetectors and performance comparisons are made with the incoherent PPM method.

Polarization shift keying (PolSK) has been a candidate for optical communications for more than a decade, and Chapter 8 is dedicated to incoherent transceiver structures for the only optical modulation which takes advantage of the vector property of the lightwave in conjunction with OCDMA. The system has been accurately analysed taking into account the presence of optical amplified spontaneous emission (ASE) noise, optical receiver thermal and shot noise, and more importantly multiple access interference. The application of optical tapped-delay lines in the receiver configuration as the CDMA-decoder is presented. Furthermore, the two-dimensional optical modulation scheme deploying hybrid frequency and polarization shift keying (F-PolSK) is also included for the first time in optical CDMA concepts to increase the security and enhance the capacity.

Chapter 9 begins with challenges in access networks and existing solutions with their limitations in terms of scalability, performance and capacity. This chapter covers most relevant and applicable aspects of OCDMA in optical networking. Passive optical network (PON) as a strong candidate for the next generation access networks is discussed, considering various flavours supporting asynchronous transfer mode (ATM), gigabit transmission (GPON) and Ethernet (EPON). Since OCDMA supports random access protocols, the need for medium access control (MAC) is also described here. The configurations for optical network units (ONU) and terminals (ONT) are also analysed. The network scalability in terms of fibre link distances as well as BER, considering the main performance degrading issues are also analysed. This chapter also explores the Internet protocol (IP) traffic transmission and compatibility scheme over OCDMA network as well as a network node structure to map the IP traffic into an equivalent or compatible OCDMA scheme over an optical infrastructure. The overall IP-over-OCDMA performance is analysed in terms of user channel utilization factor in the network. To take full advantage of OCDMA networking, most recent random access protocols, including pure selfish, threshold algorithm and overlap section algorithms, are analysed alongside the practical network characteristics. Various techniques of multiprotocol label switching (MPLS) in the optical domain under different access methods such as TDMA, WDM and mainly OCDMA are discussed fundamentally, as well as generalized MPLS (GMPLS).

As observed, serious transitions have been recently proposed for gradual migration from WDMA to OCDMA technology due to its promising potential in optical networking. By considering OCDMA at the heart of optical networks, the guaranteed quality-of-services (QoS) and services differentiation (DiffServ) will be the next step in establishing optical networks. Hence, Chapter 10 is dedicated to demonstrating the capability of service-oriented OCDMA networks. Different services with different qualities and data-rates can be simply implemented in OCDMA by variable-weight and variable-length spreading codes respectively. Accordingly, various families of reconfigurable spreading codes and their impact on the service-oriented architectures are examined in detail in this final chapter.

This book is referenced throughout by end-of-chapter references which provide a guide for further reading and indicate a source for those equations and/or expressions which have been quoted without derivation. The principal readers of this book are expected to be undergraduate and postgraduate students who would like to consolidate their knowledge of optical networking, specifically OCDMA technologies. In addition, researchers and practicing engineers are potential readers who need to equip themselves with the foundations of understanding and using the continuing innovations in this technology. Finally, readers are expected to have the fundamental knowledge of communications theory, optical communications, computer and communications networking, statistical analysis and mathematics.

# Acknowledgements

Dr. Hooshang Ghafouri-Shiraz owes a particular debt of gratitude to many of his former Ph.D. students for their excellent research works on optical CDMA. We also express our utmost gratitude to the authors of numerous papers, articles and books which we have referenced while preparing this book, and especially to those authors and publishers who have kindly granted permission for reproduction of some of their diagrams.

Dr. Hooshang Ghafouri-Shiraz
Dr. M. Massoud Karbassian

# 1

# Introduction to Optical Communications

## 1.1 Evolution of Lightwave Technology

The invention of the solid state ruby laser (acronym for light amplification by stimulated emission of radiation) in May 1960 [1] and the He-Ne gas laser [2] in December 1960 has led to some wide-ranging and very significant scientific and technological progress. This so-called 'discovery of the century', followed by the first use of semiconductor lasers [3–5] in communications, heralded the start of optical communications. The laser provided a powerful coherent light source together with the possibility of modulation at high frequency, and this opened up a new portion of the electromagnetic spectrum with frequencies many times higher than those commonly available in radio communication systems. In addition the narrow beam divergence of the laser made enhanced free-space optical transmission a practical possibility.

Since optical frequencies are of the order of 100 THz, and information capacity increases directly with frequency bandwidth, the laser potentially offers a few order of magnitude increase in available bandwidth compared with microwave systems. Thus, by using only a small portion of the available frequency spectrum, a single laser could, in principle, carry millions of telephone conversations or TV channels.

With the potential of such wideband transmission capabilities in mind, a number of experiments [6] using atmospheric optical channels were carried out in the early 1960s. These experiments showed the feasibility of modulating a coherent optical carrier wave at very high frequencies. However, the high cost of development for all the necessary components, and the limitations imposed on the atmospheric channel by rain, fog, snow and dust make such high-speed systems economically unattractive. However, numerous developments of free-space optical channel systems operating at baseband frequencies were in progress for earth-to-space communications [7, 8].

It soon became apparent that some form of optical waveguide was required. By 1963, bundles of several hundred glass fibres were already being used for small-scale

*Optical CDMA Networks: Principles, Analysis and Applications*, First Edition. Hooshang Ghafouri-Shiraz and M. Massoud Karbassian.
© 2012 John Wiley & Sons, Ltd. Published 2012 by John Wiley & Sons, Ltd.

illumination, but these early fibres had very high attenuations and so their use as a transmission medium for optical communications was not considerable. Optical fibres can provide a much more reliable and versatile optical channel than the atmosphere. Initially, the extremely large losses of more than 1000 dB/km observed in even the best optical fibres made them appear impractical. In fact, to compete with existing coaxial cable transmission lines, the glass fibre attenuation had to be reduced to less than 20 dB/km. It was in 1966 that C.K. Kao (2009 Nobel Laureate) and G.A. Hockman [9] speculated that these high losses were as a result of impurities in the fibre material, and that the losses could be reduced to the point where optical waveguides would be a viable transmission medium. This was realized in 1970 when Kapron, Keck and Maurer [10] of the Corning Glass Works fabricated a fibre having 20 dB/km attenuation. At this attenuation, repeater spacing for optical fibre links become comparable to those of copper systems, thereby making lightwave technology an engineering reality. A whole new era of optical fibre communications was thus launched.

The ensuing development of optical fibre transmission systems grew from the combination of semiconductor technology, which provided the necessary light sources and photodetectors, and optical waveguide technology upon which the optical fibre is based. The result was a transmission link that had certain inherent advantages over conventional copper systems in telecommunications applications. For example, optical fibres have lower transmission losses and wider bandwidths as compared to copper wires.

This means that, with optical fibre cable systems, more data can be sent through one fibre, over longer distances, thereby decreasing the number of channels and reducing the number of repeaters needed over these distances. In addition, the low weight and the small hair-sized dimensions of fibres offer a distinct advantage over heavy, bulky wire cables in crowded underground city ducts. The low weight and small size are also of importance in aircraft where small lightweight cables are advantageous, and in tactical military applications where large amounts of cable must be unreeled and retrieved rapidly.

An especially important feature of optical fibres relates to their dielectric nature. This provides optical waveguides with immunity to electromagnetic interference, such as inductive pick-up from signal-carrying wires and from lightning, and freedom from electromagnetic pulse effects, the latter being of particular interest for military applications. Furthermore, ground loops are no longer an issue, fibre-to-fibre crosstalk is very low and a high degree of data security is afforded since the optical signal is well confined within the waveguide. Of additional importance is the advantage that silica is the principal material of which optical fibres are made. This material is abundant and inexpensive since the main source of silica is sand.

The recognition of optical fibre advantages in the early 1970s created a flurry of activity in all areas related to optical fibre transmission systems. This development led to the first laboratory demonstration of optical communication with glass fibre in the early 1970s. Such a progress resulted in significant technological advances in optical sources, fibres, photodetectors and fibre cable connectors. Since then, research on material for optical fibre transmission has made dramatic progress. Fibre loss was reduced from 1000 dB/km to 20 dB/km in 1970 and to 4 dB/km in 1973. By using longer-wavelength transmission the optical fibre losses were further reduced to 2 dB/km in 1974, 0.5 dB/km in 1976 and 0.2 dB/km by 1979.

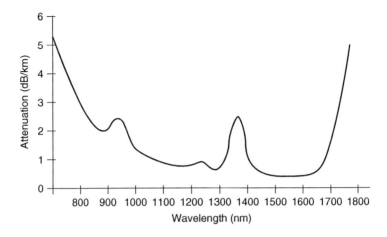

**Figure 1.1**   Optical signal attenuation against wavelength

Also, a study of the spectral response of glass fibres showed the presence of low-loss transmissions at 850 nm, 1300 nm and 1550 nm as shown in Figure 1.1. Although the early optical links used the 850 nm window, the longer wavelength windows exhibit lower losses, typically 0.5 dB/km at 1300 nm and 0.22 dB/km at 1550 nm. As a result, most modern links use 1300–1550 nm wavelength light sources.

New types of fibre materials have also been investigated [11–14] for use in the 3 μm–5 μm wavelength bands. It was found that fluoride glasses have extremely low transmission losses at mid-infrared wavelengths (i.e. $0.2\,\mu m < \lambda < 8\,\mu m$) with the lowest loss being around 2.25 μm. The material that has been concentrated on is a heavy-metal fluoride glass which uses $ZrF_4$ as the major component. Although this glass potentially offers intrinsic minimum losses of 1.01–0.001 dB/km, fabricating long lengths of these fibres is difficult. Firstly, ultra-pure materials must be used to reach this low level. Secondly, fluoride glass is prone to devitrification. Fibre-making techniques have to take this into account to avoid the formation of microcrystallites, which have a drastic effect on scattering losses.

## 1.2   Laser Technologies

The prospects for lower fibre losses at longer wavelengths led to intensive research on lasers and photodetectors. The advent of the semiconductor laser in 1962 meant that a fast light source was available. The material used was gallium arsenide (GaAs) which emits light at a wavelength of 870 nm. With the discovery of the 850 nm window, the wavelength of emission was reduced by doping the GaAs with aluminium (Al). Later modifications included different laser structures to increase device efficiency and lifetime. As the longer wavelength windows exhibit lower losses, various materials – in particular InGaAsP/InP – were also investigated to produce devices for operation at 1300 nm and 1550 nm. These efforts have also been successful. Commercial systems that used 1300 and 1550 nm technologies appeared in early and mid 1980s, respectively. Semiconductor sources are now available which emit at any one of the above wavelengths, with modulation speeds of several Gbits/s being routinely achieved.

The field of semiconductor lasers which has already reached a considerable level of development has recently undergone more changes. Initially, semiconductor laser wavelengths were at the infrared end of the spectrum. However, by shortening the wavelengths it becomes possible to concentrate the light in a smaller area thus increasing the energy density and producing a light source that is suited to optical data processing devices such as optical disk and optical printers. As a result, research was carried out into reducing the wavelength of the semiconductor laser to within the visible spectrum. At present, laser diodes emitting at 780 nm are available for use as light sources in digital audio disks and related devices. In addition, high-power semiconductor lasers (higher than 40 mW) emitting at 980 nm and 1480 nm are used as pump sources for erbium-doped fibre amplifiers [15]. At longer wavelengths (i.e. 1000–1600 nm) both bulk Fabry–Perot (FP) and distributed feedback (DFB) lasers [16] and more recently multiple quantum well lasers have been fabricated and used successfully in long-haul fibre communication systems. Currently new optical sources, termed strained bulk and quantum well lasers, which offer better performance have been investigated intensively [17].

According to the results of a recent survey, laser techniques are used in a wide range of industrial fields such as measuring technologies, communications and data processing. Among laser light sources in general, there has been a notable increase in the use of semiconductor and carbonic acid gas lasers. New light sources that are currently at the research and development stage include the excimer laser, synchrotron-generated light, the free electron laser and strained bulk and quantum well lasers. Currently the most promising technological development in the field of optoelectronics is the excimer laser. This laser produces wavelength at the ultraviolet end of the spectrum and it has the advantages of short wavelength and high output power. As a result, the excimer laser promises a whole range of technological applications in areas of the semiconductor and chemical industries.

Today, optoelectronics covers an extensive range of applications. Light-based technologies are now applied not only to optical communications but also to industrial processing, medicine, measurement and data processing. At present, there is a growth in the number of companies offering a range of devices and systems. They cover the manufacture of materials and components for lasers, optical fibres, optical ICs and the production of optical measuring instruments. Optical processing and communication devices and optical disks are based largely on laser light technologies and they may also include related optical technologies developed from laser techniques.

## 1.3   Optical Fibre Communication Systems

By 1980, advances in optical technologies had led to the development and worldwide installation of practical and economically feasible optical fibre communication systems that carry live telephone, cable TV and various types of telecommunications traffic. These installations all operate as baseband systems in which the data is sent by simply turning the transmitter on and off. Despite their apparent simplicity such systems have already offered very good solutions to some vexing problems in conventional applications.

Data communication capacity and the distances over which the data can be transmitted have continued to increase. Commercial long-distance fibre communication systems being installed in the mid and late 80s use the 'second-generation' 1300 nm technology and transmit data at 256 Mbps on each of several fibres over distances of 50 km between repeaters. The first commercial long-distance optical fibre telephone system was put into service by the American Telephone and Telegraph Company (AT&T) in 1985 [18]. Today, most major cities in the USA, Europe, Japan, South Korea and some other countries are linked by optical fibre systems. For example, in Japan, Nippon Telegraph and Telephone public corporation (NTT) has played an important role in the research and development of optical communications. NTT has completed development of small and medium-capacity (100 Mbps to 1 Gbps) as well as large-capacity (10, 40 and 100 Gbps) inland trunk and submarine non-repeater systems. In the small- and medium-capacity systems graded index multimode fibres – and in the large-capacity systems single-mode fibres – have been used. In other countries, optical communication systems have been introduced in on-land trunk network systems. In USA, between 1981 and 1985, 6, 45, 90 and 430 Mbps systems, and in the UK, 8, 34 and 140 Mbps, and in France 34 and 140 Mbps have been used successfully.

The capacity of a communication system is often measured through the bit rate-distance product $B \times L$, where $B$ is the bit rate and $L$ is the repeater spacing. This product has increased through technological advances since 1850 from 10 bps.km to beyond 1 Tbs.km [19]. Also, the progress in the performance of lightwave systems has indeed been rapid, as evidenced by the several-orders-of-magnitude improvement in the bit rate-distance product over the recent period [19].

Optical transmission can be classified into short-distance and long-distance categories depending on whether the optical signal is transmitted over relatively short or long distances compared with typical intercity distances of 50–100 km. Short-haul optical systems cover intercity and local loop traffic, and typically operate at low bit rates over distances of less than 20 km. While, long-haul fibre communication systems operate at high bit rates over long distances. A typical optical fibre system link is shown in Figure 1.2. It comprises three main parts:

- An optical transmitter consisting of an optical light source and its associated drive circuitry;
- A transmission medium which is an optical fibre cable; and
- An optical receiver consisting of a photodetector which is followed by an amplifier and a signal demodulation circuit.

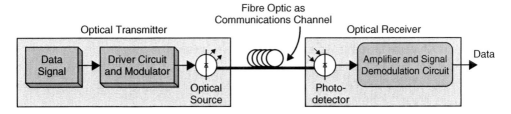

**Figure 1.2**   Optical fibre communications system

The function of an optical transmitter is to convert the electrical signal into optical form and launch the resulting optical signal into the optical fibre. Both semiconductor laser diodes (LD) and light emitting diodes (LED) are used as optical sources because of their compatibility with optical fibres. The optical signal is produced by modulating the LD or LED optical carrier wave. Modulation can be performed either by an external modulator or by direct modulation. In the latter method the input signal is applied to the driver circuit of the optical source like the one in Figure 1.2 and the signal is modulated by varying the injection current of the semiconductor optical source. In the 0.7 μm to 0.9 μm wavelength bands the light source materials are generally alloys of GaAlAs. At longer wavelengths ranging from 1.1 μm to 1.6 μm an InGaAsP alloy is the principal optical source material.

The optical signal is usually launched into the optical fibre via a micro-lens so as to maximize the coupling efficiency between the optical source and the optical fibre [20]. Often the launched power $P_0$, is expressed in dBm units (i.e. 1 mW is 0 dBm) and is an important design parameter as it determines how much fibre loss can be tolerated. Typical values of $P_0$ for LEDs are less than $-10$ dBm and for LDs are in the 0 to 20 dBm range. The LD or LED beam (i.e. the optical carrier) may be modulated using either an analogue or a digital information signal. Analogue modulation involves the variation of the light emitted from the optical source in a continuous manner. With digital modulation, however, discrete changes in the light intensity are obtained similar to amplitude-shift keying (ASK) or on-off modulation (OOK) pulses. Although it is often easy to implement analogue modulation with an optical fibre communication system, it is less efficient and requires a far higher signal-to-noise ratio at the receiver than digital modulation. Also the linearity needed for analogue modulation is not always provided by semiconductor optical sources, especially at high modulation frequencies. So analogue optical fibre communication links are generally limited to shorter distances and lower bandwidths than digital links.

The receiver in an optical fibre communication system essentially consists of the photodetector followed by an amplifier with possibly additional signal processing circuits. Therefore, the receiver initially converts the optical signal, incident on the detector into an electrical signal, which is then amplified before further processing to extract the information originally carried by the optical signal, as shown in Figure 1.2. When an optical signal is launched into the optical fibre it becomes progressively attenuated and distorted with increasing distance because of scattering, absorption and dispersion mechanisms in the fibre waveguide. At the receiver the attenuated and distorted modulated optical power emerging from the fibre will be detected by a photodetector (PD) and converted into an electrical current output (i.e. photocurrent) and processed. There are two principal photodetectors used in a fibre-optic link, PIN and avalanche photodiodes (APD). Both devices exhibit high efficiency and response speed. For applications in which a low-power optical signal is received, an APD is normally used since it has a greater sensitivity owing to an inherent internal gain mechanism. Silicon PDs are used in the 0.7 μm to 0.9 μm region whereas Germanium or InGaAs are the prime material candidates in the 1.1 μm to 1.6 μm region.

The design of the optical receiver is inherently more complex than that of the transmitter since it has to both amplify and reshape the degraded signal received by the PD. The principal figure of merit for a receiver's sensitivity is the minimum optical power necessary at the desired datarate to attain either a given error probability for digital systems or a specified signal-to-noise ratio for analogue systems. The ability of a receiver to achieve a certain performance level depends on the PD type, the noise performance in the system and the characteristics of the successive amplification stages at the receiver.

In addition, there has been an ongoing research in soliton transmission [21]. A soliton is a non-dispersive pulse that makes use of nonlinearity in a fibre to cancel out chromatic dispersion effects. Researchers at NTT have transmitted solitons at 20 Gbps over 70 km of optical fibre having on the average 3.6 ps/km/nm of dispersion [22]. In 1992, AT&T reported 32 Gbps optical soliton data transmission over 90 km [23], and also KDD reported transmission of 5 Gbps modulated optical solitons over 3000 km of optical fibre using erbium-doped fibre amplifiers [24]. Much longer transmission distances have also been reported using dispersion-shifted fibre [25].

## 1.4 Lightwave Technology in Future

Beyond the technologies on the immediate horizon and the systems proposed or envisaged, several other opportunities can be dimly seen and still more are expected. In other words, despite this rapid growth and the many successful applications, lightwave technology is not even close to being mature. New fibre materials and structures can greatly increase the available bandwidth and decrease the attenuation. Coherent communication techniques can improve receiver sensitivity and make more sophisticated and powerful modulation methods available to increase the system throughput such as multidimensional and multilevel modulation techniques.

An important objective is the perfection of all-optical networks, which include all-optical signal processing, switches, repeaters and network units. Applications include local area networks (LANs), subscriber loops and TV distribution [26].

## 1.5 Optical Lightwave Spectrum

Light is electromagnetic energy which travels in a wavefront. Thus, it can be identified within the general electromagnetic spectrum of alternating waves. Light as an electromagnetic wave is characterized by a combination of time varying electric $E$ and magnetic $H$ fields propagating through space as seen in Figure 1.3. The frequency of oscillation of the fields, $f$ and their wavelength, $\lambda$ are related:

$$f = \frac{c}{n\lambda} \tag{1.1}$$

where $c = 3 \times 10^8$ m/s is the light velocity in vacuum and $n$ is the refractive index of the medium given by:

$$n = \sqrt{\varepsilon_r \cdot \mu_r} \tag{1.2}$$

where $\varepsilon_r, \mu_r$ are the relative permeability and relative permittivity of the medium, respectively. The electric and magnetic fields oscillate perpendicularly to one another and also to the direction of propagation as illustrated in Figure 1.3. The simplest waves are sinusoidal waves, which can be expressed as:

$$E(z,t) = E_o \cos(\omega t - kz + \Phi) \tag{1.3}$$

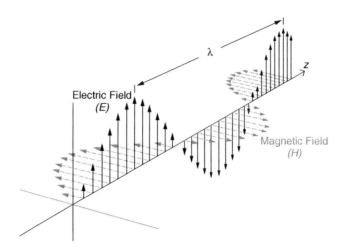

**Figure 1.3** $E$ and $H$ propagating in orthogonal planes, perpendicular to the direction of propagation

where $E(z, t)$ is the value of the electric field at the point $z$ and at time $t$, $E_o$ is the amplitude of the wave, $\omega = 2\pi f$ is the angular frequency, $k = 2\pi / \lambda$ is the wave number and $\Phi$ is the phase constant. The term $(\omega t - kz + \Phi)$ is the phase of the wave. Equation (1.3) describes a perfectly monochromatic plane wave of infinite extent propagating in the positive $z$ direction.

To obtain a better understanding of the wave characteristics, Equation (1.3) can be represented diagrammatically at one instance in time and for one point in space as shown in Figure 1.4(a). It is shown as the variation of the electric field, $E$ with distance at $t = 0$, assuming $\Phi = 0$. This spatial variation of the electric field is given by:

$$E = E_o \cos (kz) \tag{1.4}$$

Similarly Figure 1.4(b) depicts the variation of the electric field as a function of time at some specific location in space (e.g. $z = 0$) which is:

$$E = E_o \cos (\omega t) = E_o \cos (2\pi f t) = E_o \cos \left( \frac{2\pi}{T} t \right) \tag{1.5}$$

where $T = 1/f$ is the period. The frequencies and the wavelengths of the electromagnetic waves as seen in Equation (1.3) are related by Equation (1.1).

The term 'optical' usually refers to frequencies in the infrared, visible and ultraviolet portions of the electromagnetic spectrum. Figure 1.5 shows the electromagnetic spectrum range of optical frequencies that primarily interest us. In this region it is customary to specify the band of interest in terms of wavelength instead of frequency as in the radio region. The optical spectrum ranges from about $0.2\,\mu m$ (the far ultraviolet) to about $100\,\mu m$ (the far infrared). Visible wavelengths extend from $0.4\,\mu m$ (colour blue) to $0.7\,\mu m$ (colour red) as shown in Figure 1.5. Optical fibres are not very good transmitters of light in the visible region. They attenuate the waves to such an extent that only short transmission links are practical. Losses in the ultraviolet are even greater. In the infrared, however, there

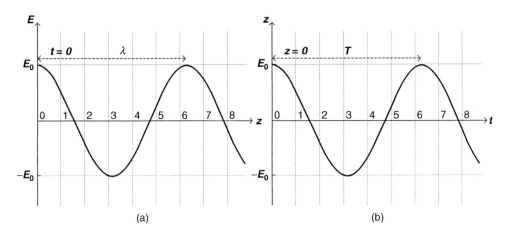

**Figure 1.4**   Electric field $E$ plotted as a function of (a) spatial coordinate $z$ (b) time

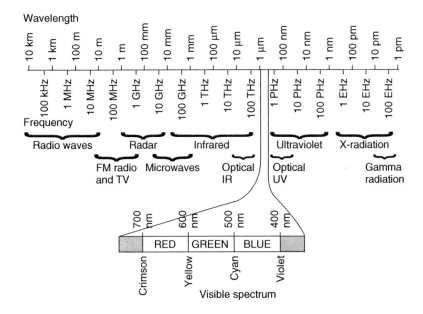

**Figure 1.5**   Electromagnetic spectrum

are two regions in which glass is a very efficient transmitter. This occurs at wavelengths close to $0.85\,\mu m$ and in the region from 1.1 to $1.6\,\mu m$ as shown from Figure 1.1.

## 1.6   Optical Fibre Transmission

As the transmission medium in a communication system, optical fibres have the potential for being used wherever twisted wire pairs or coaxial cables are employed. This is due to their lowloss and widebandwidth as apparent from Figure 1.1. Optical fibres have a flat transfer function far beyond 100 MHz and have very low loss compared with wire pairs or coaxial cables. In short, communication using an optical carrier waveguide along a glass fibre has the following extremely attractive features [11, 28–31]:

- Enormous potential bandwidth
- Small size and weight
- Electrical isolation
- Immunity to interference and crosstalk
- Signal security
- Low transmission loss
- Ruggedness and flexibility
- System reliability and ease of maintenance
- Potential low cost

The fundamental principles underlying these enhanced performance characteristics, and their practical realizations are considered as a whole medium in the systematic and networking techniques in the following chapters. It is assumed that the readers of this book have general understanding of the basic principles and properties of light as well as fibre-optics.

The transport network, also known as the *first-mile* network, connects the service provider central offices to businesses and residential subscribers. This network is also referred to in the literatures as the 'subscriber access network', the 'local loop' or even sometimes the 'last-mile'. Residential subscribers demand first-mile access solutions that have high bandwidth, offer media-rich Internet services, and are comparable in price with existing networks. Similarly, corporate users demand broadband infrastructure through which they can connect their local area networks (LAN) to the Internet backbone.

## 1.7   Multiple Access Techniques

In order to make full use of the available bandwidth in optical fibres and to satisfy the bandwidth demand in future information networks, it is necessary to multiplex low-rate data streams onto optical fibre to increase the total throughput. There is a need for technologies that allow multiple users to share the same frequency, especially as wireless telecommunications continues to increase in popularity. Currently, there are three common types of multiple access systems:

- Wavelength division multiple access (WDMA)
- Time division multiple access (TDMA)
- Code division multiple access (CDMA)

This section reviews the basic multiple access techniques in the optical domain and introduces the current state of optical access techniques.

### 1.7.1   Wavelength Division Multiple Access (WDMA)

In WDMA systems, each channel occupies a narrow optical bandwidth ($\geq 100$ GHz) around a centre wavelength or frequency [32].

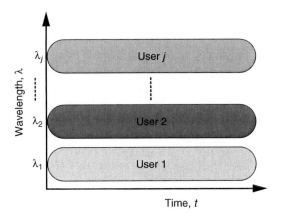

**Figure 1.6**   Resource sharing based on the WDMA technique

The modulation format and speed at each wavelength can be independent of those of other channels as shown in Figure 1.6.

Arrayed or tuneable lasers will be needed for WDMA applications [33]. Because each channel is transmitted at a different wavelength, they can be selected using an optical filter [34]. Tuneable filters can be realized using acousto-optics [35], liquid crystal [36] or fibre Bragg grating [37]. To increase the capacity of the fibre link using WDMA we need to use more carriers or wavelengths, and this requires optical amplifiers [38] and filters to operate over extended wavelength ranges. Due to the greater number of channels and larger optical power the increased nonlinear effects in fibres causes optical crosstalk such as four-wave mixing [39] over wide spectral ranges. Another approach to increase the capacity of WDMA links is to use dense WDM (DWDM) which will have to operate with reduced channel spacing, and this is based on the ITU-T recommendation G.692 [40] defining 43 wavelength channels from 1530 to 1565 nm, with a spacing of 100 GHz as shown in Figure 1.7.

This requires a sharp optical filter with linear phase response, wavelength stable components and optical amplifiers with flat gain over wide bandwidths. Also, optical fibres must support hundreds of channels without distortion or crosstalk. With respect to channel

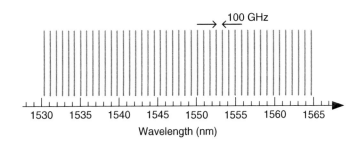

**Figure 1.7**   The ITU-T DWDM spacing

switching, wavelength routing is the next switching dimension for DWDM, with interferometric crosstalk being an essential issue in the implementation of cross-connects based on space and wavelength [41]. Hence, the extent of wavelength routing that is realizable places fundamental limits on network flexibility, which in turn determines switch size, implementation complexity and costs.

## 1.7.2   Time Division Multiple Access (TDMA)

In TDMA system, each channel occupies a pre-assigned time-slot, which interleaves with the time-slots of other channels as shown in Figure 1.8.

Synchronous digital hierarchy/synchronous optical network (SDH/SONET) is the current transmission and multiplexing standard for high-speed signals, which is based on time division multiplexing [42]. Optical TDMA (OTDMA) networks can be based on a broadcast topology or incorporate optical switching [43]. In broadcast networks, there is no routing or switching within the network. Switching occurs only at the periphery of the network by means of tuneable transmitters and receivers. The switch-based networks perform switching functions optically within the network in order to provide packet-switched services at very high bit rates [44]. In an electrically time-multiplexed system, multiplexing is carried out in the electrical domain, before the electrical-to-optical (E/O) conversion and demultiplexing is performed after optical-to-electrical (O/E) conversion. Major electronic bottlenecks occur in the multiplexer E/O, and the demultiplexer O/E, where electronics must operate at the full multiplexed bit rate. As an alternate solution, the bottleneck of O/E/O conversion moved further in processing rather than right after the mux/demux via optically time division multiplexed technique where the optical signals remain in optical domain and push the E/O and O/E to the transmitters and receivers respectively instead of at the mux/demux stage. OTDMA systems offer a large number of node addresses; however, the performance of OTDMA systems is ultimately limited by

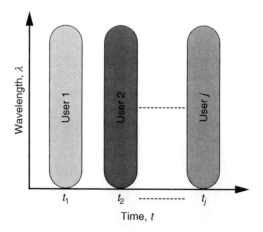

**Figure 1.8**   Resource sharing based on the TDMA technique

the time-serial nature of the technology. OTDMA systems also require strong centralized control to allocate time-slots and to manage the network operation.

### 1.7.3   Code Division Multiple Access (CDMA)

CDMA is one of a family of transmission techniques generically called spread spectrum, explained in the following section. In this technique, the network resources are shared among users that are assigned a code instead of a time-slot as in TDMA or a wavelength as in WDMA. Then, different users are capable of accessing the resources using the same channel at the same time, as shown in the Figure 1.9.

The concepts of spread spectrum, i.e. CDMA, seem to contradict normal intuition, since in most communications systems we try to maximize the amount of useful signal we can fit into a minimal bandwidth. In CDMA we transmit multiple signals over the same frequency band, using the same modulation techniques at the same time [45]. Traditional thinking would suggest that communication would not be possible in this environment. The following effects of spreading are worth mentioning.

#### 1.7.3.1   Capacity gain

Using the Shannon–Hartly law for the capacity of a band-limited channel, it is easy to see that, for a given signal power, the wider the bandwidth used, the greater the channel capacity. So if we broaden the spectrum of a given signal we get an increase in channel capacity and/or an improvement in the signal-to-noise ratio (SNR). The Shannon–Hartly law gives the capacity of a band-limited communications channel in the presence of Gaussian noise (most communications channel has Gaussian noise) [46].

$$Capacity = B \cdot \log\left(1 + \frac{P_s}{2BN_o}\right) \tag{1.6}$$

where $P_s$ represents signal power, $N_o$ noise power and $B$ available bandwidth.

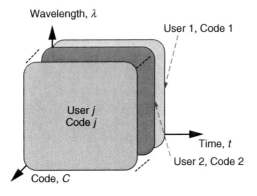

**Figure 1.9**   Resource sharing based on the CDMA technique

It is easy to see that with $P_s$ and $N_o$ held constant, capacity increases as bandwidth increases (though not quite as fast). Thus, for a given channel capacity, the required power decreases as utilized bandwidth increases. The wider the bandwidth the lower the power that we need to use for a given capacity.

### 1.7.3.2  Security

Spread spectrum was invented by military communications people for the purpose of battlefield communications. Spread spectrum signals have an excellent rejection of intentional jamming (jammer power must be very great to be successful). In addition, the direct sequence (DS) technique results in a signal which is very hard to distinguish from background noise unless you know the particular random code sequence used to generate the signal. Thus, not only are DS signals hard to jam, but they are extremely difficult to decode (unless you have the key) and quite hard to intercept even if all you need to know is when something is being transmitted.

## 1.8  Spread Spectrum Communications Techniques

Spread spectrum communication (SSC) involves spreading the desired signal over a bandwidth much larger than the minimum bandwidth necessary to send the signal. It has become very popular in the realm of personal communications [47]. Spread spectrum methods can be combined with CDMA methods to create multiuser communications systems with very good interference performance.

This section covers the details behind the method of SSC, as well as analysing two main types of SS systems, namely direct-sequence spread spectrum (DS-SS) and frequency-hopping spread spectrum (FH-SS).

As stated above, spread spectrum systems afford protection against jamming and interference from other users in the same band, as well as noise, by spreading the signal to be transmitted and performing the reverse de-spread operation on the received signal at the receiver. This de-spreading operation in turn spreads those signals which are not properly spread when transmitted, decreasing the effect that spurious signals will have on the desired signal. Spread spectrum systems can be thought of having two general properties: first, they spread the desired signal over a bandwidth much larger than the minimum bandwidth needed to send the signal, and second, this spreading is carried out using a pseudo-random noise (PN) sequence. In general, we will see that the increase in bandwidth above the minimum bandwidth in a spread spectrum system can be thought of as applying gain to the desired signal with respect to the undesirable signals. We can now define the processing gain $G_P$ as [48]:

$$G_P = \frac{BW_{car.}}{BW_{data}} \tag{1.7}$$

where $BW_{car.}$ is the bandwidth that the signal has been increased to (i.e. carrier bandwidth), and $BW_{data}$ is the minimum bandwidth necessary to transmit the data signal. Processing gain can be thought of as the improvement over conventional communication schemes due to the spreading of the signal. Often, a better measure of this gain is given by the jamming margin, $M_j$:

$$M_j(dB) = G_P(dB) - SNR_{min} \tag{1.8}$$

This indicates the amount of interference protection offered before the signal is corrupted. The spreading function is achieved through the use of a PN sequence. The data signal is combined with the PN sequence such that each data bit is encoded with several if not all the bits in the PN sequence. In order to achieve the same data rate as was desired before spreading, the new data must be sent at a rate equal to the original rate multiplied by the number of PN sequence bits (i.e. code-length) used to encode each bit of data. This increase in bandwidth is the 'processing gain', which is a measure of the noise and interference immunity of this method of transmission.

To see how the spreading process helps protect the signal from outside interference, the types of possible interference are introduced: (i) noise, and (ii) interference from other users of the same frequency band. Noise can be considered as background additive white Gaussian noise (AWGN) having power spectral density of $N_0$. Since the noise is white (i.e. includes all frequencies) the spreading of the bandwidth does not have much of an effect here. The noise power is constant over the entire bandwidth, thus increasing the bandwidth in fact lets more noise into the system, which might be seen as unfavourable. However, we will see that this is not really a problem. The major source of signal corruption comes from multiple user or multiple access interference (MAI). The technique of CDMA is to combat this type of interference.

In a wireless communications network, all the signals propagate through the air by electromagnetic waves, thus there is no way to ensure that a given user can receive their desired signal. In fact, the user receives all the signals that are sent in that band.

### 1.8.1   Direct-Sequence Spread Spectrum (DS-SS)

DS-SS is the most common version of spread spectrum in use today due to its simplicity and ease of implementation. In DS-SS, the carrier (data signal) is modulated by the PN sequence, which is of a much higher frequency than the desired data rate.

Let $f$ be the frequency of the data signal, with an appropriate pulse period $T = 1/f$. Let the PN sequence be transmitted at a rate $f_c$ so that the increase in the data rate is $f_c/f$. The frequency $f_c$ is known as the chip-rate, with each individual bit in the modulating sequence known as a 'chip'.

Thus the width of each pulse in the modulating sequence is $T_c$, the chip duration. Figure 1.10 illustrates the encoded signal, the data signal for one pulse width, and the PN sequence over the same time [49].

As a result, the frequency domain will look like the signal in Figure 1.11. Assuming that the data signal is $D(t)$, transmitted at frequency $f$, and the PN sequence is $PN(t)$ at frequency $f_c$, then the transmitted signal is:

$$S(t) = D(t) \cdot PN(t) \tag{1.9}$$

The PN sequence is designed such that it has very good correlation properties, for example:

$$R_{PN}(\tau) = \begin{cases} 1 & \tau = 0, N, 2N \\ 1/N & \text{otherwise} \end{cases} \tag{1.10}$$

where $N$ is the length of the PN sequence (code-length).

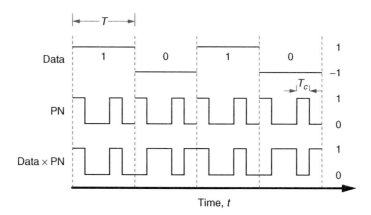

**Figure 1.10**   DS-SS signalling format

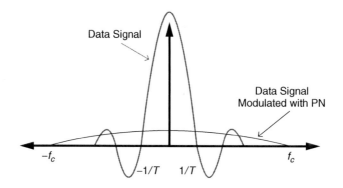

**Figure 1.11**   Data signal and DS-SS modulated data signal in the frequency domain

Therefore, when the signal is correlated with the PN sequence at the receiver, the received signal will be recovered exactly (assuming that there is synchronization between the sent and received PN sequences) as:

$$S(t) \cdot PN(t) = D(t) \cdot PN(t) \cdot PN(t) = D(t) \tag{1.11}$$

Here we recall the correlation functions. Two signals are correlated by multiplying each other bit by bit in discrete format. When a signal is correlated by its own shifted version, for example, let $x_i$ be the sequence code, $x_{i+\tau}$ be the signal shifted by time $\tau$, and $n$ be the PN code-length, then the auto-correlation function is:

$$R_{XX}(\tau) = \sum_{i=0}^{n-1} x_i x_{i+\tau} \tag{1.12}$$

When two signals $x_i$ and $y_i$ are correlated the cross-correlation function is:

$$R_{XY}(\tau) = \sum_{i=0}^{n-1} x_i y_{i+\tau} \tag{1.13}$$

where $y_{i+\tau}$ is the signal shifted by time $\tau$. It should be noted that two signals are 'orthogonal' when the cross-correlation value between them is zero. If we allow both noise and a jamming signal $J(t)$ with finite power distributed evenly across the frequency band, the received signal at the input to the receiver, $Y(t)$ is:

$$Y(t) = D(t) \cdot PN(t) + J(t) + N(t) \tag{1.14}$$

Now, this received signal is correlated with the associated PN sequence at the receiver. The result of correlating $Y(t)$ and the PN sequence is the product of PN (i.e. $PN(t)$) with $D(t) \cdot PN(t)$, $J(t)$ and $N(t)$. Thus, the product of $D(t) \cdot PN(t) \cdot PN(t)$ equals $D(t)$ which is the data signal, since the $PN(t) \cdot PN(t)$ represents the auto-correlation and de-spreads the data signal back to the original frequency of $f = 1/T$. While, the product of $J(t) \cdot PN(t)$ and $N(t) \cdot PN(t)$ denotes the further spreading the jamming and noise signals respectively. Therefore, the jamming and noise signals power reduces within the carrier frequency of $f_c$. Using a matched filter at the receiver to pass the data signal results in a decrease in the jamming and noise power by a factor of the processing gain $G_P$ i.e. $G_P = \frac{BW_{Car.}}{BW_{data}} \approx f_c/f$.

So we see that the data signal has been made immune to the effect of a malicious third-party jammer as well. As stated earlier, even though a factor of $f_c/f$ more noise was let into the system by the increased bandwidth, the effect of that noise was also reduced by $G_P$ due to the processing gain of the system, and thus the effect of AWGN has not been increased by this DS-SS system [48].

### 1.8.2   CDMA and DS-SS

In a CDMA system, each user is identified by its own spreading code, and in order to prevent users from interfering with each other, these codes are designed to be orthogonal to each other (ideally, the cross-correlation function between any two of these codes is zero). In practice, perfect orthogonality is difficult to achieve, but for now we assume perfect orthogonality in order to easily understand the concept of CDMA theory. Each user's signal is being encoded with not only a PN sequence, but also its own orthogonal code. Therefore, the transmitted signal $S(t)$ is:

$$S_i(t) = D_i(t) \cdot PN(t) \cdot C_i(t) = D_i(t) \cdot P_i(t) \tag{1.15}$$

where $C_i(t)$ denotes the CDMA code of the $i^{th}$ user whose data signal is $D_i(t)$, and $P_i(t)$ denotes the combination of the PN sequence and the orthogonal code for the $i^{th}$ user. Ideally, this allows a large number of users to use the same bandwidth; that is, now not only we do have the intentional interference rejection properties but also we have a

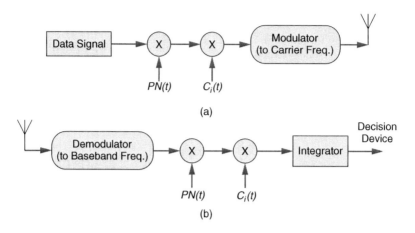

**Figure 1.12**   DS-SS basic transceiver (a) transmitter (b) receiver

multiuser interference rejection. Assume that there are $N$ users with $N$ orthogonal codes in this system, all using the same frequency band. Thus the $i^{th}$ receiver's signal is:

$$Y_i(t) = D_i(t) \cdot P(t) + \sum_{k=1, k \neq i}^{N} D_k(t + \theta) \cdot P_k(t + \theta) \qquad (1.16)$$

where $\theta$ is a random delay. When this is correlated with the PN sequence and the $i^{th}$ orthogonal code, the result will become zero (i.e. orthogonality), and only the signal due to the desired transmission will remain. The basic transmitter and receiver structures for DS-SS are shown in Figure 1.12. The transmitter only multiplies the data signal with the PN sequence and the CDMA code, and then modulates the resulting signal to the carrier frequency, and the receiver just performs the reverse operation and integrates the received signal. However, all this assumes perfect synchronization between transmitter and receiver.

## 1.8.3   Frequency-Hopping Spread Spectrum (FH-SS)

In FH-SS, the signal itself is not spread across the entire large bandwidth, rather the wide bandwidth is divided into $N$ sub-bands, and the signal hops from one band to the other in a pseudo-random manner instead. The centre frequency of the signal changes from one sub-band to another, as shown in Figure 1.13. As one can see, the large frequency bandwidth $N \cdot f_b$ at $f_c$ is divided into $N$ sub-bands of width $f_b$. The bandwidth $f_b$ must be enough to transmit the data signal $D(t)$, and, at a predetermined time interval, the centre frequency of the data signal changes from one sub-band to another pseudo-randomly [47].

As Figure 1.13 shows, the data signal hops from band $N$ at $f_c + (N/2)f_b$ to band 2 at $f_c - ((N/2) - 1)f_b$ and to band $N - 2$, and so on. Usually, the width of each sub-band is set so that the amount of signal that overlaps with adjacent sub-bands is minimal, and is thus approximately the bandwidth of the original data signal.

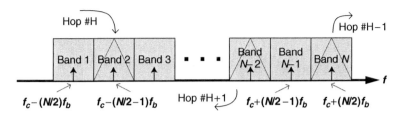

**Figure 1.13**  FH-SS signalling format

Two different kinds of FH-SS are commonly used: slow FH and fast FH. In slow FH-SS, several bits are sent for each hop, so the signal stays in a particular sub-band for a long time relative to the data-rate. In fast FH-SS, the signal switches sub-bands several times for each bit transmitted, so the signal stays in a sub-band for a very short time relative to the data-rate. It should be noted that slow FH is not really a spread spectrum technique, since this does not really spread the system and also, because the time spent in one sub-band is very large, the corresponding width of the band can be small, thus possibly violating the first principle of a spread spectrum system, namely that the spread bandwidth must be much greater than the non-spread bandwidth.

In fast FH, the performance of the system with respect to AWGN is not affected as in the direct sequence (DS). The noise power seen at the receiver is approximately the same as that in the un-hopped case, since each sub-band is approximately the same size as the original data signal's bandwidth. Here, if we again assume that the jamming signal $J(t)$ is distributed uniformly over the entire band, it is clear that the only portion of the jamming signal that affects the data is the part within the bandwidth of $f_b$, and thus the jamming signal is reduced by the factor of the processing gain $G_P$ which is defined as:

$$G_P = \frac{BW_{Car.}}{BW_{data}} = \frac{N \cdot f_b}{f_b} = N \qquad (1.17)$$

Thus in FH, the protection afforded is equal to the number of frequency bands. In case of interference in certain frequency bands, the probability of a bit being in error is then given by $P_{BE} = J/N$, where $J$ is the number of channels interfered, and $N$ is the total number of frequencies available to the hopping.

However, FH allows us to very simply decrease the bit-error rate (BER). If we choose to have a large number of chips per bit (i.e. a chip represents a hop), then we can use a simple majority function to determine what the transmitted bit was. We assume that the number of available hop channels is larger than the number of channels being interfered. If the simple majority function is being used, then the formula for the error rate becomes:

$$P_E = \sum_{x=r}^{c} \binom{c}{x} p^x \cdot q^{c-x} \qquad (1.18)$$

where $\binom{c}{r} = c!/[(c-r)! \cdot r!]$ and it reads the combination of $r$ out of $c$ which is the number of chips per bit (hops per bit), $r$ is the number of chip errors necessary to cause a bit error, $p$ is the probability of one bit error (i.e. $P_{BE} = J/N$), and $q$ is the

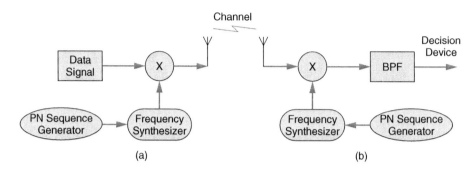

**Figure 1.14**   FH-SS basic transceiver (a) transmitter (b) receiver

probability of no error for a chip, or $1 - p$. By increasing the number of chips per bit from one to three and assuming $r = 2$ and $p = 0.01$, for example, we find that the error rate will be:

$$P_E = \sum_{x=r}^{c} \binom{c}{x} p^x \cdot q^{c-x} = \binom{3}{2} (p^2 - p^3) + \binom{3}{3} p^3$$

$$= 3p^2 - 2p^3 = 3 \times 10^{-4} - 2 \times 10^{-6} \approx 3 \times 10^{-4} \qquad (1.19)$$

Thus by just increasing the hopping rate from once per bit to three times per bit, the bit-error rate can be decreased dramatically.

The PN sequence is used here to determine the hopping sequence. So in order to transmit the signal, the data signal must be modulated up to the centre frequency of the band determined by the PN sequence. The structure of the transmitter is as shown in Figure 1.14. The data signal is modulated by the PN sequence-generator frequency by the frequency synthesizer and transmitted. The frequency synthesizer demodulates the signal down to a baseband frequency, and then the signal is filtered – thus only the desired data signal is passed through – and finally the signal is decoded. Again, to get multiple users using the same wide frequency band, CDMA techniques must be used.

### 1.8.4   CDMA and FH-SS

The method of CDMA used in this case is to provide each user with an orthogonal hop sequence, i.e. no two users occupy the same sub-band at the same time. In this way, multiple users can be accommodated without any chance of interfering with one another, since ideally only one user will be in a frequency sub-band during a given hop, and thus the receiver, due to its band-pass filter, will be able to detect the intended signal. Thus the transceiver given in Figure 1.14 needs only to be modified to incorporate an orthogonal code sequence in determining the centre frequency of the current hop (i.e. use the orthogonal code combined with the PN sequence as the input to the frequency synthesizer), and the system can support multiple users.

## 1.9  Motivations for Optical CDMA Communications

Multiple access techniques are required to meet the demand for high-speed and large-capacity communications in optical networks, which allow multiple users to share the fibre bandwidth. Three major multiple access techniques have been looked at above: each user is allocated a specific time-slot in time-division multiple-access (TDMA), or a specific frequency (wavelength) slot in wavelength division multiple-access (WDMA). Both techniques have been extensively explored and utilized in optical communication systems [32–34, 36, 40, 50–58]. The alternative, optical code-division multiple-access (OCDMA) [59–80], is receiving increasing attention due to its potential for enhanced information security, simplified and decentralized network control, improved spectral efficiency and increased flexibility in the granularity of bandwidth that can be provided. In optical CDMA (OCDMA), different users whose signals may be overlapped both in time and frequency share a common communications medium; multiple-access is achieved by assigning unlike minimally interfering code sequences to different transmitters, which must subsequently be detected in the presence of multiple access interference (MAI) from other users.

CDMA is derived from radio frequency (RF) spread spectrum communications, originally developed for military applications due to an inherent low probability of intercept and to immunity to interference, and more recently for commercial cellular radio applications such as third-generation (3G) and beyond [47–49, 81]. CDMA is now becoming the dominant multiple-access technique in RF wireless networks. In contrast, the need to perform encoding and decoding for optical CDMA (OCDMA) poses one immediate challenge because of both the optical carrier frequency and the much higher bit rate of multi-gigabit/s per user, which already approaches the limit of electronic processing. Furthermore, the challenges for OCDMA come from critical requirements which are routinely needed in optical communication systems. Therefore, innovative all-optical processing technologies are also wanted. These requirements include: extreme high quality-of-service (QoS), i.e. BER at $10^{-9}$ or below; large capacity, i.e. hundreds of users; enhanced bandwidth of $100^{+}$ Gbps; and higher scalability at long distances such as tens or hundreds of kilometres for LANs and metropolitan area networks (MAN). Significant progress of OCDMA research has been achieved worldwide in recent years, including several different OCDMA schemes that have been proposed based on different choices of sources [54, 66, 70, 82], coding schemes [37, 59, 67] and detections [76, 80, 83–87] which will be discussed in detail in this book through the following chapters.

OCDMA techniques may be classified according to the choice of coherent versus incoherent processing; coherent versus incoherent broadband optical source; and encoding methods, i.e. time versus frequency domain and/or amplitude versus phase spectra. To increase the available coding space, time-wavelength (two-dimensional) coding schemes have been proposed [88–91] where each code chip corresponds to a specific time position and wavelength position within a bit, as determined by a code matrix. However, this brings considerable complexity to the system implementations. Various spreading coding techniques will be discussed in detail in Chapter 2.

CDMA is well suited for bursty network environments, and the asynchronous nature of data transmission can simplify and decentralize network management and control. However, due to complex system requirements, full asynchronism is very difficult to implement in practice while real-time simultaneous MAI suppression, due to imperfect spreading codes, is still a hot research topic and under investigation [70, 72, 75, 79, 86, 92–96]. On the other hand, the synchronous scheme takes advantage of accommodating higher numbers of subscribers due to the time-shift property of the codes, and although the synchronization is difficult some methods have been proposed and are in use [97–99].

Several challenging research topics are still missing for practical OCDMA realization and development. These include the high co-channel interference (i.e. MAI) naturally present in almost all forms of OCDMA; low network capacity in terms of number of concurrent users; and codes that can support various traffic demands in terms of bandwidth and BER performance.

The proposed improvement in channel utilization and relatively low technical complexity and ease of implementation will have a direct impact on the current state of OCDMA networking, as will be discussed in Chapter 9 concerning scalability and compatibility of OCDMA with Internet protocol (IP) and passive optical networks (PON). A few significant CDMA properties of which simple, very high speed and cost-effective optical transport network can take advantage are as follows [71].

- *Statistical allocation of network capacity*: Any particular CDMA receiver experiences other users' signals as noise. This means that you can continue adding channels until the signal-to-noise ratio (SNR) becomes low enough that you start getting bit errors. A link can be allocated as many active connections such that the total data traffic stays below the channel capacity. For example, if there are a few hundred voice channels over CDMA, and the average power is the channel limit, then many more voice connections can be managed than with TDMA or WDMA methods. This can also be applied to data traffic on a LAN or access network where the traffic is inherently bursty in nature, as with Internet protocol (IP).

- *No guard time or guard bands*: In a TDMA system when multiple users share the same channel there must be a way to ensure that they do not transmit at the same time and overlap each other's signals. Since there is no really accurate clock recovery, a length of time must be allowed between the end of one user's transmission and the beginning of the next. As the speed gets higher, this guard time comes to dominate the system throughput. In a CDMA network, the stations simply transmit when they are ready. Also, in a WDMA system, unused frequency space is allocated between bands due to frequency overlapping avoidance through filtering process. These guard bands again represent wasted bandwidth.

- *Easier system management*: The users must have frequencies and/or time-slots assigned to them through some central administration processes in WDMA and TDMA systems. All you need with CDMA is for communicating stations to have the assigned code.

## 1.10 Access Networks Challenges

High capacity backbone networks have been highlighted. The high capacity standard OC-192 backbone networks are currently provided by operators with 10 Gbps data rate. State-of-the-art access network technologies such as digital subscriber line (DSL) provide up to a few Mbps, at best, of downstream bandwidth to customers, depending on various flavours. Therefore, the access network in fact challenges the current technology of networking on how to transfer a huge amount of traffic from backbone (i.e. core network) to access network (i.e. end-users) through wide bandwidth, high data rate services, such as video-on-demand (VoD) [100].

On the other hand, DSL technologies are highly distance-dependent in that the distance of any DSL subscriber to a central office must be less than 6 km due to the signal attenuation and distortion. And, usually, service providers are reluctant to service to distances more than 4 km. However, different DSL flavours such as very high bit rate DSL (VDSL) can support up to 50 Mbps of downstream bandwidth which is emerging. It should be noted that when the data rate goes high on DSL technologies the distances also shrink: for example, the maximum distance which VDSL is supported is only 500 metres [101].

Alternatively, community access television (CATV) networks are commonly used for broadband access [43]. By dedicating some radio frequency (RF) channels in coaxial cable for data, CATV networks provide Internet services as well as TV channels. Bearing in mind that CATV networks are mainly focused on broadcast services (like TV), they hardly fit well in distributed access bandwidth. This could be the reason why CATV end-users become frustrated at higher loads. There is now a strong feeling that (i) faster (ii) more reliable and (iii) higher scalable access networks are needed for next-generation ultra-high-speed crystal-clear broadband applications (e.g. high-definition IPTV).

The most promising proposed solution is to bring fibre closer to the home (i.e. end-users) in the next-generation access networks. The proposed models of FTTx [102, 103] including: fibre-to-the-home (FTTH), fibre-to-the-curb (FTTC), fibre-to-the-building (FTTB), etc. potentially offer extensive access bandwidth to their subscribers. FTTx aims to take over technologies such as VDSL by providing fibre directly to the homes, or very close by. FTTx platforms are mainly based on the passive optical network (PON) architectures which will be studied in detail in Chapter 9.

## 1.11  Summary

This chapter reviews the historical evolution of optical communications over fibre-optic in the past 50 years since laser discovery. The need for high-speed high-bandwidth reliable and scalable communications networks has led the research communities and industry to develop transatlantic and transpacific fibre-optic communication links. The carry-on motivations bring the optical fibre to homes and premises to serve the end-users with higher quality services such as multimedia-on-demand and high-definition holographic telepresence.

To take advantages of extremely wide bandwidth offered by fibre-optics, several multiple access techniques are utilized to share the communication channel efficiently. Among them, the most popular ones, wavelength-division multiple access (WDMA), time-division multiple access (TDMA) and the more promising code-division multiple access (CDMA) techniques, have already been introduced. Spread-spectrum communication methods including direct-sequence spread spectrum and frequency-hopping spread spectrum have also been studied as a prerequisite to understanding CDMA procedure. Chapter 2 is dedicated to introducing various types of spreading code designs deployed in the optical domain for OCDMA, while later chapters are more dedicated to systematic and networking aspects of OCDMA for various applications depending on the specifications.

# References

1. Maiman, T.H. (1960) Stimulated optical radiation in ruby. *Nature*, **187**, 493–494.
2. Javan, A., Bennett Jr., W.R. and Herriot, D.R. (1961) Population inversion and continuous optical MASER oscillation in a gas discharge containing a He-Ne mixture. *Physical Review Letters*, **6** (3), 106–110.
3. Nathan, M.I. *et al*. (1962) Stimulated emission of radiation from GaAs p-n junction. *Applied Physics Letters*, **1** (3), 62–64.
4. Quist, T.M. *et al*. (1962) Semiconductor MASER of GaAs. *Applied Physics Letters*, **1** (4), 91–92.
5. Holonyak Jr., N. and Bevacqua, S.F. (1962) Coherent (visible) light emission from GaAS1-xPx Junctions. *Applied Physics Letters*, **1** (4), 82–83.
6. Kompfner, R. (1965) Optical communications. *Science*, **150**, 149–155.
7. Kraemer, A.R. (1977) Free-space optical communications. *Signal*, **32**, 26–32.
8. Staff, L.F. (1980) Blue-green laser links to subs. *Laser Focus*, **16**, 14–18.
9. Kao, K.C. and Hockman, G.A. (1986) Dielectric fibre surface waveguides for optical frequencies. *Proc. IEE*, **133** (3), 1151–1158.
10. Kapron, F.P., Keck, D.B. and Maurer, R.D. (1970) Radiation losses in glass optical waveguides. *Applied Physics Letters*, **17** (10), 423–425.
11. Cherin, A.H. (1983) *An introduction to optical fibres*. McGraw-Hill, USA.
12. Tran, D.C., Sigel Jr., G.H. and Bendow, B. (1984) Heavy metal fluoride glasses and fibres: A review. *J. Lightw. Technol.*, **2** (10), 566–586.
13. Lucas, J. (1989) Review: fluoride glasses. *J. Materials Science*, **24** (1), 1–13.
14. Folweiler, R.C. (1989) Fluoride glasses. *GTE J. Science and Tech.*, **3** (1), 25–37.
15. Ghafouri-Shiraz, H. and Shum, P. (1992) Simulation of the Erbium-doped fibre amplifier characteristics. *J. Microw. & Opt. Tech. Let.*, **5** (4), 191–194.
16. Ghafouri-Shiraz, H. and Lo, B.S.K. (1996) *Distributed feedback laser diodes: principles and physical modelling*. John Wiley and Sons, UK.
17. Ghafouri-Shiraz, H. and Tsuji, S. (1994) Strain effects on refractive index and confinement factor of InxGa(1-x)As laser diodes. *J. Microw. & Opt. Tech. Let. (Special Issue)*, **7** (3), 113–119.
18. O'Neill, E.F. (1985) *A history of science and engineering in Bell systems*. AT&T Bell Lab.
19. Agrawal, G.P. (1992) *Fibre-optic communication systems*. John Wiley and Sons, USA.
20. Kotsas, A., Ghafouri-Shiraz, H. and Maclean, T.S.M. (1991) Microlens fabrication on single-mode fibres for efficient coupling from laser diodes. *J. Optical and Quantum Electronics*, **23** (3), 367–378.
21. Hasegawa, A. (1989) *Optical soliton in fibres*. Springer-Verlag, Germany.
22. Watsuki, K.T. *et al*. (1990) 20 Gb/s optical soliton data transmission over 70 km using distributed fibre amplifiers. *IEEE Photonics Tech. Letters*, **2** (12), 905–907.
23. Andrekson, P.A. *et al*. (1992) 32 Gb/s optical data transmission over 90 km. *IEEE Photonics Tech. Letters*, **4** (1), 76–79.
24. Taga, H. *et al*. (1992) 5 Gbits/s optical soliton transmission experiment over 3000 km employing 91 cascaded Er-doped fibre amplifier repeaters. *Electronics Letters*, **28** (24), 2247–2248.
25. Nakazawa, M. *et al*. (1991) 10 Gbit/s soliton data transmission over one million kilometers. *Electronics Letters*, **27** (14), 1270–1272.
26. Horimatsu, T. and Sasaki, M. (1989) OEIC technology and its application to subscriber loops. *J. Lightw. Technol.*, **7** (11), 1612–1622.
27. Electromagnetic Spectrum. Available from: http://www.lbl.gov/MicroWorlds/ALSTool/EMSpec/EMSpec2.html (accessed 22 August 2011).
28. Senior, J.M. (1992) *Optical Fibre Communications Principles and Practice*. Prentice-Hall (Second Edition), Prentice-Hall Europe.
29. Born, M., Wolf, E. and Bhatia, A.B. (1999) *Principles of optics*. 7th ed. Cambridge University Press, New York, USA.

30. Zanger, H. and Zanger, C. (1992) *Fiber optics communication and other applications*. Macmillan, New York, USA.
31. Palais, J.C. (1988) *Fiber optic communications*. Prentice-Hall International Editions, New York.
32. Borella, M.S. *et al*. (1997) Optical components for WDM lightwave networks. In: *Proc. of IEEE*, **85** (8).
33. Lee, T.P. and Zah, C.E. (1989) Wavelength-tunable and single-frequency lasers for photonic communications networks. *IEEE Comm. Mag.*, **27** (10), 42–52.
34. Kobrinski, H. and Cheung, K.W. (1994) Wavelength-tunable optical filters: applications and technologies. *IEEE Comm. Mag.*, **32** (12), 50–54.
35. Baron, J.E. *et al*. (1989) Multiple channel operation of an integrated acousto-optic tunable filter. *Electronics Letters*, **25** (6), 375–376.
36. Sneh, A. and Johnson, K.M. (1994) High-speed tunable liquid crystal optical filter for WDM systems. In: *Proc. IEEE/LEOS Summer Topical Meetings on Optical Networks and Their Enabling Technologies*.
37. Ito, M. *et al*. (1995) Fabrication and application of fiber Bragg gratings review. *J. Optoelectron. Devices Technol.*, **10** (3), 119–130.
38. Morkel, P.R. *et al*. (1991) Erbium-doped fiber amplifier with flattened gain spectrum. *IEEE Photonics Tech. Letters*, **3** (2), 118–120.
39. Chraplyvy, A.R. *et al*. (1994) Reduction of four-wave mixing crosstalk in WDM systems using unequally spaced channels. *IEEE Photonics Tech. Letters*, **6** (6), 754–756.
40. Brackett, C.A. (1990) Dense wavelength division multiplexing networks: principle and applications. *IEEE J. on Selected Areas in Comm.*, **8** (8), 948–964.
41. Hinton, H.S. (1990) Photonic switching fabrics. *IEEE Comm. Mag.*, **28** (4), 71–89.
42. Perros, H.G. (2005) *Connection-oriented networks: SONET/SDH, ATM, MPLS and optical networks*. John Wiley & Sons, Chichester, England.
43. Ramaswami, R. and Sivarajan, K.N. (1998) *Optical Networks: a practical perspective*. Morgan Kaufmann.
44. Mukherjee, B., Yao, S. and Dixit, S. (2000) Advances in photonic packet switching: an overview. *IEEE Comm. Mag.*, **38** (2), 84–94.
45. Ilyas, M. and Moftah, H.T. (2003) *Handbook of optical communication networks*. CRC Press, Florida, USA.
46. Proakis, J.G. (1995) *Digital communications*. McGraw-Hill, New York, USA.
47. Prasad, R. (1996) *CDMA for wireless personal communications*. Artech House publisher, Boston, USA.
48. Meel, I.J. (1999) *Spread spectrum – introduction and application*. Siruis Communication, Malaysia.
49. Viterbi, A.J. (1995) *CDMA, principles of spreading spectrum communication*. Addison Wesley, Boston, USA.
50. Eisenstein, G., Tucker, R.S. and Korotky, S.K. (1988) Optical time-division multiplexing for very high bit rate transmission. *J. Lightw. Technol.*, **6** (11), 1737–1749.
51. Fujiwara, M. *et al*. (2002) Novel polarization scrambling technique for carrier-distributed WDM networks. In: *ECOC*.
52. Xu, R., Gong, Q. and Ya, P. (2001) A novel IP with MPLS over WDM-based broadband wavelength switched IP network. *J. Lightw. Technol.*, **19** (5), 596–602.
53. Iwatsuki, K., Kani, J.I. and Suzuki, H. (2004) Access and metro networks based on WDM technologies. *J. Lightw. Technol.*, **22** (11), 2623–2630.
54. Tsang, W.T. *et al*. (1993) Control of lasing wavelength in distributed feedback lasers by angling the active stripe with respect to the grating. *IEEE Photonics Tech. Letters*, **5** (9), 978–980.
55. Assi, C., Ye, Y. and Dixit, S. (2003) Dynamic bandwidth allocation for quality of service over Ethernet PON. *IEEE J. on Selected Areas in Comm.*, **21** (11), 1467–1477.
56. Killat, U. (1996) *Access to B-ISDN via PON – ATM communication in practice*. Wiley Teubner Communications, Chichester, England.
57. Lam, C.F. (2007) *Passive optical network: principles and practice*. Academic Press, Elsevier, USA.
58. Kramer, G. (2005) *Ethernet passive optical network*. McGraw-Hill, New York, USA.
59. Azizoghlu, M., Salehi, J.A. and Li, Y. (1992) Optical CDMA via temporal codes. *IEEE Trans on Comm.*, **40** (8), 1162–1170.
60. Heritage, J.P., Salehi, J.A. and Weiner, A.M. (1990) Coherent ultrashort light pulse code-division multiple access communication systems. *J. Lightw. Technol.*, **8** (3), 478–491.
61. Salehi, J.A. (1989) Code division multiple-access techniques in optical fiber networks – part I: fundamental principles. *IEEE Trans. on Comm.*, **37** (8), 824–833.

62. Salehi, J.A. and Brackett, C.A. (1989) Code division multiple-access technique in optical fiber networks – part II: system performance analysis. *IEEE Trans. on Comm.*, **37** (8), 834–842.

63. Kwong, W.C., Perrier, P.A. and Prucnal, P.R. (1991) Performance comparison of asynchronous and synchronous code-division multiple-access techniques for fiber-optic local area networks. *IEEE Trans. on Comm.*, **39** (11), 1625–1634.

64. Wei, Z. and Ghafouri-Shiraz, H. (2002) Proposal of a novel code for spectral amplitude coding optical CDMA systems *IEEE Photonics Tech. Letters*, **14** (3), 414–416.

65. Smith, E.D.J., Blaikie, R.J. and Taylor, D.P. (1998) Performance enhancement of spectral-amplitude-coding optical CDMA using pulse position modulation. *IEEE Trans. on Comm.*, **46** (9), 1176–1185.

66. Wei, Z., Ghafouri-Shiraz, H. and Shalaby, H.M.H. (2001) Performance analysis of optical spectral-amplitude-coding CDMA systems using super-fluorescent fiber source. *IEEE Photonics Tech. Letters*, **13** (8), 887–889.

67. Kavehrad, M. and Zaccarin, D. (1995) Optical code division-multiplexed systems based on spectral encoding of noncoherent sources. *J. Lightw. Technol.*, **13** (3), 534–545.

68. Wei, Z. and Ghafouri-Shiraz, H. (2002) IP transmission over spectral-amplitude-coding CDMA links. *J. Microw. & Opt. Tech. Let.*, **33** (2), 140–142.

69. Wei, Z. and Ghafouri-Shiraz, H. (2002) IP routing by an optical spectral-amplitude-coding CDMA network. *IEE Proc. Communications*, **149** (5), 265–269.

70. Cooper, A.B. *et al.* (2007) High spectral efficiency phase diversity coherent optical CDMA with low MAI. In: *Lasers and Electro-Optics (CLEO)*.

71. Prucnal, P.R. (2005) *Optical code division multiple access: fundamentals and Applications*. CRC Taylor & Francis Group.

72. Shalaby, H.M.H. (1995) Synchronous fiber-optic CDMA systems with interference estimators. *J. Lightw. Technol.*, **17** (11), 2268–2275.

73. Shalaby, H.M.H. (1995) Performance analysis of optical synchronous CDMA communication systems with PPM signaling. *IEEE Trans. on Comm.*, **43** (2/3/4), 624–634.

74. Shalaby, H.M.H. (1999) A performance analysis of optical overlapping PPM-CDMA communication systems. *J. Lightw. Technol.*, **19** (2), 426–433.

75. Lee, T.S., Shalaby, H.M.H. and Ghafouri-Shiraz, H. (2001) Interference reduction in synchronous fiber optical PPM-CDMA systems *J. Microw. & Opt. Tech. Let.*, **30** (3), 202–205.

76. Shalaby, H.M.H. (1999) Direct-detection optical overlapping PPM-CDMA communication systems with double optical hard-limiters. *J. Lightw. Technol.*, **17** (7), 1158–1165.

77. Hamarsheh, M.M.N., Shalaby, H.M.H. and Abdullah, M.K. (2005) Design and analysis of dynamic code division multiple access communication system based on tunable optical filter. *J. Lightw. Technol.*, **23** (12), 3959–3965.

78. Shalaby, H.M.H. (2002) Complexities, error probabilities and capacities of optical OOK-CDMA communication systems. *IEEE Trans on Comm.*, **50** (12), 2009–2017.

79. Shalaby, H.M.H. (1998) Cochannel interference reduction in optical PPM-CDMA systems. *IEEE Trans. on Comm.*, **46** (6), 799–805.

80. Shalaby, H.M.H. (1998) Chip-level detection in optical code division multiple access. *J. Lightw. Technol.*, **16** (6), 1077–1087.

81. Buehrer, R.M. (2006) *Code division multiple access (CDMA)*. Morgan & Claypool Publishers, Colorado, USA.

82. Huang, W., Andonovic, I. and Tur, M. (1998) Decision-directed PLL for coherent optical pulse CDMA system in the presence of multiuser interference, laser phase noise and shot noise. *J. Lightw. Technol.*, **16** (10), 1786–1794.

83. Liu, X. *et al.* (2004) Tolerance in-band coherent crosstalk of differetial phase-shift-keyed signal with balanced detection and FEC. *IEEE Photonics Tech. Letters*, **16** (4), 1209–1211.

84. Betti, S., Marchis, G.D. and Iannone, E. (1992) Polarization modulated direct detection optical transmission systems. *J. Lightw. Technol.*, **10** (12), 1985–1997.

85. Ohtsuki, T. (1999) Performance analysis of direct-detection optical CDMA systems with optical hard-limiter using equal-weight orthogonal signaling. *IEICE Trans. on Comm.*, **E82-B** (3), 512–520.

86. Wang, X. *et al.* (2006) Demonstration of DPSK-OCDMA with balanced detection to improve MAI and beat noise tolerance in OCDMA systems. In: *OFC*.

87. Benedetto, S. *et al.* (1994) Coherent and direct-detection polarization modulation system experiment. In: *ECOC*, Firenze, Italy.

88. Griner, U.N. and Arnon, S. (2004) A novel bipolar wavelength-time coding scheme for optical CDMA systems. *IEEE Photonics Tech. Letters*, **16** (1), 332–334.
89. Gu, F. and Wu, J. (2005) Construction of two-dimensional wavelength/time optical orthogonal codes using difference family. *J. Lightw. Technol.*, **23** (11), 3642–3652.
90. Teixeira, A.L.J. *et al.* (2001) All-optical time-wavelength code router for optical CDMA networks. In: *LEOS, The 14th Annual Meeting of the IEEE*.
91. Liang, W. *et al.* (2008) A new family of 2D variable-weight optical orthogonal codes for OCDMA systems supporting multiple QoS and analysis of its performance. *Photonic Network Communications*, **16** (1), 53–60.
92. Liu, M.Y. and Tsao, H.W. (2000) Cochannel interference cancellation via employing a reference correlator for synchronous optical CDMA system. *J. Microw. & Opt. Tech. Let.*, **25** (6), 390–392.
93. Yamamoto, F. and Sugie, T. (2000) Reduction of optical beat interference in passive optical networks using CDMA technique. *IEEE Photonics Tech. Letters*, **12** (12), 1710–1712.
94. Gamachi, Y. *et al.* (2000) An optical synchronous M-ary FSK/CDMA system using interference canceller. *J. Electro. & Comm. in Japan*, **83** (9), 20–32.
95. Yang, C.C. (2007) Optical CDMA passive optical network using prime code with interference elimination. *IEEE Photonics Tech. Letters*, **19** (7), 516–518.
96. Lin, C.L. and Wu, J. (2000) Channel interference reduction using random Manchester codes for both synchronous and asynchronous fiber-optic CDMA systems. *J. Lightw. Technol.*, **18** (1), 26–33.
97. Mustapha, M.M. and Ormondroyd, R.F. (2000) Dual-threshold sequential detection code synchronization for an optical CDMA network in the presence of multiuser interference. *J. Lightw. Technol.*, **18** (12), 1742–1748.
98. Yang, G.-C. (1994) Performance analysis for synchronization and system on CDMA optical fiber networks. *IEICE Trans. on Comm.*, **E77B** (10), 1238–1248.
99. Keshavarzian, A. and Salehi, J.A. (2005) Multiple-shift code acquisition of optical orthogonal codes in optical CDMA systems. *IEEE Trans on Comm.*, **53** (4), 687–697.
100. Sivalingam, K.M. and Subramanian, S. (2005) *Emerging optical network technologies*. Springer Science+Business Media Inc, Boston, USA.
101. Goralski, W. (1998) *ADSL and DSL Technologies*. McGraw-Hill, USA.
102. Ohara, K. (2003) Traffic analysis of Ethernet-PON in FTTH trial service. In: *OFC*.
103. Kitayama, K., Wang, X. and Wada, N. (2006) OCDMA over WDM PON – solution path to gigabit symetric FTTH. *J. Lightw. Technol.*, **24** (4), 1654–1662.

# 2

# Optical Spreading Codes

## 2.1 Introduction

In an optical code-division multiple access (OCDMA) system, the main purpose is to recognize the intended user in the presence of other users, since all users share the same channel. Another aim is to accommodate more possible subscribers in the system. In this case, the optical sequence codes with the best orthogonal characteristic should be employed. In other words, spreading code sequences are selected for the features of maximum auto-correlation and minimum cross-correlation in order to optimize the differentiation between the correct signal and the interference.

This chapter reviews fundamental spreading codes used in the application of OCDMA communications including coherent/incoherent, asynchronous/synchronous, temporal/spectral and one-dimensional/multidimensional schemes.

Basic bipolar codes including $m$-sequences, Gold sequences, Hadamard and Walsh-Hadamard codes will be studied in terms of construction and properties. The bipolar codes have been borrowed from wireless CDMA concept. Although, due to the differences in the signal carries of RF and lightwave as well as channel characteristics of air-interfaces in RF communications and fibre-optics in optical communications, unipolar codes are mostly acceptable in the OCDMA communications and networks. However, recent progresses in the optical coherent modulations and coding, bipolar codes are also employed in coherent OCDMA systems.

Optical orthogonal codes (OOC) as one of the successful codes in the asynchronous OCDMA systems will be reviewed as well as various prime code families suitable for synchronous OCDMA schemes.

Since OCDMA provides multiple users sharing single channel, the most degrading issue will be mixing the intended user's signal with other users' signals in the same channel. Accordingly, channel coding such as Manchester encoding and Turbo codes seem necessary to elevate the overall system performance. Here in this chapter, we will also study various channel coding as forward error correction (FEC) techniques including convolutional codes, Manchester and differential Manchester encodings, and Turbo codes

*Optical CDMA Networks: Principles, Analysis and Applications*, First Edition. Hooshang Ghafouri-Shiraz and M. Massoud Karbassian.
© 2012 John Wiley & Sons, Ltd. Published 2012 by John Wiley & Sons, Ltd.

algorithms. Turbo-coded OCDMA will be analysed here whereas Manchester encoded OCDMA systems will be extensively studied and analysed in Chapter 5.

## 2.2 Bipolar Codes

In this section we will introduce the bipolar codes that can be applied to the coherent and the incoherent OCDMA systems using differential detection such as balanced detection. We are primarily concerned with the maximal length sequences–so-called $m$-sequences, Gold sequences and Walsh–Hadamard codes. Bipolar codes comprise $(-1, 1)$ sequences whereas unipolar codes comprise $(1, 0)$ sequences

### 2.2.1  m-Sequence Codes

The $m$-sequence comprises a pseudo-random sequence generated by a feedback shift register which has the maximal length, referred to as a maximal linear feedback shift register sequence [1]. The period of the $m$-sequence code is determined by the number of shift-registers and linear feedback logic gates. If an $s$-stage shift register is used, the $m$-sequence period will be $n = 2^s - 1$. The linear feedback logic is also determined by the following function, $f(x)$, which is a primitive polynomial function of degree $s$. That is:

$$f(x) = \sum_{i=0}^{s} c_i x^i \tag{2.1}$$

The above polynomial function must satisfy the following properties [1]:

1. It is an irreducible polynomial, i.e. it cannot be decomposed into smaller factors.
2. $x^n + 1$ is divisible by $f(x)$ when $n = 2^s - 1$.
3. $x^q + 1$ is not divisible by $f(x)$ when $q < n$.

As an example, let us find the $m$-sequences of primitive polynomials of degree $s = 4$ which have $m$-sequence period of $2^4 - 1 = 15$. By breaking down the polynomial $x^{15} + 1$ we have:

$$x^{15} + 1 = (x^4 + x^3 + x^2 + x + 1)(x^4 + x^3 + 1)(x^4 + x + 1)(x^2 + x + 1)(x + 1) \tag{2.2}$$

In Equation (2.2) terms $(x^2 + x + 1)$ and $(x + 1)$ are not primitive polynomials as they are not the polynomials of degree $s = 4$. Also, since the expression $(x^4 + x^3 + x^2 + x + 1)$ is a factor of $(x^5 + 1)$, it does not satisfy the above-mentioned third property, so it is not a primitive polynomial either. On the other hand, terms $(x^4 + x + 1)$ and $(x^4 + x^3 + 1)$ satisfy all of the above three properties and accordingly they are primitive polynomials.

Figure 2.1 shows an $s$-stage linear feedback shift-register where $a_{s-i}$ denoted the state of the $(s - i)^{th}$ stage shift-register and $c_i (i = 0, 1, \ldots, s)$ denotes the link state of the feedback line in the shift-register; meaning that $c_i = 0$ represents the disconnected line and $c_i = 1$ represents the connected line.

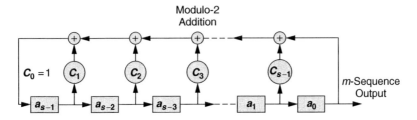

**Figure 2.1**  An $s$-stage linear feedback shift-register

The logic behind this expression can be written as:

$$a_s = c_1 a_{s-1} \oplus c_2 a_{s-2} \oplus \ldots \oplus c_s a_0 \tag{2.3}$$

where $\oplus$ represents modulo-2 addition, which is an exclusive-OR logic. By moving the $a_s$ to the right-hand side of Equation (2.3) and substituting $a_s = c_0 a_s$ where $c_0 = 1$ as shown in Figure 2.1, the Equation (2.3) can be expressed as:

$$\sum_{i=0}^{s} c_i a_{s-i} = 0 \tag{2.4}$$

where $c_i (i = 0, 1, \ldots, s)$ is the coefficient of the primitive polynomials of Equation (2.1). By substituting $c_i$ into Equation (2.3), after being obtained (see the example), we can find the linear feedback logic. Now the $m$-sequence codes can be generated by the $s$-stage linear shift-register.

**Example:** Let us generate an $m$-sequence with the period $n = 2^4 - 1 = 15$ from the primitive polynomial $x^4 + x + 1$. This implies that the coefficients $c_4 = c_1 = c_0 = 1$ and $c_3 = c_2 = 0$. Accordingly, the linear feedback logic is obtained as:

$$a_4 = a_3 \oplus a_0 \tag{2.5}$$

Figure 2.2 illustrates the resulting four-stage maximal length linear feedback shift-register.

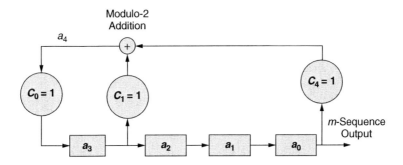

**Figure 2.2**  Four-stage maximal length linear feedback shift-register

Assuming the initial state of the shift-register is set to '1000' i.e. $a_3 = 1, a_2 = 0$, $a_1 = 0, a_0 = 0$, then after four clocks the same sequence will come out as 1000; then the fifth sequence comes out as a result of initial $a_4 = a_3 \oplus a_0 = 1 \oplus 0 = 1$ at the first clock, and the sixth bit is generated as a result of $a_4 = a_3 \oplus a_0 = 1 \oplus 0 = 1$ at the second clock, etc. The maximal length linear feedback shift-register sequence of '010110101111000' is thus produced after 15 clock times, representing the period. Therefore, we can generate $m$-sequence codes as long as a primitive polynomial is available. The $m$-sequence cardinality (i.e. number of available codes) with period of $n = 2^s - 1$ is equal to the number of the $s$-stage primitive polynomials, hence the number of $s$-stage primitive polynomials is [2]:

$$|C| = \frac{\xi(2^s - 1)}{s} \tag{2.6}$$

where $\xi(2^s - 1)$ is the Euler number denoting the number of positive integers including 1 relatively prime to and less than $2^s - 1$, i.e.

$$\xi(2^s - 1) = \begin{cases} 2^s - 2 & \text{if } 2^s - 1 \text{ is a prime number} \\ (q_1 - 1)(q_2 - 2) & \text{if } 2^s - 1 = q_1 q_2 \end{cases} \tag{2.7}$$

where $q_1, q_2$ are prime numbers. More examples and tables including the different coefficients for $m$-sequence generation are discussed in [1]. Properties of $m$-sequences are as follow:

- The code-length (i.e. period) of $m$-sequence codes is $2^s - 1$.
- The appearance of '0' and '1' in the codes is approximately equiprobable. Note that in an $m$-sequence code the number of '1's is only one more than that of '0's.
- Subsequences of '1's and '0's in an $m$-sequence is called a run, thus there are $2^s - 1$ runs in an $m$-sequence. The number of runs for length one is $(2^s - 1)/2$ and the number of runs for length two is $(2^s - 1)/4$ and the number of runs for length three is $(2^s - 1)/8$ and so on. Finally, there will be only one run for length $s$ and consecutive zero run for lengths $-1$.
- The cyclic auto-correlation function of $m$-sequences when '0' and '1' elements are replaced with '1' and '$-1$' respectively is expressed as:

$$R_{XX}(\tau) = \sum_{i=0}^{n-1} x_i x_{i+\tau} = \begin{cases} 2^s - 1 & \tau = 0 \\ -1 & 0 < \tau \leq n - 1 \end{cases} \tag{2.8}$$

where $X = (x_0, x_1, x_2, \ldots, x_{n-1})$ is a codeword of $m$-sequence and $n = 2^s - 1$. Accordingly, the cross-correlation function of two $m$-sequence codes $X = (x_0, x_1, x_2, \ldots, x_{n-1})$ and $Y = (y_0, y_1, y_2, \ldots, y_{n-1})$ is:

$$R_{XY}(\tau) = \sum_{i=0}^{n-1} x_i y_{i+\tau} \quad 0 \leq \tau \leq n - 1 \tag{2.9}$$

Figures 2.3 and 2.4 illustrate the auto-correlation and cross-correlation values of two $m$-sequence codes of $(1, -1, -1, -1, 1, -1, -1, 1, 1, -1, 1, -1, 1, 1, 1)$ and $(1, -1, -1, -1, 1, 1, 1, 1, -1, 1, -1, 1, 1, -1, -1)$. It can be seen that the ratio of the maximum values of cross-correlation and auto-correlation is $7/15 = 46\%$.

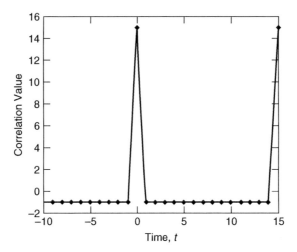

**Figure 2.3**  Auto-correlation function for $m$-sequence $(1, -1, -1, -1, 1, -1, -1, 1, 1, -1, 1, -1,$
$1, 1, 1)$

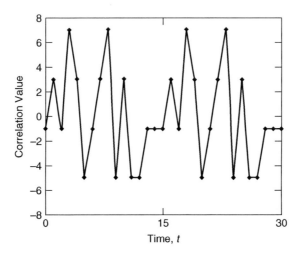

**Figure 2.4**  Cross-correlation function for $m$-sequence $(1, -1, -1, -1, 1, -1, -1, 1, 1, -1, 1, -1,$
$1, 1, 1)$ and $(1, -1, -1, -1, 1, 1, 1, 1, 1, -1, 1, -1, 1, 1, -1, -1)$

## 2.2.2   Gold Sequences

Gold codes are generated by modulo-2 addition of two $m$-sequence codes bit-by-bit under
the synchronizing clock [1–3]. Figure 2.5 depicts the combination of two shift-registers
producing $m$-sequences. Note that two $m$-sequences must have same length and rate,
resulting in a Gold code of the same length and rate. Although Gold codes are generated
by two $m$-sequence codes, they are not maximal length linear shift-register sequences
because they do not satisfy the primitive polynomial properties discussed in $m$-sequence
generation.

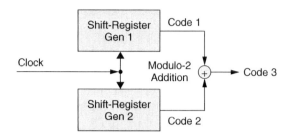

**Figure 2.5**  Gold code generator

By varying the relative shifts of two $m$-sequence codes of $2^s - 1$ long, $2^s - 1$ Gold sequences (i.e. cardinality) can be produced. Considering two original $m$-sequence codes as Gold code seeds, the Gold code cardinality will be:

$$|C| = (2^s - 1) + 2 = 2^s + 1 \tag{2.10}$$

Assuming $X = (x_0, x_1, x_2, \ldots, x_{n-1})$ and $Y = (y_0, y_1, y_2, \ldots, y_{n-1})$ as two $m$-sequence codes generating Gold codes, the resulting codewords will then be:

$$X, Y, X \oplus Y, X \oplus Y_1, X \oplus Y_2, \ldots, X \oplus Y_{(n-1)} \tag{2.11}$$

where $Y_i$ denotes the $i^{th}$ cyclic right-shift of codeword $Y$ and $\oplus$ represents modulo-2 addition, equivalent to exclusive-OR logic.

Gold codes improve the cyclic shift of $m$-sequence codes and expand to more codewords as well as improving the correlation values. Gold codes have three auto-correlation and cross-correlation $R_{XY}(\tau)$ values of $(-t_s - 1, -1, t_s - 1)$ where $t_s = 2^{\lfloor (s+2/2)-1 \rfloor}$ and $\lfloor x \rfloor$ denotes the largest integer of less than or equal to $x$ [1]. The cross-correlation values of Gold codes do not exceed the maximal value of the cross-correlation function between the two preferred $m$-sequence codes generating the Gold codes. However, the auto-correlation function of Gold codes, unlike $m$-sequence codes, has side lobes. Figure 2.6 illustrates the auto-correlation functions of one sequence of $(-1, -1, -1, -1, 1, -1, 1, -1, 1, -1, 1, 1, 1, 1, -1, -1, -1, -1, -1, 1, -1, -1, -1, -1, 1, 1, -1, -1, -1, 1, 1)$ as a Gold code. The ratio of the maximum cross-correlation value, shown in Figure 2.7, versus the peak of auto-correlation here is $9/31 = 29\%$.

### 2.2.3  Walsh–Hadamard Codes

Walsh–Hadamard codes, also called Walsh codes, comprise the row vectors of arranged Hadamard matrices [1]. This code family is also bipolar, meaning that the elements are $(-1, 1)$. The procedure of Walsh–Hadamard code generation is described in the following:

$$H(i+1) = \begin{bmatrix} H(i) & H(i) \\ H(i) & -H(i) \end{bmatrix} \tag{2.12}$$

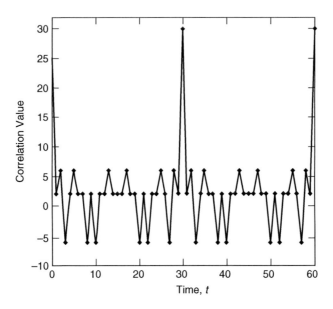

**Figure 2.6** Auto-correlation function for Gold codes $(-1, -1, -1, -1, 1, -1, 1, -1, 1, -1, 1, 1, 1, 1, 1, -1, -1, -1, -1, -1, 1, -1, -1, -1, -1, 1, 1, 1, -1, -1, -1, 1, 1)$

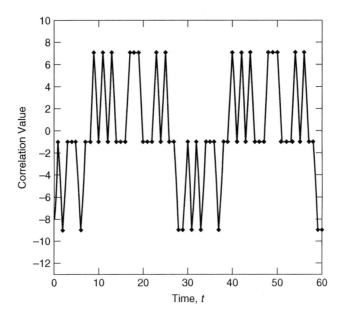

**Figure 2.7** Cross-correlation function of Gold codes $(-1, -1, -1, -1, -1, -1, -1, 1, 1, 1, 1, -1, 1, 1, -1, 1, 1, 1, 1, 1, -1, 1, 1, 1, 1, -1, -1, -1, -1, 1, -1)$ and $(-1, -1, -1, -1, 1, -1, 1, -1, 1, 1, -1, 1, 1, 1, 1, -1, -1, -1, -1, 1, -1, 1, -1, -1, -1, -1, 1, 1, 1, -1, -1, -1, 1, 1)$

where $i = (0, 1, 2, \ldots)$ and $H(0) = +1$. Having this recursive expression, we then have:

$$H(1) = \begin{bmatrix} 1 & 1 \\ 1 & -1 \end{bmatrix} \tag{2.13}$$

$$H(2) = \begin{bmatrix} 1 & 1 & 1 & 1 \\ 1 & -1 & 1 & -1 \\ 1 & 1 & -1 & -1 \\ 1 & -1 & -1 & 1 \end{bmatrix} \tag{2.14}$$

and so on. The matrix $H(i)$ is a $2^i \times 2^i$ square matrix comprising the $(-1, 1)$ elements. $H(1)$ or simply the $H$ matrix is called the Hadamard matrix. $H_{2^i}$ is also equal to the $H(i)$ of dimension $2^i \times 2^i$ when $i = (1, 2, 3, \ldots)$.

$$H_{2^1} = H_2 = H(1) = \begin{bmatrix} 1 & 1 \\ 1 & -1 \end{bmatrix}; \quad H_{2^i} = \underbrace{H_2 \otimes H_2 \otimes \ldots \ldots \otimes H_2}_{i \ times} \tag{2.15}$$

where $\otimes$ denotes the Kronecker product of two matrices, for example:

$$X = \begin{bmatrix} x_{11} & x_{12} & \cdots & x_{1n} \\ \vdots & & \ddots & \vdots \\ x_{m1} & x_{m2} & \cdots & x_{mn} \end{bmatrix} \quad \text{and} \quad Y = \begin{bmatrix} y_{11} & y_{12} & \cdots & y_{1l} \\ \vdots & & \ddots & \vdots \\ y_{k1} & y_{k2} & \cdots & y_{kl} \end{bmatrix} \tag{2.16}$$

$$X \otimes Y = \begin{bmatrix} x_{11}Y & x_{12}Y & \cdots & x_{1n}Y \\ \vdots & & \ddots & \vdots \\ x_{m1}Y & x_{m2}Y & \cdots & x_{mn}Y \end{bmatrix} \quad \text{and} \quad Y \otimes X \begin{bmatrix} x_{11}X & x_{12}X & \cdots & x_{1n}X \\ \vdots & & \ddots & \vdots \\ x_{m1}X & x_{m2}X & \cdots & x_{mn}X \end{bmatrix} \tag{2.17}$$

$X \otimes Y$ and/or $Y \otimes X$ are matrices with $(m \cdot k) \times (n \cdot l)$ dimensions. Accordingly, $H(i)$ is also called the Hadamard matrix. Now, each row of the $H_{2^i}$ matrix is referred to as the Hadamard code. This is also referred to as the Walsh–Hadamard code family.

The in-phase correlation function of Walsh–Hadamard codes is defined and obtained by a Walsh matrix of $2^i \times 2^i$ dimensions as:

$$R_{XX}(\tau) = (H(i)_j)(H(i)_k)^T = \begin{cases} 2^i & j = k; \ \text{Auto–correlation} \\ 0 & j \neq k; \ \text{Cross–correlation} \end{cases} \tag{2.18}$$

For example, $H(2)_1 = [1 \ -1 \ -1 \ 1]$ and $H(2)_2 = [1 \ 1 \ 1 \ 1]$, thus:

$$(H(2)_1)(H(2)_1)^T = [1 \ -1 \ -1 \ 1] \begin{bmatrix} 1 \\ -1 \\ -1 \\ 1 \end{bmatrix} = 4 \tag{2.19}$$

$$(H(2)_1)(H(2)_2)^T = [1 \ -1 \ -1 \ 1] \begin{bmatrix} 1 \\ 1 \\ 1 \\ 1 \end{bmatrix} = 0. \tag{2.20}$$

## 2.3   Unipolar Codes: Optical Orthogonal Codes

An important type of temporal code is optical orthogonal code (OOC), proposed for intensity modulation and direct-detection (IM-DD) OCDMA systems [4–11]. These are very sparse codes, meaning that the code weight is very low, thus limiting the efficiency in practical coding and decoding. Moreover, the number of codes that can be supported is very low as compared to a codeset of the same length used in radio communications like the previously introduced Walsh–Hadamard codes. To get more codes we need to increase the length of the code, demanding the use of very short pulse optical sources having a pulse width much smaller than the bit duration.

The required temporal OCDMA codes must satisfy the following conditions:

1. The peak auto-correlation function of the code should be maximized.
2. The cross-correlation between any two codes should be minimized.
3. The side lobes of the auto-correlation function of the code should be minimized.

Conditions (1) and (2) insure that the multiple access interference (MAI) is minimized, while condition (3) simplifies the synchronization process at the receiver.

The correlation $R_{C_i C_j}(\tau)$ of two signature signals $C_i(t)$ and $C_j(t)$ is defined as:

$$R_{C_i C_j}(\tau) = \int_{-\infty}^{+\infty} C_i(t)C_j(t+\tau)dt \qquad (2.21)$$

where $i,j = (1,2,\ldots)$, and the signature signal $C_k(t)$ is defined as:

$$C_k(t) = \sum_{n=-\infty}^{+\infty} C_k(n)u_{T_c}(t - nT_c) \qquad (2.22)$$

where $k = (1,2,\ldots)$ and $C_k(n) \in \{1,0\}$ is the periodic sequence of period $N$ and chip duration $T_c$. The discrete correlation function of any two code sequences $C_i(n)$ and $C_j(n)$ is then given by:

$$R_{C_i C_j}(m) = \sum_{n=0}^{N-1} C_i(n)C_j(n+m) \qquad (2.23)$$

where $m = (\ldots, -1, 0, 1, \ldots)$. The sum in the argument of $C_j(n+m)$ is calculated modulo $N$; here in this chapter we present this operation from now on as $[x]_y$ read as $x$ modulo $y$. In the discrete form, the above conditions are rewritten as:

1. The number of ones in the zero-shift discrete auto-correlation function should be maximized.
2. The number of coincidences of the non-zero shift discrete auto-correlation function should be minimized.
3. The number of coincidences of the discrete cross-correlation function should be minimized.

An OOC is usually represented by a quadruple, $(N, w, \lambda_a, \lambda_c)$ where $N$ is the code-length; $w$ is the code-weight (i.e. the number of ones); $\lambda_a$ is the upper-bound on the auto-correlation value for non-zero shift and $\lambda_c$ is the upper-bound on cross-correlation values. The conditions for OOCs are then:

$$R_{C_i C_j}(m) = \sum_{n=0}^{N-1} C_i(n) C_j(n+m) \leq \lambda_c \quad \text{for } \forall m \tag{2.24}$$

and

$$R_{C_i C_j}(m) = \sum_{n=0}^{N-1} C_i(n) C_j(n+m) \leq \lambda_a \quad \text{for } [m]_n \neq 0 \tag{2.25}$$

There is a special case of $\lambda_a = \lambda_c = \lambda$ that the OOC is represented by $(N, w, \lambda)$ as optimal OOC. $|C|$ denotes the cardinality of the OOC family, i.e. the size of the codeset which refers to the number of codewords contained in the codeset. The largest possible size of the set with conditions of $(N, w, \lambda)$ denotes $\phi(N, w, \lambda)$. By the aid of Johnson bound [5], it is known that $\phi(N, w, \lambda)$ should satisfy [4]:

$$\phi(N, w, \lambda) \leq \frac{(N-1)(N-2)\dots(N-\lambda)}{w(w-1)(w-2)\dots(w-\lambda)} \tag{2.26}$$

In case of $\lambda_a = \lambda_c = 1$, i.e. strict OOC, it is shown that the $|C|$ is upper-bounded by:

$$|C| \leq \left\lfloor \frac{(N-1)}{w(w-1)} \right\rfloor \tag{2.27}$$

where $\lfloor x \rfloor$ denotes the largest integer of less than or equal to $x$.

An example of a strict OOC$(N = 13, w = 3, \lambda = 1)$ codeset is $C_i \in \{1100100000000, 1010000100000\}$. It is clear that the auto-correlation is thus equal to the code-weight of $w = 3$, and the non-zero shift auto-correlation and the cross-correlation is less than or equal to $\lambda = 1$. The same codeset can be represented using the set notation of $C_i \in \{(1, 2, 5), (1, 3, 8)\} \mod(13)$, where the elements in the codeset represent the position of the pulses, i.e. 1s, in the code sequence of code-length 13.

Assuming an OCMDA system with OOC coding and an avalanche photodetector (APD), the compound effect of APD noise, thermal noise and MAI was evaluated in [12]. The complex statistics of the APD, described in [13], was not used but instead a simplifying Gaussian approximation was considered in [14]. It was shown that when noise effects are considered, the performance of OCDMA based on OOCs can be two orders of magnitude worse than that of the ideal case. Also the improvement in bit-error rate (BER) by using hard-limiters is not significant because the MAI during the zero data bit transmission cannot be completely suppressed as in the noise-free case.

The main drawback of OOCs is the limited number of users for a reasonable code-length and code-weight; therefore, two-dimensional OOC codes that use the wavelength-time dimensions were proposed and their performance analysis and construction methods were thoroughly investigated [7, 11]. Here in this chapter we also review a few of them.

The OOC construction methodologies seem dazzling and complicated, including projective geometry, greedy algorithm, iterative construction, algebraic coding and the combinatorial method [4]. Generally, the major advantages of combinatorial construction are to generate the optimal codeset with as low as possible cross-correlation providing the reasonable cardinality under the predefined algorithm, and greatly simplify the processes of generating signature codes without any further optimal code extraction. However, the increasing complexity of hardware design not only gives the practical difficulties but also becomes incredibly expensive. The OOC $(N, w, \lambda)$, when $\lambda$ denotes the largest cross-correlation value, reads the user-defined code-length $N$ and code-weight $w$ and then generates the corresponding sequence index terms of $\{a, b, c, d\}$ according to the cardinality of:

$$\phi(N, w, 1) = |C|_{(n,w,1)} = \left\lfloor \frac{N-1}{w(w-1)} \right\rfloor \tag{2.28}$$

Figure 2.8 shows the construction model of OOC based on the sequence index terms $\{a, b, c, d\}$ for example.

### *Example: OOC (N, 3, 1)*

According to the length constraint of OOC, the code generation of the OOC$(N, 3, 1)$ codeset can be segmented into two functional parts. When $N \leq 49$, the look-up table of sequence index is given in Table 2.1 determining the position of the ones in the sequences [4].

For example, $N = 31$ shows that $\phi(31, 3, 1) = \left\lfloor \frac{31-1}{3(3-1)} \right\rfloor = 5$ and sequence index $\{1, 4, 16\}$ means three ones are situated in first, fourth and sixteenth chips respectively: Table 2.2 shows some examples.

If the length of signature code $N > 49$, the combinatorial method can generate the optimal sequence indices. However, there is a limitation of the user-defined input $N$ rather than $\mod(N, 6) = 2$; note that here $w(w-1) = 6$ where $w = 3$. There are four cases to

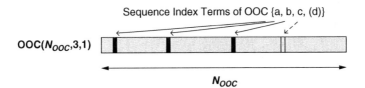

Sequence Index Terms of OOC {a, b, c, (d)}

OOC($N_{OOC}$,3,1)

$N_{OOC}$

**Figure 2.8**  Construction model of OOC family

**Table 2.1**  OOC $(N, 3, 1)$ sequence indexes for various lengths

| $N$ | Sequence Index, when $N \leq 49$ |
|---|---|
| 7 | $\{1, 2, 4\}$ |
| 13 | $\{1, 2, 5\}\{1, 3, 8\}$ |
| 19 | $\{1, 2, 6\}\{1, 3, 9\}\{1, 4, 11\}$ |
| 25 | $\{1, 2, 7\}\{1, 3, 10\}\{1, 4, 12\}\{1, 5, 14\}$ |
| 31 | $\{1, 2, 8\}\{1, 3, 12\}\{1, 4, 16\}\{1, 5, 15\}\{1, 6, 14\}$ |
| 37 | $\{1, 2, 12\}\{1, 3, 10\}\{1, 4, 18\}\{1, 5, 13\}\{1, 6, 19\}\{1, 7, 13\}$ |
| 43 | $\{1, 2, 20\}\{1, 3, 23\}\{1, 4, 16\}\{1, 5, 14\}\{1, 6, 17\}\{1, 7, 15\}\{1, 8, 19\}$ |

**Table 2.2**  OOC (31, 3, 1) sequences

| Index | Sequence Codes |
| --- | --- |
| $\{1,2,8\}$ | 1 1 0 0 0 0 0 1 0 0 0 0 0 0 0 0 0 0 0 0 0 0 0 0 0 0 0 0 0 0 0 |
| $\{1,3,12\}$ | 1 0 1 0 0 0 0 0 0 0 0 1 0 0 0 0 0 0 0 0 0 0 0 0 0 0 0 0 0 0 0 |
| $\{1,4,16\}$ | 1 0 0 1 0 0 0 0 0 0 0 0 0 0 0 1 0 0 0 0 0 0 0 0 0 0 0 0 0 0 0 |
| $\{1,5,15\}$ | 1 0 0 0 1 0 0 0 0 0 0 0 0 0 1 0 0 0 0 0 0 0 0 0 0 0 0 0 0 0 0 |
| $\{1,6,14\}$ | 1 0 0 0 0 1 0 0 0 0 0 0 0 1 0 0 0 0 0 0 0 0 0 0 0 0 0 0 0 0 0 |

calculate for the sequence indices based on the value of mod($t$,4) and $N = 6t + r$, where $r$ is the remainder, i.e. $1 < r < 6$. In each case, the additional codeset is summarized and padded into the main group [4]:

**Case 1:** mod$(t,4) = 0, t = 4k \geq 8$
$\{1, 4k + i + 1, 8k\}$ $(1 \leq i \leq 2k - 1)$
$\{1, 8k + i, 12k - i + 1\}$ $(1 \leq i \leq k)$
$\{1, 9k + i + 2, 18k - i + 1\}$ $(1 \leq i \leq k - 2)$
And additional sequence codes are:
$\{1, 6k + 1, 10k + 1\}, \{1, 9k + 1, 9k + 2\}, \{1, 10k + 2, 12k + 1\}$

**Case 2:** mod$(t,4) = 1, t = 4k + 1 \geq 9$
$\{1, t + i + 1, 2t + 2 - i\}$ $(1 \leq i \leq 2k)$
$\{1, 2t + i + 1, 3t - i + 1\}$ $(1 \leq i \leq k)$
$\{1, 2t + k + 3 + i, 3t - k - i + 1\}$ $(1 \leq i \leq k - 2)$
And additional sequence codes are:
$\{1, t + 2k + 2, 2t + 2k + 2\}, \{1, 2t + k + 2, 2t + k + 3\}, \{1, 2t + 2k + 3, 3t + 1\}$

**Case 3:** mod$(t,4) = 2, t = 4k + 2 \geq 6$ and mod$(k,6) \neq 2$
$\{1, t + i + 1, 2t\}$ $(1 \leq i \leq t/2 - 1)$
$\{1, 2t - i + 2, 3t - i + 1\}$ $(1 \leq i \leq k)$
$\{1, 2t + k + i + 2, 3t - k - i + 1\}$ $(1 \leq i \leq k - 1)$
And additional sequence codes are:
$\{1, 3k/2 + 1, 3k/2 + 1\}, \{1, 2t + k + 1, 2t + k + 2\}, \{1, 5t/2 + 2, 3t + 2\}$

**Case 4:** mod$(t,4) = 3$ and mod$(r,6) \neq 2$
$\{1, t + i + 1, 2t + 2 - i\}$ $(1 \leq i \leq 2k + 1)$
$\{1, 2t + i + 1, 3t - i + 1\}$ $(1 \leq i \leq k + 1)$
$\{1, 2t + k + 4 + i, 3t - k - i\}$ $(1 \leq i \leq k - 2)$
And additional sequence codes are:
$\{1, t + 2k + 3, 2t + 2k + 3\}, \{1, 2t + k + 3, 3t + 4\}, \{1, 2t + 2k + 4, 3t + 2\}$

For example, for $\phi(50, 3, 1) = \left\lfloor \frac{50-1}{3(3-1)} \right\rfloor = 8$ the first, fourth and eighth sequences have the indices of $\{1, 13, 21\}, \{1, 10, 16\}$ and $\{1, 18, 23\}$, respectively, as shown in Table 2.3.

As seen, the code construction is complicated, and there are many different algorithms for code generation in the literature [4–11]. Basically, as long as the codeset correlation constraints are met, the algorithm that can generate unipolar sequences can be assumed to be an OOC codeset. OOC sequences are well-established in the asynchronous

**Table 2.3** OOC (50, 3, 1) code sequences

| No. | Index | Sequence Codes |
|-----|-------|----------------|
| 1 | {1, 13, 21} | 1 0 0 0 0 0 0 0 0 0 0 0 1 0 0 0 0 0 0 0 1 0 0 0 0 0 0 0 0 0 0 0 0<br>0 0 0 0 0 0 0 0 0 0 0 0 0 0 0 0 0 |
| 4 | {1, 10, 16} | 1 0 0 0 0 0 0 0 0 1 0 0 0 0 0 1 0 0 0 0 0 0 0 0 0 0 0 0 0 0 0 0 0<br>0 0 0 0 0 0 0 0 0 0 0 0 0 0 0 0 0 |
| 8 | {1, 18, 23} | 1 0 0 0 0 0 0 0 0 0 0 0 0 0 0 0 0 1 0 0 0 0 1 0 0 0 0 0 0 0 0 0 0<br>0 0 0 0 0 0 0 0 0 0 0 0 0 0 0 0 0 |

OCDMA due to their correlation properties. However, limited number of users will be accommodated under this scheme and complex medium access control is required as will be discussed in Chapter 9.

Although synchronization is beyond the scope of this book, it is important to point briefly to some of the works done on the topic. In the above, asynchronous operation was assumed, but synchronization of OOC systems will be a major requirement to introduce burst and packet-based systems. Also, performance degradation of OOC systems will be severe if synchronization is not maintained. A simple synchronization method was considered in [15] and more recently a multiple search method that reduces the mean synchronization time was studied and analysed in [16].

## 2.4   Unipolar Codes: Prime Code Families

From the practical point of view, the two previously mentioned primary goals of OCDMA system must be achieved: the receiver should correctly recognize the desired users' signals among the interfering signals, and more possible subscribers should be accommodated in the system. Therefore, according to the three desig conditions described in Section 2.3, the above two goals of OCDMA transceivers should be accomplished by employing suitable optical code sequences with the best orthogonal characteristics.

In the last few decades, various optical spreading sequences for OCDMA networks have been investigated and experimented with [4, 6, 11, 17–24]. In this section we focus on prime code families including prime codes (PC), modified prime codes (MPC), new-MPC (n-MPC), group padded MPC (GPMPC) and transposed MPC (T-MPC). Their construction methods and properties are studied in details in the following sections.

### 2.4.1   Prime Codes

Prime sequences were first proposed in [23], and were previously developed for an optical fibre network using asynchronous OCDMA with relaxed requirements. Compared with OOC, the generation process of prime codes is relatively simple. The construction of prime codes is divided into two steps.

First, the aim is to build a set of prime sequences $S_x = (S_{x_0}, S_{x_1}, \ldots, S_{x_j}, \ldots, S_{x_{P-1}}) S_{x_j}$, from the Galois Field, $GF(P) = (0, 1, 2, \ldots, j, \ldots, P-1)$ where $P \geq 3$ is a prime number. This $S_x$ sequence could be obtained by multiplying every single element of $x, j \in GF(P) \leftrightarrow x, j \in \{0, 1 \ 2, \ldots, P-1\}$ modulo-$P$ as in Equation (2.29). Hence, the number of $P$ distinct prime sequences $S_x$ will be derived. The elements of prime sequences $S_{x_j}$ are given by:

$$S_{xj} = (x \times j)\mathrm{mod}(P) \tag{2.29}$$

where $x, j \in \{0, 1\ 2, \ldots, P - 1\}$.

Second, each prime sequence $S_x$ is mapped into binary sequences $C_x = (C_{x_0}, C_{x_1}, \ldots, C_{x_j}, \ldots, C_{x_{p^2-1}})$ with code-length $P^2$ according to the following rule:

$$C_{x_i} = \begin{cases} 1 & i = S_{x_j} + jP,\ j = 0, 1, \ldots, P - 1 \\ 0 & \text{or else} \end{cases} \tag{2.30}$$

Therefore, one set of prime code sequences could be accomplished by using the above approach.

An example of a prime code codeset where $P = 5$ is listed in Table 2.4 to clearly interpret the development process of the sequences.

It is apparent from Table 2.4 that the codeset with code-length $P^2$ and code-weight $P$ has $P$ distinct sequences.

The auto- and cross-correlation functions for any pair of code sequences $C_n$ and $C_m$ with discrete format are:

$$C_n \cdot C_m = \begin{cases} P & m = n;\ \text{Auto–Correlation} \\ 1 & m \neq n;\ \text{Cross–Correlation} \end{cases} \tag{2.31}$$

where $m, n \in \{1\ 2, \ldots, P\}$.

It can be easily seen from the above correlation function that the peak value of auto-correlation is $P$, which happens when $m = n$. Also, the cross-correlation value '1' occurs at each synchronized time $T$.

Figure 2.9 displays the auto-correlation values of prime code, $S_3$ and the peak values equals 5, as expected, at every synchronized time $T$.

Meanwhile, the cross-correlation values of prime codes $S_3$ and $S_1$ for the same data stream ($P = 5$) is illustrated in the Figure 2.10. The peak value of cross-correlation function is bounded by '1' at each synchronized time $T$ when the signal follows the data stream 10101 as illustrated in the figures. As seen from the Figures 2.9 and 2.10, and also throughout this chapter, since the data bits are encoded (i.e. multiplied by the spreading codes) throughout the bit duration $T$, the final result becomes available at the end of the bit duration as the synchronization occurs. Thus, here we see the actual value of the bits

**Table 2.4** Prime code sequences when $P = 5$

| Groups $x$ | $i$ 0 1 2 3 4 | Sequence | PC Sequences |
|---|---|---|---|
| 0 | 0 0 0 0 0 | $S_0$ | $C_0 = 10000\ 10000\ 10000\ 10000\ 10000$ |
| 1 | 0 1 2 3 4 | $S_1$ | $C_1 = 10000\ 01000\ 00100\ 00010\ 00001$ |
| 2 | 0 2 4 1 3 | $S_2$ | $C_2 = 10000\ 00100\ 00001\ 01000\ 00010$ |
| 3 | 0 3 1 4 2 | $S_3$ | $C_3 = 10000\ 00010\ 01000\ 00001\ 00100$ |
| 4 | 0 4 3 2 1 | $S_4$ | $C_4 = 10000\ 00001\ 00010\ 00100\ 01000$ |

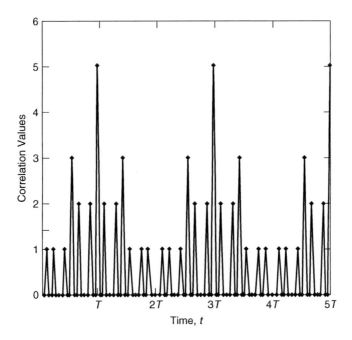

**Figure 2.9**  Auto-correlation values of prime codes of $S_3$, where $P = 5$ following the data stream of 10101

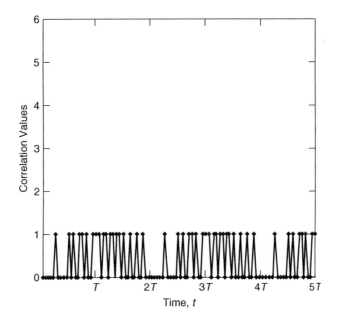

**Figure 2.10**  Cross-correlation values of prime codes of $S_3$ and $S_1$, where $P = 5$ following the data stream of 10101

(zero or one) at every synchronized time $T$, depending on the correlation value too, where other values within the bit duration, i.e. $nT < t < (n + 1)T$ are not considered.

The major drawback associated with prime code sequences is the limited number of available codes. Therefore, the corresponding codeset cardinality to support subscribers would be insufficient in the probable network design based on prime codes.

## 2.4.2  Modified Prime Codes (MPC)

In order to overcome the limitation of prime codes, a modified version of prime codes, named the modified prime code (MPC), has been introduced [23]. These optical sequences have the ability to support more users simultaneously transmitted in the system with the lower multiple access interference (MAI).

An MPC codeset can be achieved through $P - 1$ times shifting of the pervious prime codes. Therefore, the available number of signature code sequences can be extended to $P^2$ with $P$ groups of $P$ codes, where $P \geq 3$ is a prime number [25].

Firstly, the original prime code generator $S_{xj}$ is left (or right) rotated. Then, the new time-shifted sequence $S_{xt} = (S_{x_{t0}}, S_{x_{t1}}, \ldots, S_{x_{tj}}, \ldots, S_{x_{t(P-1)}})$ is obtained in terms of the following function, where $t$ denotes the number of times as $S_x$ has been left (or right) rotated [23]. Hence, this method can result in a significant increase in the number of possible subscribers. The same mapping technique is used to produce binary sequences, as:

$$C_{x_i} = \begin{cases} 1 & i = S_{x_{ij}} + jP, \ j = 0, 1, \ldots, P - 1 \\ 0 & \text{or else} \end{cases} \tag{2.32}$$

Finally, by applying this method an MPC codeset is generated. Similarly, one example of MPC is exhibited in Table 2.5 which implies that the MPC sequences with code-length $P^2$ and code-weight $P$ has $P^2$ distinct code sequences. Hence, in the OCDMA system using MPC the number of possible subscribers is significantly extended up to $P^2$ which is a factor of $P$ larger than that of prime codes.

The auto- and cross-correlation functions for any pair of MPC codes $C_n$ and $C_m$ with discrete format are as follow [23]:

$$C_n \cdot C_m = \begin{cases} P & m = n \\ 0 & m \neq n, \ m \text{ and } n \text{ share same group} \\ 1 & m \neq n, \ m \text{ and } n \text{ share different groups} \end{cases} \tag{2.33}$$

where $m, n \in \{1, 2, \ldots, P^2\}$. Therefore, it should be noted that the cross-correlation value of two codes within the same group is strictly orthogonal. However, for sequences located in two different groups, the cross-correlation value is bounded by a one. Additionally, the auto-correlation peak value equals $P$.

Figure 2.11 illustrates the auto-correlation property of MPC sequence $S_{2,1}$ and the maximum peak value equals 5 at each synchronized time $T$ when the sequences follow the data stream 10101.

Similarly, the cross-correlation values of MPC sequences placed in the same group, e.g. $S_{2,0}$ and $S_{2,1}$, for the same data stream is displayed in Figure 2.12. As the figure shows, the value of cross-correlation function is equal to zero at each synchronized time $T$.

**Table 2.5** MPC sequences where $P = 5$

| Group $x$ | $i$<br>0 1 2 3 4 | Sequence | MPC Sequences |
|---|---|---|---|
| | 0 0 0 0 0 | $S_{0,0}$ | $C_{0,0} = 10000\ 10000\ 10000\ 10000\ 10000$ |
| | 4 4 4 4 4 | $S_{0,1}$ | $C_{0,1} = 00001\ 00001\ 00001\ 00001\ 00001$ |
| 0 | 3 3 3 3 3 | $S_{0,2}$ | $C_{0,2} = 00010\ 00010\ 00010\ 00010\ 00010$ |
| | 2 2 2 2 2 | $S_{0,3}$ | $C_{0,3} = 00100\ 00100\ 00100\ 00100\ 00100$ |
| | 1 1 1 1 1 | $S_{0,4}$ | $C_{0,4} = 01000\ 01000\ 01000\ 01000\ 01000$ |
| | 0 1 2 3 4 | $S_{1,0}$ | $C_{1,0} = 10000\ 01000\ 00100\ 00010\ 00001$ |
| | 1 2 3 4 0 | $S_{1,1}$ | $C_{1,1} = 01000\ 00100\ 00010\ 0000110000$ |
| 1 | 2 3 4 0 1 | $S_{1,2}$ | $C_{1,2} = 00100\ 00010\ 00001\ 10000\ 01000$ |
| | 3 4 0 1 2 | $S_{1,3}$ | $C_{1,3} = 00010\ 00001\ 10000\ 01000\ 00100$ |
| | 4 0 1 2 3 | $S_{1,4}$ | $C_{1,4} = 00001\ 10000\ 01000\ 00100\ 00010$ |
| | 0 2 4 1 3 | $S_{2,0}$ | $C_{2,0} = 10000\ 00100\ 00001\ 01000\ 00010$ |
| | 2 4 1 3 0 | $S_{2,1}$ | $C_{2,1} = 00100\ 00001\ 01000\ 00010\ 10000$ |
| 2 | 4 1 3 0 2 | $S_{2,2}$ | $C_{2,2} = 00001\ 01000\ 00010\ 10000\ 00100$ |
| | 1 3 0 2 4 | $S_{2,3}$ | $C_{2,3} = 01000\ 00010\ 10000\ 00100\ 00001$ |
| | 3 0 2 4 1 | $S_{2,4}$ | $C_{2,4} = 00010\ 10000\ 00100\ 00001\ 01000$ |
| | 0 3 1 4 2 | $S_{3,0}$ | $C_{3,0} = 10000\ 00010\ 01000\ 00001\ 00100$ |
| | 3 1 4 2 0 | $S_{3,1}$ | $C_{3,1} = 00010\ 01000\ 00001\ 00100\ 10000$ |
| 3 | 1 4 2 0 3 | $S_{3,2}$ | $C_{3,2} = 01000\ 00001\ 00100\ 10000\ 00010$ |
| | 4 2 0 3 1 | $S_{3,3}$ | $C_{3,3} = 00001\ 00100\ 10000\ 00010\ 01000$ |
| | 2 0 3 1 4 | $S_{3,4}$ | $C_{3,4} = 00100\ 10000\ 00010\ 01000\ 00001$ |
| | 0 4 3 2 1 | $S_{4,0}$ | $C_{4,0} = 10000\ 00001\ 00010\ 00100\ 01000$ |
| | 4 3 2 1 0 | $S_{4,1}$ | $C_{4,1} = 00001\ 00010\ 00100\ 01000\ 10000$ |
| 4 | 3 2 1 0 4 | $S_{4,2}$ | $C_{4,2} = 00010\ 00100\ 01000\ 10000\ 00001$ |
| | 2 1 0 4 3 | $S_{4,3}$ | $C_{4,3} = 00100\ 01000\ 10000\ 00001\ 00010$ |
| | 1 0 4 3 2 | $S_{4,4}$ | $C_{4,4} = 01000\ 10000\ 00001\ 00010\ 00100$ |

Finally, Figure 2.13 shows the cross-correlation values of $S_{1,0}$ and $S_{2,1}$ which are the MPC sequences within two different groups. In this case, the cross-correlation value is '1' at each synchronized time $T$ which clearly indicates a reduction in the cross-correlation value between these two code sequences and hence a lowering of the multiple access interference.

MPC sequences with a relative large number of available code sequences and lower cross-correlation values are identified as a suitable candidate, in particular for synchronous OCDMA due to their excellent in-phase correlation properties, to accommodate a greater number of subscribers under the same bandwidth expansion. On the other hand, as noticed, code-length plays a significant role in raising the system performance in terms of multiple access interference (MAI), bit-error rate (BER) and correlation properties. Therefore, investigating longer codes that maintain the desired properties can be beneficial, though at the cost of lower bit rate in the time-spreading schemes as will be studied in Chapters 3 and 5.

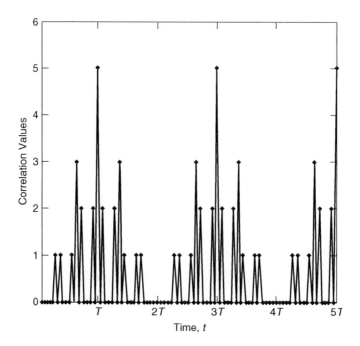

**Figure 2.11**  Auto-correlation values of MPC sequence of $S_{2,1}$, where $P = 5$ following the data stream of 10101

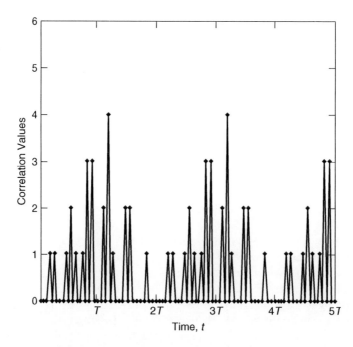

**Figure 2.12**  Cross-correlation values of MPC sequences of $S_{2,0}$ and $S_{2,1}$, where $P = 5$ within the same group, following the data stream of 10101

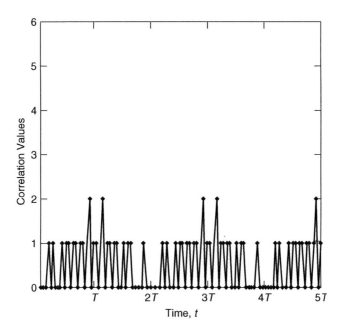

**Figure 2.13** Cross-correlation values of MPC sequences of $S_{1,0}$ and $S_{2,1}$, where $P = 5$ within the different groups, following data stream of 10101

### 2.4.3  The New-Modified Prime Code (n-MPC)

New-modified prime codes (n-MPC) have been introduced in [26, 27]. They are generated through repeating the last sequence-stream of the previous MPC sequence and rotating in the same group with the aid of a subsequence of length $P$. This code family, like the MPC codeset, has $P$ groups, each of which has $P$ codes. The length of each code is $P^2 + P$ and the weight is $P + 1$, where $P \geq 3$ is a prime number. Hence, the total number of available sequences is $P^2$. Table 2.6 shows an example of an n-MPC for $P = 5$.

The auto- and cross-correlation functions for any pair of n-MPC codes $C_n$ and $C_m$ with discrete format are as follow [26, 27]:

$$C_n \cdot C_m = \begin{cases} P + 1 & m = n \\ 0 & m \neq n, \ m \ and \ n \ share \ same \ group \\ \leq 2 & m \neq n, \ m \ and \ n \ share \ different \ groups \end{cases} \qquad (2.34)$$

where $m, n \in \{1, 2, \ldots, P^2\}$.

The auto- and cross-correlation properties of the n-MPC for the data stream 11010 and $P = 5$ are illustrated in Figures 2.14 to 2.16, respectively. Figure 2.14 depicts the auto-correlation values of n-MPC sequence, $S_{2,3}$, which has the value of $P + 1 = 6$ at each synchronized time $T$.

**Table 2.6**   n-MPC sequences where $P = 5$

| $x$ | $i$<br>0 1 2 3 4 | Seq. | MPC Sequences | Padded<br>Seq. |
|---|---|---|---|---|
| | 0 0 0 0 0 | $S_{0,0}$ | $C_{0,0} = 10000\ 10000\ 10000\ 10000\ \underline{\mathit{10000}}$ | ***01000*** |
| | 4 4 4 4 4 | $S_{0,1}$ | $C_{0,1} = 00001\ 00001\ 00001\ 00001\ 00001$ | *10000* |
| 0 | 3 3 3 3 3 | $S_{0,2}$ | $C_{0,2} = 00010\ 00010\ 00010\ 00010\ 00010$ | 00001 |
| | 2 2 2 2 2 | $S_{0,3}$ | $C_{0,3} = 00100\ 00100\ 00100\ 00100\ 00100$ | 00010 |
| | 1 1 1 1 1 | $S_{0,4}$ | $C_{0,4} = 01000\ 01000\ 01000\ 01000\ \mathbf{\mathit{01000}}$ | 00100 |
| | 0 1 2 3 4 | $S_{1,0}$ | $C_{1,0} = 10000\ 01000\ 00100\ 00010\ 00001$ | 00010 |
| | 1 2 3 4 0 | $S_{1,1}$ | $C_{1,1} = 01000\ 00100\ 00010\ 00001\ 10000$ | 00001 |
| 1 | 2 3 4 0 1 | $S_{1,2}$ | $C_{1,2} = 00100\ 00010\ 00001\ 10000\ 01000$ | 10000 |
| | 3 4 0 1 2 | $S_{1,3}$ | $C_{1,3} = 00010\ 00001\ 10000\ 01000\ 00100$ | 01000 |
| | 4 0 1 2 3 | $S_{1,4}$ | $C_{1,4} = 00001\ 10000\ 01000\ 00100\ 00010$ | 00100 |
| | 0 2 4 1 3 | $S_{2,0}$ | $C_{2,0} = 10000\ 00100\ 00001\ 01000\ 00010$ | 01000 |
| | 2 4 1 3 0 | $S_{2,1}$ | $C_{2,1} = 00100\ 00001\ 01000\ 00010\ 10000$ | 00010 |
| 2 | 4 1 3 0 2 | $S_{2,2}$ | $C_{2,2} = 00001\ 01000\ 00010\ 10000\ 00100$ | 10000 |
| | 1 3 0 2 4 | $S_{2,3}$ | $C_{2,3} = 01000\ 00010\ 10000\ 00100\ 00001$ | 00100 |
| | 3 0 2 4 1 | $S_{2,4}$ | $C_{2,4} = 00010\ 10000\ 00100\ 00001\ 01000$ | 00001 |
| | 0 3 1 4 2 | $S_{3,0}$ | $C_{3,0} = 10000\ 00010\ 01000\ 00001\ 00100$ | 00001 |
| | 3 1 4 2 0 | $S_{3,1}$ | $C_{3,1} = 00010\ 01000\ 00001\ 00100\ 10000$ | 00100 |
| 3 | 1 4 2 0 3 | $S_{3,2}$ | $C_{3,2} = 01000\ 00001\ 00100\ 10000\ 00010$ | 10000 |
| | 4 2 0 3 1 | $S_{3,3}$ | $C_{3,3} = 00001\ 00100\ 10000\ 00010\ 01000$ | 00010 |
| | 2 0 3 1 4 | $S_{3,4}$ | $C_{3,4} = 00100\ 10000\ 00010\ 01000\ 00001$ | 01000 |
| | 0 4 3 2 1 | $S_{4,0}$ | $C_{4,0} = 10000\ 00001\ 00010\ 00100\ 01000$ | 00100 |
| | 4 3 2 1 0 | $S_{4,1}$ | $C_{4,1} = 00001\ 00010\ 00100\ 01000\ 10000$ | 01000 |
| 4 | 3 2 1 0 4 | $S_{4,2}$ | $C_{4,2} = 00010\ 00100\ 01000\ 10000\ 00001$ | 10000 |
| | 2 1 0 4 3 | $S_{4,3}$ | $C_{4,3} = 00100\ 01000\ 10000\ 00001\ 00010$ | 00001 |
| | 1 0 4 3 2 | $S_{4,4}$ | $C_{4,4} = 01000\ 10000\ 00001\ 00010\ 00100$ | 00010 |

Meanwhile, Figure 2.15 plots the cross-correlation values of $S_{1,1}$ and $S_{1,3}$ n-MPC sequences which are within the same group which are zero at every synchronized time $T$ and Figure 2.16 shows the same for $S_{3,2}$ and $S_{2,2}$ n-MPC sequences which are from different groups. As this figure shows the cross-correlation values are less than or equal to two at every synchronous instants of $T$.

The conclusion that we can draw from the above analysis is that increasing the code-length can enhance the correlation properties and hence improve the OCDMA system performance in the detection process. However, the code-length is also a trade-off between the network performance and the throughput due to longer code and lower data rate in the time-spreading scheme. Therefore, as we will discuss later in this chapter, the code-length cannot increase infinitely because of the power and processing time restrictions of both encoder/decoder.

In principle, the major degrading factors which affect the overall performance of OCDMA networks are the auto-correlation, $\lambda_a$, cross-correlation, $\lambda_c$ and multiple access

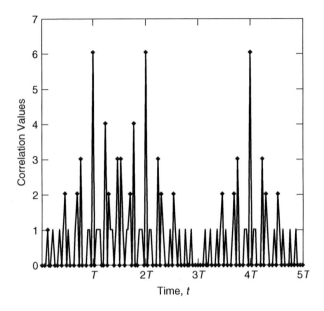

**Figure 2.14**   Auto-correlation values of n-MPC sequence of $S_{2,3}$, where $P = 5$ following the data stream of 11010

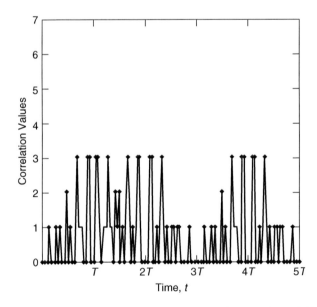

**Figure 2.15**   Cross-correlation values of n-MPC sequences $S_{1,1}$ and $S_{1,3}$ within the same group where $P = 5$, following the data stream of 11010

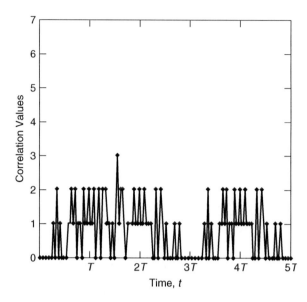

**Figure 2.16**  Cross-correlation values of n-MPC sequences of $S_{3,2}$ and $S_{2,2}$, within the different groups where $P = 5$, following the data stream of 11010

interference (MAI) constraints. Note that the cross-correlation value implies the interference between users. Accordingly, to reduce the effect of MAI, codes with minimum cross-correlation value $\lambda_c$ are preferred for OCDMA communications. On the other hand, the auto-correlation value implies the signal power and the distinction in the presence of other users. Thus, codes with maximum auto-correlation value $\lambda_a$ are desirable too. These issues limit the capacity of OCDMA networks for a given power budget and system cost.

To compare OOCs ($\lambda_c = 1$) with prime code families $\lambda_c \leq 2$, prime codes are still preferable. To compare OOCs ($\lambda_c = 1$) with prime code families $\lambda_c \leq 2$, prime codes are still preferable, due to the fact that prime codes encoder/decoder tuneability can be simply implemented by tuneable optical tapped-delay lines (OTDL) for time-spreading applications for example. Also, since the prime codes enjoy the equal-length 'subsequences' in each codeword, both encoding power and transmitter cost during the encoding/decoding process will be reduced. To further reduce the cost and power, low-weight prime codes which have symmetrically distributed 'pulses' in any codeword can be considered and constructed [28].

## 2.4.4  Padded Modified Prime Codes

In the past two decades much effort has been devoted to investigating new optical address codes which have optimum orthogonal properties or ideal-in-phase cross-correlation value. The motivations in constructing new optical code families for the OCDMA applications come from the purpose of correctly distinguishing the desired users' signals in the presence of other users' interfering signals. Therefore, another optical spreading code family based on prime codes has been introduced as padded modified prime code (PMPC) in [24, 29].

**Table 2.7** PMPC sequences where $P = 5$

| $x$ | $i$<br>0 1 2 3 4 | Seq. | MPC Sequences | Padded<br>Seq. |
|---|---|---|---|---|
| **0** | 0 0 0 0 0 | $S_{0,0}$ | $C_{0,0}$ = 10000 10000 10000 10000 10000 | **10000** |
|  | 4 4 4 4 4 | $S_{0,1}$ | $C_{0,1}$ = 00001 00001 00001 00001 00001 | **10000** |
|  | 3 3 3 3 3 | $S_{0,2}$ | $C_{0,2}$ = 00010 00010 00010 00010 00010 | **10000** |
|  | 2 2 2 2 2 | $S_{0,3}$ | $C_{0,3}$ = 00100 00100 00100 00100 00100 | **10000** |
|  | 1 1 1 1 1 | $S_{0,4}$ | $C_{0,4}$ = 01000 01000 01000 01000 01000 | **10000** |
| 1 | 0 1 2 3 4 | $S_{1,0}$ | $C_{1,0}$ = 10000 01000 00100 00010 00001 | 01000 |
|  | 1 2 3 4 0 | $S_{1,1}$ | $C_{1,1}$ = 01000 00100 00010 00001 10000 | 01000 |
|  | 2 3 4 0 1 | $S_{1,2}$ | $C_{1,2}$ = 00100 00010 00001 10000 01000 | 01000 |
|  | 3 4 0 1 2 | $S_{1,3}$ | $C_{1,3}$ = 00010 00001 10000 01000 00100 | 01000 |
|  | 4 0 1 2 3 | $S_{1,4}$ | $C_{1,4}$ = 00001 10000 01000 00100 00010 | 01000 |
| 2 | 0 2 4 1 3 | $S_{2,0}$ | $C_{2,0}$ = 10000 00100 00001 01000 00010 | 00100 |
|  | 2 4 1 3 0 | $S_{2,1}$ | $C_{2,1}$ = 00100 00001 01000 00010 10000 | 00100 |
|  | 4 1 3 0 2 | $S_{2,2}$ | $C_{2,2}$ = 00001 01000 00010 10000 00100 | 00100 |
|  | 1 3 0 2 4 | $S_{2,3}$ | $C_{2,3}$ = 01000 00010 10000 00100 00001 | 00100 |
|  | 3 0 2 4 1 | $S_{2,4}$ | $C_{2,4}$ = 00010 10000 00100 00001 01000 | 00100 |
| 3 | 0 3 1 4 2 | $S_{3,0}$ | $C_{3,0}$ = 10000 00010 01000 00001 00100 | 00010 |
|  | 3 1 4 2 0 | $S_{3,1}$ | $C_{3,1}$ = 00010 01000 00001 00100 10000 | 00010 |
|  | 1 4 2 0 3 | $S_{3,2}$ | $C_{3,2}$ = 01000 00001 00100 10000 00010 | 00010 |
|  | 4 2 0 3 1 | $S_{3,3}$ | $C_{3,3}$ = 00001 00100 10000 00010 01000 | 00010 |
|  | 2 0 3 1 4 | $S_{3,4}$ | $C_{3,4}$ = 00100 10000 00010 01000 00001 | 00010 |
| 4 | 0 4 3 2 1 | $S_{4,0}$ | $C_{4,0}$ = 10000 00001 00010 00100 01000 | 00001 |
|  | 4 3 2 1 0 | $S_{4,1}$ | $C_{4,1}$ = 00001 00010 00100 01000 10000 | 00001 |
|  | 3 2 1 0 4 | $S_{4,2}$ | $C_{4,2}$ = 00010 00100 01000 10000 00001 | 00001 |
|  | 2 1 0 4 3 | $S_{4,3}$ | $C_{4,3}$ = 00100 01000 10000 00001 00010 | 00001 |
|  | 1 0 4 3 2 | $S_{4,4}$ | $C_{4,4}$ = 01000 10000 00001 00010 00100 | 00001 |

PMPC is also constructed through regularly padding the MPC sequences with the aid of a subsequence that is $P$ long. Therefore, the code-length will extend to $P^2 + P$ and the code-weight also jumps up to $P + 1$, where $P \geq 3$ is a prime number. To generate the PMPC, the sequence-stream representing each group is padded to the MPC sequences of the same group. Table 2.7 illustrates an example of PMPC generation where $P = 5$.

The auto- and cross-correlation functions for any pair of PMPC codes $C_n$ and $C_m$ with discrete format can be expressed as [24, 29]:

$$C_n \cdot C_m = \begin{cases} P + 1 & m = n; \text{Auto–Correlation} \\ 1 & m \neq n; \text{Cross–Correlation} \end{cases} \qquad (2.35)$$

where $m, n \in \{1, 2, \ldots, P^2\}$.

The above correlation function shows that the PMPC sequences have a uniform cross-correlation property that is equal to one whereas the auto-correlation value is now $P + 1$. The correlation values of PMPC sequences are plotted in Figures 2.17 and 2.18.

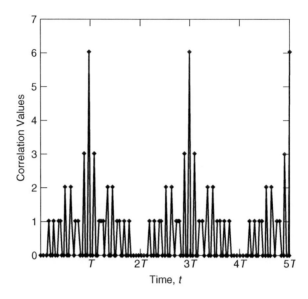

**Figure 2.17**   Auto-correlation values of PMPC sequence of $S_{1,1}$, where $P = 5$ following the data stream of 10101

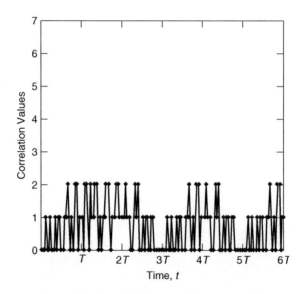

**Figure 2.18**   Cross-correlation values of PMPC sequences $S_{1,1}$ and $S_{2,3}$ where $P = 5$ following the data stream of 10101

Figure 2.17 plots the auto-correlation value of PMPC sequence $S_{1,1}$. It is apparent that the highest peak value is six as expected following the data stream 10101. Meanwhile, Figure 2.18 demonstrates the cross-correlation values of $S_{1,1}$ and $S_{2,3}$ of PMPC sequences for the same data stream. However, as the figure clearly shows, in most cases the cross-correlation values are equal to one. Consequently, PMPC sequences with low cross-correlation values are appropriate for direct-detect and spectral-amplitude-coding (SAC) OCDMA systems to limit the multiple-user interference (MUI), and as will be discussed in Section 2.5 and Chapter 4, codes with the uniform minimum ideal in-phase cross-correlation value of one are very desirable for SAC-OCDMA to cancel the MUI.

## 2.4.5   Group Padded Modified Prime Code (GPMPC)

Group padded modified prime code (GPMPC) is simply generated by padding the two last sequence-streams of the previous MPC sequence where rotating in the same group. It should be noted that the padding order should be maintained during the procedure, otherwise the cross-correlation value will increase. Finally, the two sequences are padded into each MPC sequence and consequently the code enlarges by $2P$ as compared to MPC sequences and by $P$ as compared to n-MPC and PMPC. This implies an increase in the chip-rate (processing gain in spreading) which makes this prime code family more secure, i.e. less or no interception, and also permits the OCDMA system to mitigate the MUI as will be explained in detail in Chapter 5. It is important to note that the padded sequences cannot only be the two final sequence-streams of the MPC but they can also be any two sequence-streams of the MPC sequence. This is due to the uniqueness of each MPC sequence-stream that makes each code unique. This prime code family also has $P$ groups, each of which has $P$ codes. The length of each code is $P^2 + 2P$ and the weight is $P + 2$ with the total number of available sequences of $P^2$.

Table 2.8 shows an example of the GPMPC for $P = 5$. Table 2.8 shows that each code consists of two parts, MPC and group sequence-stream (GSS) parts. For example, for code $S_{0,0}$ the MPC part is '10000 10000 10000 10000 10000' and its GSS part is '01000 01000' which are the two last sequence-stream of $S_{0,4}$ in the same group. Similarly, by padding the two last sequence-streams of $S_{0,0}$ MPC which are '10000 10000' to $S_{0,1}$ MPC sequence, GPMPC of $S_{0,1}$ is generated.

The auto- and cross-correlation functions for any pair of GPMPC codes $C_n$ and $C_m$ with discrete format are seen as:

$$C_n \cdot C_m = \begin{cases} P + 2 & m = n \\ 0 & m \neq n, \ m \text{ and } n \text{ share same group} \\ \leq 2 & m \neq n, \ m \text{ and } n \text{ share different groups} \end{cases} \qquad (2.36)$$

where $m, n \in \{1, 2, \ldots, P^2\}$.

Figure 2.19 plots the auto-correlation function for $S_{4,0}$ GPMPC sequence, indicating $P + 2 = 5 + 2 = 7$ as the in-phase maximum auto-correlation value when following data stream of 11010. Figure 2.20 shows the correlation function, Equation (2.36), when codes $S_{1,0}$ and $S_{1,2}$ hit each other at each synchronous time $T$ (i.e. in-phase cross-correlation) and generate a zero value as they share the same group.

The cross-correlation values of $S_{3,2}$ and $S_{4,0}$ are depicted in Figure 2.21. It can be seen that the in-phase cross-correlation value at every synchronized time $T$ becomes one as the

**Table 2.8**  GPMPC sequences where $P = 5$

| $x$ | $i$<br>0 1 2 3 4 | Seq. | MPC Part | GSS Part |
|---|---|---|---|---|
|   | 0 0 0 0 0 | $S_{0,0}$ | $C_{0,0} = 10000\ 10000\ 10000\ \mathbf{10000\ 10000}$ | ***01000 01000*** |
|   | 4 4 4 4 4 | $S_{0,1}$ | $C_{0,1} = 00001\ 00001\ 00001\ 00001\ 00001$ | **10000 10000** |
| 0 | 3 3 3 3 3 | $S_{0,2}$ | $C_{0,2} = 00010\ 00010\ 00010\ 00010\ 00010$ | 00001 00001 |
|   | 2 2 2 2 2 | $S_{0,3}$ | $C_{0,3} = 00100\ 00100\ 00100\ 00100\ 00100$ | 00010 00010 |
|   | 1 1 1 1 1 | $S_{0,4}$ | $C_{0,4} = 01000\ 01000\ 01000\ \mathit{01000\ 01000}$ | 00100 00100 |
|   | 0 1 2 3 4 | $S_{1,0}$ | $C_{1,0} = 10000\ 01000\ 00100\ 00010\ 00001$ | 00100 00010 |
|   | 1 2 3 4 0 | $S_{1,1}$ | $C_{1,1} = 01000\ 00100\ 00010\ 00001\ 10000$ | 00010 00001 |
| 1 | 2 3 4 0 1 | $S_{1,2}$ | $C_{1,2} = 00100\ 00010\ 00001\ 10000\ 01000$ | 00001 10000 |
|   | 3 4 0 1 2 | $S_{1,3}$ | $C_{1,3} = 00010\ 00001\ 10000\ 01000\ 00100$ | 10000 01000 |
|   | 4 0 1 2 3 | $S_{1,4}$ | $C_{1,4} = 00001\ 10000\ 01000\ 00100\ 00010$ | 01000 00100 |
|   | 0 2 4 1 3 | $S_{2,0}$ | $C_{2,0} = 10000\ 00100\ 00001\ 01000\ 00010$ | 00001 01000 |
|   | 2 4 1 3 0 | $S_{2,1}$ | $C_{2,1} = 00100\ 00001\ 01000\ 00010\ 10000$ | 01000 00010 |
| 2 | 4 1 3 0 2 | $S_{2,2}$ | $C_{2,2} = 00001\ 01000\ 00010\ 10000\ 00100$ | 00010 10000 |
|   | 1 3 0 2 4 | $S_{2,3}$ | $C_{2,3} = 01000\ 00010\ 10000\ 00100\ 00001$ | 10000 00100 |
|   | 3 0 2 4 1 | $S_{2,4}$ | $C_{2,4} = 00010\ 10000\ 00100\ 00001\ 01000$ | 00100 00001 |
|   | 0 3 1 4 2 | $S_{3,0}$ | $C_{3,0} = 10000\ 00010\ 01000\ 00001\ 00100$ | 01000 00001 |
|   | 3 1 4 2 0 | $S_{3,1}$ | $C_{3,1} = 00010\ 01000\ 00001\ 00100\ 10000$ | 00001 00100 |
| 3 | 1 4 2 0 3 | $S_{3,2}$ | $C_{3,2} = 01000\ 00001\ 00100\ 10000\ 00010$ | 00100 10000 |
|   | 4 2 0 3 1 | $S_{3,3}$ | $C_{3,3} = 00001\ 00100\ 10000\ 00010\ 01000$ | 10000 00010 |
|   | 2 0 3 1 4 | $S_{3,4}$ | $C_{3,4} = 00100\ 10000\ 00010\ 01000\ 00001$ | 00010 01000 |
|   | 0 4 3 2 1 | $S_{4,0}$ | $C_{4,0} = 10000\ 00001\ 00010\ 00100\ 01000$ | 00010 00100 |
|   | 4 3 2 1 0 | $S_{4,1}$ | $C_{4,1} = 00001\ 00010\ 00100\ 01000\ 10000$ | 00100 01000 |
| 4 | 3 2 1 0 4 | $S_{4,2}$ | $C_{4,2} = 00010\ 00100\ 01000\ 10000\ 00001$ | 01000 10000 |
|   | 2 1 0 4 3 | $S_{4,3}$ | $C_{4,3} = 00100\ 01000\ 10000\ 00001\ 00010$ | 10000 00001 |
|   | 1 0 4 3 2 | $S_{4,4}$ | $C_{4,4} = 01000\ 10000\ 00001\ 00010\ 00100$ | 00001 00010 |

sequences are not sharing the same group as seen from Table 2.8. Note that the maximum in-phase cross-correlation value of two codes from different groups does not exceed two as also shown in Figure 2.22, for another example of $S_{3,3}$ and $S_{4,4}$ following the data stream 11010. As mentioned, the correlation values at the synchronized bit duration $T$ are considerable, and the values within the bit duration i.e. $nT < t < (n + 1)T$ are negligible.

The GPMPC expands the code-length by retaining the excellent correlation properties of the prime code families. This attribute helps the OCDMA stay more secure as well. It has been noted that the code-length is an important feature of the spreading codeset that can raise the system performance by reducing multiple access interference (MAI) and consequently the error rate. On the other hand, Figure 2.19 shows, the auto-correlation peak has increased which accordingly enhances the difference between the auto- and cross-correlation values of GPMPC sequences as compared with those of the other prime code families. This feature assists the detection process significantly and reduces MAI remarkably.

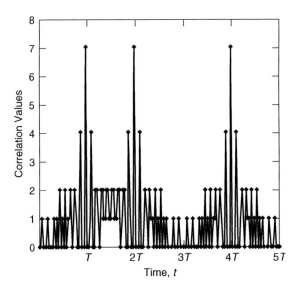

**Figure 2.19** Auto-correlation values of GPMPC sequence of $S_{4,0}$ where $P = 5$ following the data stream of 11010

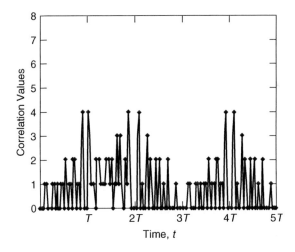

**Figure 2.20** Cross-correlation values of GPMPC sequences $S_{1,0}$ and $S_{1,2}$ within the same group where $P = 5$, following the data stream of 11010

### 2.4.6 Transposed Modified Prime Codes

As observed with the previously introduced prime code families, the codes are focused on the code-length increment to optimize their correlation function. However, the code-length cannot be increased indefinitely and practical implementations must be considered at the same time as the code design process. Another important parameter of code design

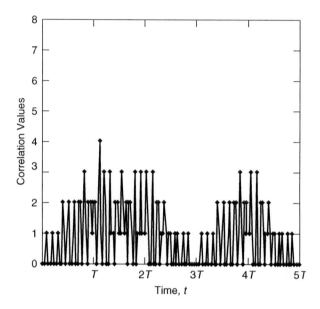

**Figure 2.21** Cross-correlation values of GPMPC sequences of $S_{3,2}$ and $S_{4,0}$, within the different groups where $P = 5$, following the data stream of 11010

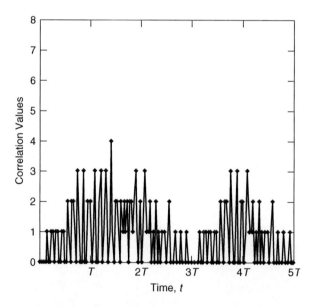

**Figure 2.22** Cross-correlation values of GPMPC sequences of $S_{3,3}$ and $S_{4,4}$, within the different groups where $P = 5$, following the data stream of 11010

is to maximize the number of available sequences (cardinality) in the code family. Here, a novel code design technique is introduced as it optimally increases the code cardinality of prime codes. The technique has been applied to full-padded modified prime code to generate novel transposed-MPC (T-MPC). Its construction is as follows [30].

First, full-padded MPC sequences are generated $P$ times, concatenating the MPC sequences with the last sequence-streams of the previous MPC sequence in the same group. In every step, one column is produced and padded to MPC sequences similar to the group-padded MPC introduced in Section 2.4.5, and the process continues with new padded MPC. This implies that the last sequence-stream of the previously padded MPC sequence is again padded to the next newly padded MPC sequence. Note that the padding process rotates in the same group, meaning that the last sequence-stream of the last MPC sequences is padded to the first MPC sequence in the same group and not continuing throughout the codeset. This padding procedure repeats only $P - 1$ times, because the $P^{th}$ time will reproduce the last sequence-stream column of MPC itself. Table 2.9 illustrates an example of full-padded MPC when $P = 3$. According to Table 2.9, for example $C_{1,2}$ consists of 00100 00010 00001 10000 01000 as MPC sequences and 10000 00001 00010 00100 01000 as full-padded sequences. It should be noted that the 10000 sequence-stream in full-padded sequences is the last sequence-stream of previous $C_{1,1}$. As we can see, the padding is performed column by column. Moreover, the next sequence-stream (i.e. 00001) in full-padded sequences is the last sequence-stream of padded $C_{1,1}$, and this diagonal padding continues.

It is noted that padded sequences are not restricted to the final sequence-stream column of MPC; they can be any sequence-stream column of MPC sequences. This is due to the uniqueness of each MPC sequence-stream that makes each padding-stream distinctive. However, it is important to maintain the column's order throughout the padding process, otherwise this would increase the cross-correlation values. Since the diagonal padding procedure only repeats $P - 1$ times, for the $P^{th}$ sequence-stream the group sequence-stream is padded to the sequences similar to padded-MPC introduced in Section 2.4.4 in the same group to finalize the full-padded MPC sequences.

Now the full-padded MPC has $2P$ code-weight, $P^2$ available sequences, with code-length of $2P^2$. The auto- and cross-correlation values for any pair of codes $m$ and $n$ is given by:

$$C_{mn} = \begin{cases} 2P & m = n \\ 1 & m \neq n; \text{ same groups} \\ \leq P & m \neq n; \text{ different groups} \end{cases} \qquad (2.37)$$

where $m, n \in \{1, 2, \ldots, P^2\}$. It is apparent that this full-padded codeset suffers from increased code-length and cross-correlation value as well as loses the orthogonal sequences to compare with existing prime code families [23, 24, 27].

Secondly, the unique lemma is now to treat the full-padded MPC sequences, in Table 2.9 for example, as a matrix in which every chip (i.e. every zero and one) representing the full-padded MPC sequences as the elements of the matrix. By applying a 'transpose function' on this matrix, a new matrix will be generated. After rearranging the chips into $P$ sequence-streams, the T-MPC codeset is generated as seen in Table 2.10.

The new values of code-weight, code-length and cardinality for the T-MPC are $P$, $P^2$ and $2P^2$, respectively. The code-length and code-weight are now as practical and

**Table 2.9**  Full-padded MPC sequences when $P = 5$

| Groups | Sequences Codes | | MPC Sequences | Full-Padded Sequences |
|---|---|---|---|---|
| | **0 0 0 0 0** | $C_{00}$ | 10000 10000 10000 10000 10000 | 01000 00100 00010 00001 10000 |
| | **4 4 4 4 4** | $C_{01}$ | 00001 00001 00001 00001 00001 | 10000 01000 00100 00010 10000 |
| 0 | **3 3 3 3 3** | $C_{02}$ | 00010 00010 00010 00010 00010 | 00001 10000 01000 00100 10000 |
| | **2 2 2 2 2** | $C_{03}$ | 00100 00100 00100 00100 00100 | 00010 00001 10000 01000 10000 |
| | **1 1 1 1 1** | $C_{04}$ | 01000 01000 01000 01000 01000 | 00100 00010 00001 10000 10000 |
| | **0 1 2 3 4** | $C_{10}$ | 10000 01000 00100 00010 00001 | 00010 00100 01000 10000 01000 |
| | **1 2 3 4 0** | $C_{11}$ | 01000 00100 00010 00001 10000 | 00001 00010 00100 01000 01000 |
| 1 | **2 3 4 0 1** | $C_{12}$ | 00100 00010 00001 10000 01000 | 10000 00001 00010 00100 01000 |
| | **3 4 0 1 2** | $C_{13}$ | 00010 00001 10000 01000 00100 | 01000 10000 00001 00010 01000 |
| | **4 0 1 2 3** | $C_{14}$ | 00001 10000 01000 00100 00010 | 00100 01000 10000 00001 01000 |
| | **0 2 4 1 3** | $C_{20}$ | 10000 00100 00001 01000 00010 | 01000 00001 00100 10000 00100 |
| | **2 4 1 3 0** | $C_{21}$ | 00100 00001 01000 00010 10000 | 00010 01000 00001 00100 00100 |
| 2 | **4 1 3 0 2** | $C_{22}$ | 00001 01000 00010 10000 00100 | 10000 00010 01000 00001 00100 |
| | **1 3 0 2 4** | $C_{23}$ | 01000 00010 10000 00100 00001 | 00100 10000 00010 01000 00100 |
| | **3 0 2 4 1** | $C_{24}$ | 00010 10000 00100 00001 01000 | 00001 00100 10000 00010 00100 |
| | **0 3 1 4 2** | $C_{30}$ | 10000 00010 01000 00001 00100 | 00001 01000 00010 10000 00010 |
| | **3 1 4 2 0** | $C_{31}$ | 00010 10000 00001 00100 10000 | 00100 00001 01000 00010 00010 |
| 3 | **1 4 2 0 3** | $C_{32}$ | 01000 00001 00100 10000 00010 | 10000 00100 00001 01000 00010 |
| | **4 2 0 3 1** | $C_{33}$ | 00001 00100 10000 00010 01000 | 00010 10000 00100 00001 00010 |
| | **2 0 3 1 4** | $C_{34}$ | 00100 10000 00010 01000 00001 | 01000 00010 10000 00100 00010 |
| | **0 4 3 2 1** | $C_{40}$ | 10000 00001 00010 00100 01000 | 00100 00010 00001 10000 00001 |
| | **4 3 2 1 0** | $C_{41}$ | 00001 00010 00100 01000 10000 | 01000 00100 00010 00001 00001 |
| 4 | **3 2 1 0 4** | $C_{42}$ | 00010 00100 01000 10000 00001 | 10000 01000 00100 00010 00001 |
| | **2 1 0 4 3** | $C_{43}$ | 00100 01000 10000 00001 00010 | 00001 10000 01000 00100 00001 |
| | **1 0 4 3 2** | $C_{44}$ | 01000 10000 00001 00010 00100 | 00010 00001 10000 01000 00001 |

efficient as the original MPC; while, greater available sequences (i.e. doubled) are provided resulting in higher throughput in the network running T-MPC.

Since the time-shift feature is no longer valid in order to generate the T-MPC, its predictability much reduces and then the system security enhances remarkably. The auto- and cross-correlation values for any pair of T-MPC codes $m$ and $n$ is given by:

$$C_{mn} = \begin{cases} P & m = n \\ 0 & m \neq n; \text{ same groups} \\ \leq 2 & m \neq n; \text{ different groups} \end{cases} \quad (2.38)$$

where $m, n \in \{1, 2, \ldots, P^2\}$. It is also noted that the reduction in code-length improves the correlation values, i.e. Equation (2.37). Based on this correlation function and the example of T-MPC in Table 2.10, the auto- and cross-correlation values of the T-MPC for the data stream 11010 are displayed graphically.

Figure 2.23 illustrates the auto-correlation values of $C_{3,2}$ at every bit synchronous position(s) $T$, which is equal to the code-length times chip duration following data stream of 10110. As expected, the peak value is $P = 5$. It can be seen in Figure 2.24 that the

**Table 2.10** Transposed-MPC sequences when $P = 5$

| Groups | Codes | T-MPC Sequences |
|---|---|---|
| | $C_{00}$ | 10000 10000 10000 10000 10000 |
| | $C_{01}$ | 00001 01000 00010 00100 00001 |
| 0 | $C_{02}$ | 00010 00100 01000 00001 00010 |
| | $C_{03}$ | 00100 00010 00001 01000 00100 |
| | $C_{04}$ | 01000 00001 00100 00010 01000 |
| | $C_{10}$ | 10000 00001 00001 00001 00001 |
| | $C_{11}$ | 00001 10000 00100 01000 00010 |
| 1 | $C_{12}$ | 00010 01000 10000 00010 00100 |
| | $C_{13}$ | 00100 00100 00010 10000 01000 |
| | $C_{14}$ | 01000 00010 01000 00100 10000 |
| | $C_{20}$ | 10000 00010 00010 00010 00010 |
| | $C_{21}$ | 00001 00001 01000 10000 00100 |
| 2 | $C_{22}$ | 00010 10000 00001 00100 01000 |
| | $C_{23}$ | 00100 01000 00100 00001 10000 |
| | $C_{24}$ | 01000 00100 10000 01000 00001 |
| | $C_{30}$ | 10000 00100 00100 00100 00100 |
| | $C_{31}$ | 00001 00010 10000 00001 01000 |
| 3 | $C_{32}$ | 00010 00001 00010 01000 10000 |
| | $C_{33}$ | 00100 10000 01000 00010 00001 |
| | $C_{34}$ | 01000 01000 00001 10000 00010 |
| | $C_{40}$ | 10000 01000 01000 01000 01000 |
| | $C_{41}$ | 00001 00100 00001 00010 10000 |
| 4 | $C_{42}$ | 00010 00010 00100 10000 00001 |
| | $C_{43}$ | 00100 00001 10000 00100 00010 |
| | $C_{44}$ | 01000 10000 00010 00001 00100 |
| | $C_{50}$ | 01000 00100 00100 00100 00100 |
| | $C_{51}$ | 10000 00010 10000 00001 01000 |
| 5 | $C_{52}$ | 00001 00001 00010 01000 10000 |
| | $C_{53}$ | 00010 10000 01000 00010 00001 |
| | $C_{54}$ | 00100 01000 00001 10000 00010 |
| | $C_{60}$ | 00100 00010 00010 00010 00010 |
| | $C_{61}$ | 01000 00001 01000 10000 00100 |
| 6 | $C_{62}$ | 10000 10000 00001 00100 01000 |
| | $C_{63}$ | 00001 01000 00100 00001 10000 |
| | $C_{64}$ | 00010 00100 10000 01000 00001 |
| | $C_{70}$ | 00010 00001 00001 00001 00001 |
| | $C_{71}$ | 00100 10000 00100 01000 00010 |
| 7 | $C_{72}$ | 01000 01000 10000 00010 00100 |
| | $C_{73}$ | 10000 00100 00010 10000 01000 |
| | $C_{74}$ | 00001 00010 01000 00100 10000 |
| | $C_{80}$ | 00001 10000 10000 10000 10000 |
| | $C_{81}$ | 00010 01000 00010 00100 00001 |
| 8 | $C_{82}$ | 00100 00100 01000 00001 00010 |
| | $C_{83}$ | 01000 00010 00001 01000 00100 |
| | $C_{84}$ | 10000 00001 00100 00010 01000 |

*(continued overleaf)*

**Table 2.10**   (*continued*)

| Groups | Codes | T-MPC Sequences |
|--------|-------|-----------------|
|        | $C_{90}$ | 11111 00000 00000 00000 00000 |
|        | $C_{91}$ | 00000 11111 00000 00000 00000 |
| 9      | $C_{92}$ | 00000 00000 11111 00000 00000 |
|        | $C_{93}$ | 00000 00000 00000 11111 00000 |
|        | $C_{94}$ | 00000 00000 00000 00000 11111 |

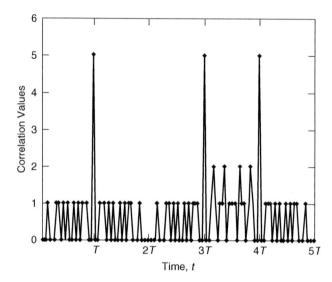

**Figure 2.23**   Auto-correlation of T-MPC of $C_{3,2}$ for the data stream of 10110

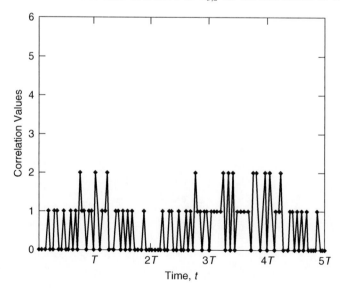

**Figure 2.24**   Cross-correlation of T-MPC of $C_{4,2}$ and $C_{4,3}$ (same group) for the data stream of 10110

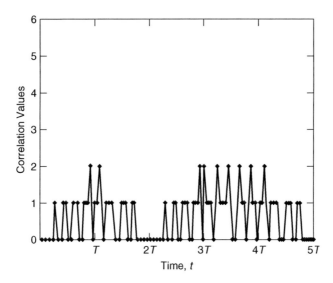

**Figure 2.25** Cross-correlation of T-MPC of $C_{0,2}$ and $C_{8,1}$ (different groups hit one) for the data stream of 10110

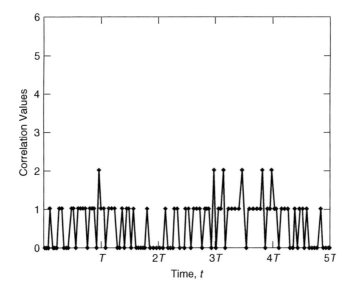

**Figure 2.26** Cross-correlation of T-MPC of $C_{5,4}$ and $C_{6,0}$ (different groups hit 'two') for the data stream of 10110

values of in-phase cross-correlation of two sequences from the same group, e.g. $C_{4,2}$ and $C_{4,3}$, hits zero, which implies perfect orthogonality. As one of the key features of T-MPC, it is worth mentioning that the number of groups containing perfect orthogonal sequences has also been doubled as well as the total number of available sequences in T-MPC as observed in Table 2.10.

Two pairs of other sequences from different groups, e.g. $C_{0,2}, C_{8,1}$ and $C_{5,4}, C_{6,0}$ are plotted in Figures 2.25 and 2.26 yielding a value of one and two respectively for the same data stream at every synchronized time $T$. It can be seen in the figures that the sequences follow the data stream 11010 as a result of CDMA encoding. If T-MPC sequences are employed in the asynchronous scheme, users would communicate in the network in different time-slots which causes out-of-phase and undesirable correlation values, as can be seen between every synchronized time. Thus, dynamically complex threshold and analysis as well as random access protocol will be required. Due to the properties of the MPC families, the codes are popular in synchronous schemes.

Here we have reviewed fundamental spreading codes based on prime code families including the recently introduced ones [30]. There are many different code algorithms with various features introduced regularly in the literature. For example, partial prime code families have also been studied and analysed [31], taking advantages of modified prime codes and low-weight code designing [28, 32, 33] to reduce the power consumption and system complexity.

## 2.5   Codes with Ideal In-Phase Cross-Correlation

Multiple-user interference (MUI) in spectral-amplitude-coding (SAC) OCDMA systems is only determined by the values of in-phase cross-correlation between the address sequences because frequency components are inherently in order [34]. One major advantage of such systems is that we can eliminate MUI when codes with fixed in-phase cross-correlation are used as address sequences. This elimination is realized by balanced detection of signals from a typical decoder and a reference decoder as will be explained in Chapter 4. Both Hadamard code and $m$-sequence are examples of these codes, which were introduced earlier in Section 2.2.

Let us define $\lambda = \sum_{i=1}^{N} x_i y_i$ as the in-phase cross-correlation of two different code sequences $X = \{x_1, x_2, \ldots, x_N\}$ and $Y = \{y_1, y_2, \ldots, y_N\}$. A code with length $N$, weight $w$ and in-phase cross-correlation $\lambda$ can be denoted by $(N, w, \lambda)$. When the cross-correlation value is fixed and equal to one, we say that the code has ideal in-phase cross-correlation, as this is the minimum value that can be achieved.

Nevertheless, in SAC systems, the inherent phase-induced intensity noise (PIIN) also severely affects the overall system performance [35]. This noise is due to spontaneous emission of the broadband source and its effect is proportional to the power of the generated photocurrent. Although MUI can be cancelled by the balanced detection where a subtraction of photocurrents is operated after photodetection, it still affects the system's performance by means of PIIN. When MUI is strong, the generated photocurrent is large and so is the resulting PIIN. Therefore, a higher signal-to-noise ratio can be expected if we reduce the total power of the MUI components without affecting the power of the effective signal (photocurrent after the subtraction operation that corresponds to the in-phase auto-correlation peak). For this purpose, in-phase cross-correlation $\lambda$ should be kept as small as possible since the power of MUI is determined by its value [36]. It is for this reason that codes with ideal in-phase cross-correlation become attractive.

Although codes with ideal cross-correlation have been studied intensively, little work has been done on codes having ideal in-phase cross-correlation.

A $(\frac{q^{m+1}-1}{q-1}, \frac{q^m-1}{q-1}, \frac{q^{m-1}-1}{q-1})$ code with low constant in-phase cross-correlation is being introduced [36]. This code is based on points and hyper-planes of the projective geometry $PG(m, q)$, where $q$ is a prime power given by $q = P^k$ with $k$ being a positive integer, $P$ being a prime number and $m$ (where $m \geq 2$) denoting the finite vector space dimension. It has been shown that performance can be improved significantly by using this code instead of Hadamard code.

In this section, we firstly review the fundamental knowledge of finite field algebra. Secondly, $(\frac{q^{m+1}-1}{q-1}, \frac{q^m-1}{q-1}, \frac{q^{m-1}-1}{q-1})$ code is introduced, together with two construction methods. Finally, we give the algebraic construction method for a series of $(P^2 + P, P + 1, 1)$ code families for each prime number $P$ (when $P \geq 3$). This code family is obtained by modifying the quadratic congruence codes introduced in [37]. Therefore, this spreading code family is called modified quadratic congruence (MQC) codes [20, 38]. In a similar method, based on the frequency-hopping codes, first introduced in [39], another code family in the form of $(q^2 + q, q + 1, 1)$ is studied in this chapter for prime power $q$. This code family not only possesses ideal in-phase cross-correlation, but also exists for a much wider number of integers than quadratic congruence codes. It should be pointed out that both code families are generated by a padding method, i.e. adding some zeros and ones to each of the original code sequences. For better understanding we have given some examples of each code. In addition, all these codes have been evaluated theoretically. Finally the spectral-amplitude-coding systems applying the codes with ideal in-phase cross-correlation are investigated and their performance is studied in detail in Chapter 4.

## 2.5.1 Finite Fields

This section presents a few elementary concepts of algebra for finite fields, needed to understand the construction methods of the codes. For a more extensive and rigorous account, interested readers are recommended to read [39, 40].

### Definitions

A finite field $GF(q)$ is a set of $q$ elements with rules for addition (subtraction) and multiplication (division) consistently defined. As a consequence, a finite field has a zero element and a unit element, both being unique. The zero element has the property $\alpha + 0 = \alpha$ and for the unit element $\alpha \cdot 1 = \alpha$ for all $\alpha \in GF(q)$.

We denote the elements of $GF(q)$ by the integers $\{0, 1, 2, \ldots, q - 1\}$ with the 'zero' and 'one' being the zero and unit elements, respectively. It should be realized that this notation is arbitrary and that the rules of addition and multiplication have not yet been specified.

A fundamental result of higher algebra is that there exist finite fields only for $q$ equal to a prime or the power of a prime number $P$, i.e. $q = P^k$ with $k$ being a positive integer. This means that $q = \{2, 3, 4, 5, 7, 9, 11, 13, \ldots\}$ are permissible but there is no finite field with, for instance, 10 elements. All fields with $q$ elements are isomorphic, which means that they differ only in the way that the elements are named. The field of $q$ elements is called the Galois field after the French mathematician Évariste Galois, who invented the notation $GF(q)$. Let us consider the following two cases.

## Case 1: $q$ is a prime number

When $q$ is equal to a prime number, the rules of addition and multiplication in $GF(q)$ are defined by modulo $q$ arithmetic. This means that the sum of, or product between, two elements is defined as the operation in the usual algebra of integer numbers with the results reduced by modulo $q$, i.e. equal to the remainder after dividing by $q$. If $q = 7$, we then have following examples as: $2 \cdot 3 = 6, 1 + 4 = 5, 3 \cdot 4 = 5$ (12mod7), $2 + 5 = 0$ (7mod7).

A non-zero element $\alpha \in GF(q)$ is said to be of order $N$ if $N$ is the lowest non-zero integer such that $\alpha^N = 1$. Since $\alpha^N$ is equal to a non-zero element and there are $q - 1$ such elements in the field, $N$ must be less than or equal to $q - 1$. An element with $N = q - 1$ is called a primitive element. For example, in $GF(7)$, the element $\alpha = 2$ has the powers $\alpha^0, \alpha^1, \alpha^2, \alpha^3, \alpha^4, \alpha^5 \ldots = 1, 2, 4, 1, 2, 4 \ldots$ The order of $\alpha = 2$ is thus $N = 3$. The element $\alpha = 3$ has the powers $\alpha^0, \alpha^1, \alpha^2, \alpha^3, \alpha^4, \alpha^5 \ldots = 1, 3, 2, 6, 4, 3, \ldots$ which shows that $\alpha = 3$ is a primitive element, i.e. of order $N = 6$. It is clear from the example that the powers of a primitive element are all the non-zero elements of a finite field.

## Case 2: $q = P^k$ where $P$ is a prime number and $k$ is a positive integer

When $q$ is a prime number, the rules for addition and multiplication are related to ordinary real number algebra in a simple way. However, when $q = P^k$ the operations are more complicated.

Consider the case $P = 2$: to define addition in this field with $q = P^k$, the elements of the field are represented as $P$-ary numbers or vectors of length $k$. As an example, for $q = P^3 = 8$, the elements are expressed as three-digit binary numbers, like $1 \equiv 001, 3 \equiv 011, 4 \equiv 100$ etc. which makes $\{0, 1, 2, \ldots, 7\}$ the octal representation of these numbers.

The addition is now defined as mod-$P$ addition of the components. For $P = 2$, this will be the binary addition, such as $1 + 1 = 0, 1 + 0 = 1, 110 + 011 = 101 \equiv 1001$, etc. To specify multiplication in the $GF(P^k)$ field, the $k$-tuples are transformed into polynomials in $Z$ of degree $k - 1$ by letting the first digit be the coefficient of $Z^{k-1}$, the second digit be the coefficient of $Z^{k-2}$, etc. For example, the triplet $(111)$ corresponds to $Z^2 + Z + 1$, $(011)$ corresponds to $Z + 1$, and so on. Addition and multiplication of polynomials are defined as in ordinary algebra using the mod-$P$ rule for the coefficients.

The multiplication rule of $GF(P^k)$ is polynomial multiplication modulo an irreducible polynomial $P(Z)$ of degree $k$. A polynomial is irreducible if it is not possible to factor it into a product of polynomials of lower degree. It thus has the features of a prime

**Table 2.11**   Power of $b = 2$

| Elements | mod-$P$ | Binary | Octal |
|----------|---------|--------|-------|
| $b^0$ | 1 | 001 | 1 |
| $b^1$ | $Z$ | 010 | 2 |
| $b^2$ | $Z^2$ | 100 | 4 |
| $b^3$ | $Z^3 = Z + 1$ | 011 | 3 |
| $b^4$ | $Z^2 + Z$ | 110 | 6 |
| $b^5$ | $Z^3 + Z^2 = Z^2 + Z + 1$ | 111 | 7 |
| $b^6$ | $Z^3 + Z^2 + Z = Z^2 + 1$ | 101 | 5 |
| $b^7$ | $Z^3 + Z = 1$ | 001 | 1 |

number in the algebra of polynomials. As an example, consider $P = 2$ and $k = 3$; thus the polynomial $P(Z) = Z^3 + Z + 1$ is irreducible. The multiplication of $5 \equiv 101$ and $3 \equiv 011$ gives $(Z^2 + 1)(Z + 1) = (Z^3 + Z^2 + Z + 1)$ mod $P(Z) = Z^2$. Now, here it is $5 \cdot 3 = 15 \equiv (1111)$ mod $(1011) = 0100 \equiv 4$ in the $GF(8)$ field.

The element $b = Z = 010$ is primitive, having the table of powers shown in Table 2.11. Multiplication in the $GF(P^k)$ field is most easily performed by using, for example, a list of the non-zero elements expressed as the power of a primitive element: $5 = b^6$ and $3 = b^3$ determines $5 \cdot 3 = b^6 \cdot b^3 = b^9 = b^7 \cdot b^2 = b^2 = 4$.

## 2.5.2 Balanced Incomplete Block Design Codes

A code with fixed in-phase cross-correlation can be constructed based on a symmetric balanced incomplete block design (BIBD) because every pair of blocks intersects in an equal number of elements [41]. We call a code acquired in this way a BIBD code. In fact, the $(\frac{q^{m+1}-1}{q-1}, \frac{q^m-1}{q-1}, \frac{q^{m-1}-1}{q-1})$ code [36] comes from a symmetric BIBD with element number $\frac{q^{m+1}-1}{q-1}$, block size $\frac{q^m-1}{q-1}$, where every pair of blocks intersects in $\frac{q^{m-1}-1}{q-1}$ elements. It has been indicated that such a symmetric BIBD exists for each prime power $P$. In this section we study the following two algebraic construction methods for the BIBD codes.

### First Construction Method

Let $\overline{\alpha} = (\alpha_0, \alpha_1, \alpha_2, \ldots, \alpha_m)$ denote a $(m + 1)$-tuple where $\alpha_j$ is an element of $GF(q)$ and $j \in \{0, 1, 2, \ldots, m\}$. Consider the tuple set $S$ where $\alpha_j$ are not all zero for each member. Given the prime power $q$ and the integer $m$, $S$ has $(q^{m+1} - 1)$ members. Let $X$ and $Y$ be two tuples in $S$; if $X = \lambda Y$ for any non-zero element $\lambda$ of $GF(q)$, $X$ is an equivalent tuple of $Y$. Because of the closure of multiplication in $GF(q)$, there always exists a group of $(q - 1)$-tuples in $S$, such that the parts of any tuple-pair in this group are equivalent each other. This tuple group is defined as an equivalence class, and there exist $v(m) = \frac{q^{m+1}-1}{q-1}$ equivalence classes in total. Based on the above definitions, a BIBD code can be constructed by using the following steps [41]:

1. Construct a tuple set $E$ where each member is chosen from a different equivalence class.
2. For each tuple $\overline{\alpha} \in E$, solve equation $\overline{\alpha} \cdot x^T = 0$, where $x = (x_0, x_1, x_2, \ldots, x_m)$ is also a member of $E$, and $x^T$ is the transpose of $x$. In doing so, $v(m-1)$ groups of roots for each $\overline{\alpha}$ are obtained.
3. Map each root group into a $q$-ary number $(x_0, x_1, x_2, \ldots, x_m)$, represent it with a decimal number between 1 and $v(m)$, and then a sequence $y(k)$ composed of $v(m-1)$ numbers is obtained for each $\overline{\alpha}$. Selecting all the tuples in $E$, we can generate $v(m)$ number sequences that constitute a symmetric BIBD.
4. Construct a binary sequence $S(i)$ based on each number sequence $y(k)$ by using the following mapping method:

$$S(i) = \begin{cases} 1 & \text{if } i = y(k) \\ 0 & \text{otherwise} \end{cases} \tag{2.39}$$

In this method, a binary code sequence can be generated for each tuple $\bar{\alpha} \in E$. Since there are $v(m)$ tuples in $E$, it results in the same number of code sequences. However, the above procedures are very complex since we have to solve $v(m)$ equations.

### Second Construction Method

According to Singer's theorem [41], a cyclic symmetric BIBD exists when the representation of the $q$-ary numbers $(x_0, x_1, x_2, \ldots, x_m)$ to decimal integers satisfy certain requirements. To simplify the first construction method, this cyclic property is used and we have a second construction method as follows:

1. Select any non-zero tuple $\bar{\alpha} \in S$ then $q^m - 1$ groups of roots can be obtained by solving equation $\bar{\alpha} \cdot x^T = 0$, where $x \in S$. These $q^m - 1$ groups of roots form $v(m-1)$ equivalence classes. Select one group of roots from each of the $v(m-1)$ equivalence classes and map it to a $q$-ary number $(x_0, x_1, x_2, \ldots, x_m)$. Thus, a set of $v(m-1)$ $q$-ary numbers is obtained which is denoted by $R$.
2. Obtain a number sequence from $R$ using $D(R) = \{i : 0 \le i \le v(m), \beta^i \in R\}$ where $\beta$ is a primitive element of $GF(q^{m+1})$. It has been indicated that the elements in this sequence constitute a difference set with size $\frac{q^m - 1}{q - 1}$ [41].
3. Generate $v(m)$ number sequences by using equation $D_k = D + k$ where $k \in \{0, 1, 2, \ldots, v(m) - 1\}$, and the $v(m)$ number sequences constitute a symmetric BIBD.
4. Construct a binary sequence $S(i)$ for each number sequence $D_k$ by the same mapping method as in the first construction method which is rewritten as:

$$S(i) = \begin{cases} 1 & \text{if } i = D_k \\ 0 & \text{otherwise} \end{cases} \tag{2.40}$$

Thus, the result is $v(m)$ code sequences. In this second method, only one equation needs to be solved. However, we need to calculate the powers of $\beta$.

### Discussions and Code Examples

Figure 2.27 shows a mapping example for *Step 4* in both introduced construction methods. It is due to this mapping method that $v(m)$ is the length of the final binary code. The design obtained in *Step 3* is a symmetric BIBD. Therefore, every pair of binary sequences intersects in exactly $\frac{q^{m-1} - 1}{q - 1}$ places [41].

When $m = 2$, a code of $(q^2 + q + 1, q + 1, 1)$ results, where it has ideal in-phase cross-correlation. As an example, Table 2.12 lists some code sequences generated by the second construction method where $q = 2, m = 2$ and the irreducible polynomial of $GF(2^3)$ is $x^3 + x + 1$. Since $q = 2$, only one non-zero member exists in $GF(2)$ and hence each

$y(k)$:   1 2 4

$S(i)$:   1 1 0 1 0 0 0

**Figure 2.27**   Example of the mapping method in BIBD code construction

**Table 2.12**   BIBD code example

| $q = 2$<br>$m = 2$ | $y(k)$ | $S(i)$ |
|---|---|---|
| *Irreducible* | (1 2 4) | 1101000 |
| *polynomials of d* | (2 3 5) | 0110100 |
| *and GF* (8) = | (3 4 6) | 0011010 |
| $x^3 + x + 1$ | (4 5 7) | 0001101 |

equivalence class contains only one group of roots. In Table 2.12, the number sequence obtained in *Step 2* is (1 2 4).

Although the construction method is simplified by Singer's theorem, root searching for equation $\bar{\alpha} \cdot x^T = 0$ is still a complex procedure. In the following, two code families will be constructed based on simple algebraic techniques. Both code families are obtained by modifying former codes with ideal cross-correlation, i.e. cross-correlation less than or equal to one. Here we should remember that the padded modified prime code (PMPC) family introduced in Section 2.4.4 also has ideal in-phase cross-correlation value.

### 2.5.3   Modified Quadratic Congruence Codes

Quadratic congruence (QC) codes with ideal cross-correlation can be constructed for each prime number, $P$ [37]. It has been shown that each sequence of QC code consists of $P$ subsequences and that each subsequence contains a single one and $(P - 1)$ zeros. By padding another similar subsequence, we can obtain a series of new $(P^2 + P, P + 1, 1)$ code families for each $P$ (when $P > 2$). These codes are referred to as modified quadratic congruence (MQC) codes.

***Code Construction Method***

The MQC of $(P^2 + P, P + 1, 1)$ family can be constructed by using the following two steps:

1. A sequence of integer numbers $Y_{d,\alpha,\beta}^{MQC}(k)$, which includes elements of a finite field $GF(P)$ over a prime number $P$, is constructed by using the following expression:

$$Y_{d,\alpha,\beta}^{MQC}(k) = \begin{cases} [d(k + \alpha)^2 + \beta]\text{mod} - P & k = 0, 1, \ldots, P - 1 \\ [\alpha + b]\text{mod} - P & k = P \end{cases} \tag{2.41}$$

where $P > 2$, $d \in \{1, 2, \ldots, P - 1\}$ and $b, \alpha, \beta \in \{0, 1, 2, \ldots, P - 1\}$. The upper part of Equation (2.41) is the original construction of quadratic congruence codes [42]; and the lower part is an added item which will make the in-phase cross-correlation exactly equal to one. Each resulted sequence $Y_{d,\alpha,\beta}^{MQC}(k)$ has $(P + 1)$ elements, and we can generate $P^2$ different sequences for each pair of fixed parameters $d$ and $b$ by changing parameters $\alpha$ and $\beta$. These $P^2$ different sequences can constitute an MQC code family. Therefore, there are, in total, $P(P - 1)$ code families when both $d$ and $b$ change.

2. A sequence of binary numbers $S_{d,\alpha,\beta}^{MQC}(i)$ based on $Y_{d,\alpha,\beta}^{MQC}(k)$ is constructed by using the following mapping method:

$$S_{d,\alpha,\beta}^{MQC}(i) = \begin{cases} 1 & if\ i = kP + Y_{d,\alpha,\beta}^{MQC}(k) \\ 0 & \text{Otherwise} \end{cases} \quad (2.42)$$

where $i \in \{0, 1, 2, \ldots, P^2 - 1\}$ and $k = \lfloor i/P \rfloor$. Here $\lfloor i/P \rfloor$ denotes the largest integer value less than or equal to $i/P$. Figure 2.28 shows a mapping example of generating binary sequence $S_{d,\alpha,\beta}^{MQC}(i)$ from the number sequence $Y_{d,\alpha,\beta}^{MQC}(k)$.

### Code Properties

In each constructed MQC family there are $P^2$ code sequences having following properties:

1. Each code sequence has $P^2 + P$ elements that can be divided into $(P + 1)$ subsequences, and each subsequence contains a single one and $(P - 1)$ zeros.
2. The in-phase cross-correlation $\lambda$ between any two sequences is exactly equal to one.

Of these two properties, the first one can be easily obtained from the mapping method as explained in *Step 2* of the construction method, and the second one comes from the following theorem:

### Definition

Let $n$ be a fixed positive integer; two integers $a$ and $b$ are said to be *congruent modulo-n*, symbolized by $a \equiv b(\text{mod-}n)$ if $n$ divides the difference $a - b$; provided that $a - b = kn$ for some integer $k$. For example $3 \equiv 24$ (mod-7).

### Theorem

Let $d$ and $b$ be constant numbers and $P(P > 2)$ be a prime number. The following congruent equation has exactly one non-congruent solution except when $\alpha_1 = \alpha_2$ and $\beta_1 = \beta_2$,

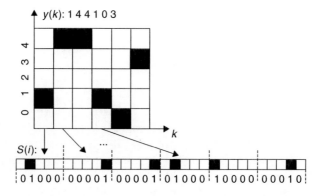

**Figure 2.28** Graphic representation of the mapping method for MQC construction

where there are $P + 1$ non-congruent solutions:

$$Y_{d,\alpha_1,\beta_1}^{MQC}(k) - Y_{d,\alpha_2,\beta_2}^{MQC}(k) = 0 \tag{2.43}$$

### Proof

It should be noted that when $\alpha_1 = \alpha_2$ and $\beta_1 = \beta_2$, the two sequences are the same. To solve Equation (2.43), let us substitute Equation (2.41) into it and consider the following two cases:

1. When $k \in \{0, 1, 2, \ldots, P - 1\}$, Equation (2.43) becomes:

$$2d(\alpha_1 - \alpha_2)k + d(\alpha_1^2 - \alpha_2^2) + \alpha_1^2 + \beta_1 - \beta_2 = 0(\text{mod} - P) \tag{2.44}$$

where $\alpha_1, \alpha_2, \beta_1$ and $\beta_2 \in \{0, 1, 2, \ldots, P - 1\}, d \in \{1, 2, \ldots, P - 1\}$. Noting that $P > 2$, by Lagrange's theorem and its corollary, we can analyse this equation as follows: when $\alpha_1 \neq \alpha_2$, there is only one solution that can be written as:

$$k = \frac{d(\alpha_2^2 - \alpha_1^2) + \beta_2 - \beta_1}{2d(\alpha_1 - \alpha_2)}(\text{mod} - P) \tag{2.45}$$

When $\alpha_1 = \alpha_2$ and $\beta_1 = \beta_2$, there are $P$ solutions for $k$ in Equation (2.45), whereas when $\alpha_1 = \alpha_2$ and $\beta_1 \neq \beta_2$, no solution can be found.
2. When $k = P$, Equation (2.43) becomes:

$$\alpha_1 - \alpha_2 = 0(\text{mod} - P) \tag{2.46}$$

When $\alpha_1 = \alpha_2$, $k = P$ is the solution.

Consequently, when $\alpha_1 = \alpha_2$ and $\beta_1 = \beta_2$, there are a total of $P + 1$ solutions of $k$ for Equation (2.43), while if either $\alpha_1 = \alpha_2$ and $\beta_1 \neq \beta_2$, or $\alpha_1 \neq \alpha_2$, Equation (2.43) has only one solution of $k$. Therefore there is exactly one solution for $k$ except if $\alpha_1 = \alpha_2$ and $\beta_1 = \beta_2$ when there are $P + 1$ non-congruent solutions.

Table 2.13 shows MQC sequences as an example, with parameters given by $P = 5, d = 1$ and $b = 2$. It is clear that the listed sequences well satisfy the two code properties.

**Table 2.13**  MQC sequences

| Parameters | $y(k)$ | $S(i)$ |
|---|---|---|
| $\alpha = 0, \beta = 0$ | (014412) | 10000 01000 00001 00001 01000 00100 |
| $\alpha = 1, \beta = 0$ | (144103) | 01000 00001 00001 01000 10000 00010 |
| $\alpha = 4, \beta = 0$ | (110441) | 01000 01000 10000 00001 00001 01000 |
| $\alpha = 1, \beta = 3$ | (422433) | 00001 00100 00100 00001 00010 00010 |
| $\alpha = 3, \beta = 4$ | (304030) | 00010 10000 00001 10000 00010 10000 |

### 2.5.4  Modified Frequency-Hopping Codes

In this section, another codeset with ideal in-phase cross-correlation from the MQC family is introduced and studied; it is usually applied to SAC-OCDMA systems, here referred to as modified frequency-hopping (MFH) codes. Although the MFH code has many advantages, it only exists for prime numbers. It can be constructed by an algebraic technique over each $GF(q)$ field for each prime power $q = P^n$ where $n$ is a positive integer. As compared with MQC the MFH codes exit for much wider integer numbers and the codeset can be selected with more flexibility on the code-length.

By recalling the modified quadratic congruence (MQC) code construction, it will be observed that there is a similar approach to constructing a family of frequency-hopping codes with ideal cross-correlation [39]. Thus, it will be seen that a set of $(q^2 + q, q + 1, 1)$ results in MFH code families as explained in the following [20, 38]:

**Codes Construction**

The MFH code family is constructed by using the following two steps:

1. Let $GF(q)$ denote a finite field of $q$ elements, and $\beta$ is a primitive element of $GF(q)$. We can construct a number sequence $Y_{a,b}(k)$ with elements of $GF(q)$ using the following expression:

$$
Y_{a,b}(k) = \begin{cases} \beta^{(a+k)} + b & k = 0, 1, 2, \ldots, q - 2 \\ b & k = q - 1 \\ a & k = q \end{cases} \tag{2.47}
$$

where $a$ and $b$ are elements of $GF(q)$ expressed by $a \in \{0, 1, \ldots, q - 2\}$ and $b \in \{0, 1, \ldots, q - 1\}$. The parameters $a$ and $b$ are fixed for each specified number of sequences, and their changes result in $q(q - 1)$ different sequences. It should be noted that the operations in Equation (2.47) are determined by $GF(q)$.

Because the values of $\beta^k$ are not equal to zero no matter what the value of $k$ is, the first $q$ elements of each sequence are all different and they can exactly constitute a whole set of $GF(q)$. Noting that $\alpha \neq q - 1$ in Equation (2.47), another $q$ sequences can be added into the code family without affecting the ideal in-phase cross-correlation of the final binary sequences. These $q$ sequences are constructed using the following expression:

$$
y(k) = \begin{cases} b & k = 0, 1, 2, \ldots, q - 1 \\ q - 1 & k = q \end{cases} \tag{2.48}
$$

Therefore, a full set of $q^2$ sequences are obtained which is denoted by $y(k)$ in the rest of this section.

2. Based on each generated number of sequences, $y(k)$, a sequence of binary sequences $S(i)$ is produced by using the same mapping method as in Equation (2.42). It is rewritten as:

$$
S(i) = \begin{cases} 1 & \text{if } i = kP + y(k) \\ 0 & \text{Otherwise} \end{cases} \tag{2.49}
$$

where $i = (0, 1, \ldots, q^2 + q - 1)$ and $k = \lfloor i/q \rfloor$.

### Code Properties

Since $q$ is a prime power, MFH code not only possesses ideal in-phase cross-correlation, but it also exists for a much wider number of integers than MQC codes. Similar to MQC codes, there are $q^2$ (1, 0) sequences in this code family with the following properties:

- Each code sequence has $q^2 + q$ elements that can be divided into $(q + 1)$ subsequences and each subsequence contains a single one and $(q - 1)$ zeros.
- In-phase cross-correlation $\lambda$ between any two sequences is always equal to one.

In the two properties, the first one can be easily obtained from the mapping method as explained in *Step 2* and the second one is proved in the following:

### Proof

Let sequences obtained from $y(k)$ and $y_{a,b}(k)$ be denoted by $A$ and $B$, respectively. It is easy to see that sequences in $A$ intersect only when $k = P$ and a sequence in $A$ intersects with a sequence in $B$ in exactly one position of $k$. Therefore, we only need to prove that there is exactly one case for all $k$ where two arbitrary sequences $y_{a_1,b_1}(k)$ and $y_{a_2,b_2}(k)$ in $B$ have the same elements where $k \in \{0, 1, 2, \ldots q\}$. This can be restated as the congruence equation:

$$y_{a_1,b_1}(k) \equiv y_{a_2,b_2}(k) \tag{2.50}$$

where it has exactly one non-congruent solution when ($\alpha_1 - \alpha_2 = 0$ AND $b_1 - b_2 = 0$) is not true. When $k \in \{0, 1, \ldots, q - 2\}$ according to Equation (2.47):

$$y_{a_1,b_1}(k) - y_{a_2,b_2}(k) = \beta^{(a_1+k)} + b_1 - (\beta^{(a_2+k)} + b_2) = 0 \tag{2.51}$$

Let $\acute{k} = k + a_1$, and the last equation can be rearranged as:

$$\beta^{\acute{k}}(1 - \beta^{(a_2-a_1)}) + (b_1 - b_2) = 0 \tag{2.52}$$

Since $a_1, a_2, b_1$ and $b_2$ are fixed, thus $(1 - \beta^{(a_2-a_1)})$ and $(b_1 - b_2)$ are constants. When $k$ changes, $\acute{k}$ also changes in $GF(q)$. As $\beta$ is a primitive element, the values of $\beta^{\acute{k}}$ are non-zero and all different. Accordingly:

> **Condition 1:** When $k = q - 1$, in Equation (2.50), $b_1 - b_2 = 0$;

> **Condition 2:** When $k = q$, in Equation (2.50), $\alpha_1 - \alpha_2 = 0$.

Now let us consider the solutions of the Equation (2.50) in all cases:

**Case 1:** When $\alpha_1 - \alpha_2 \neq 0$ and $b_1 - b_2 \neq 0$, Equation (2.52) has one solution. However, according to **Conditions 1** and **2**, it is clear that $k = q - 1$ or $k = q$ is not a solution. Therefore, exactly one solution exists.

**Case 2:** When $\alpha_1 - \alpha_2 \neq 0$ and $b_1 - b_2 = 0$, Equation (2.52) has no solution. However, $k = q - 1$ is a solution. Therefore, exactly one solution exists.

**Case 3:** When $\alpha_1 - \alpha_2 = 0$ and $b_1 - b_2 \neq 0$, Equation (2.52) has no solution. However, $k = q$ is a solution. Therefore, exactly one solution exists.

**Case 4:** When $\alpha_1 - \alpha_2 = 0$ and $b_1 - b_2 = 0$, the two sequences are same. Therefore, the number of solutions is $q + 1$.

Table 2.14 lists some MFH code sequence examples for different values of parameters $a$ and $b$ when $q$ is equal to $2^2$, i.e. $q = P^n = 4$. Here the selected primitive irreducible polynomial is written as $x^2 + x + 1$ and the primitive element $\beta$ is 10 in binary format, which can be represented as two in the decimal system. It is clearly shown that the former two properties are valid for all the listed code sequences $S(i)$.

## 2.5.5 Codes Evaluation and Comparison

Table 2.15 lists the introduced optical spreading codes with ideal in-phase cross-correlation value. MFH codes exist for each prime power the same as the BIBD codes. Although MQC codes only exist for any prime number $P$ being greater than two, which is in a smaller integer range, we can obtain many code families for each $P$. Hadamard codes (see Section 2.2.3) with length of $2^m$, where $m$ is a positive integer not less than two, is also listed in the table as a reference. Since PMPC sequences (see Section 2.4.4) and MQC codes are somehow similar in terms of code-length, code-weight, cardinality and in-phase cross-correlation value, here only MQC codes are studied as a representation of both MQC and PMPC.

To evaluate these codes, the codeset upper-bound on the size of $(N, w, \lambda)$ is analysed. Let $C = (c_{i,j})$ be an array whose rows are composed of all the sequences in a $(N, w, \lambda)$ codeset. $V(N, w, \lambda) = V$ denotes the code size $C$ in a $V \times N$ array where each of its rows

**Table 2.14** MFH sequences

| Parameters | $y(k)$ | $S(i)$ |
|---|---|---|
| $\alpha = 0, b = 0$ | 1 2 3 0 0 | 0100 0010 0001 1000 1000 |
| $\alpha = 1, b = 0$ | 2 3 1 0 1 | 0010 0001 0100 1000 0100 |
| $\alpha = 2, b = 1$ | 2 0 3 1 2 | 0010 1000 0001 0100 0010 |
| $\alpha = 0, b = 3$ | 2 1 0 3 0 | 0010 0100 1000 0001 1000 |
| $\alpha = 2, b = 2$ | 1 3 0 2 2 | 0100 0001 1000 0010 0010 |
| $b = 1$ for $y(k)$ | 1 1 1 1 3 | 0100 0100 0100 0100 0001 |

**Table 2.15** Spreading codes with ideal in-phase cross-correlation (IPC)

| Code Family | Existence | Length | Weight | Size | IPC |
|---|---|---|---|---|---|
| BIBD $(m = 2)$ | $GF(q)$ | $q^2 + q + 1$ | $q + 1$ | $q^2 + q + 1$ | 1 |
| MQC | $P > 2$ | $P^2 + P$ | $P + 1$ | $P^2$ | 1 |
| PMPC | $P > 2$ | $P^2 + P$ | $P + 1$ | $P^2$ | 1 |
| MFH | $GF(q)$ | $q^2 + q$ | $q + 1$ | $q^2$ | 1 |
| Hadamard | $m \geq 2$ | $2^m$ | $2^{m-1}$ | $2^m - 1$ | $2^{m-2}$ |

is a code sequence. Then the sum of the inner products of its rows can be expressed as:

$$\Lambda = \sum_{i=1}^{V} \sum_{j=1}^{V} \sum_{l=1}^{N} c_{i,l} \cdot c_{j,l} = V(V-1)\lambda \tag{2.53}$$

On the other hand, if we represent the number of ones in the $l^{th}$ column as $\delta_l$, the sum $\Lambda$ is also equal to $\sum_{l=1}^{N} \delta_l (\delta_l - 1)$. Note that $\sum_{l=1}^{N} \delta_l = wV$ and $\sum_{l=1}^{N} \delta_l^2$ is minimized when $\delta_l = wV/N$ for all $l$, then the inequality $(w^2 V^2/N) - wV \leq V(V-1)\lambda$ is obtained. Solving this equation for $V$ gives an upper-bound of the code cardinality as:

$$V(N, w, \lambda) \leq \frac{N(w - \lambda)}{w^2 - N\lambda} \tag{2.54}$$

When $w^2 > N\lambda$. For MQC codeset $(P^2 + P, P + 1, 1), N = P^2 + P, w = P + 1$ and $\lambda = 1$, the upper-bound of code cardinality expressed in Equation (2.55) becomes:

$$V(N, w, \lambda) \leq \frac{(P^2 + P)P}{(P + 1)^2 - (P^2 + P)} = P^2 \tag{2.55}$$

Since the MQC code has $P^2$ code sequences in each family, it is optimal theoretically. Similarly, it can be seen that both MFH and BIBD are also optimal codes. Besides ideal in-phase cross-correlation, another major advantage of both MQC and MFH lies in the first property, i.e. the elements in each sequence can be divided into subsequences and each subsequence contains only a single one. This property makes it much easier to realize the address reconfiguration in a grating-based SAC-OCDMA transmitter as will be explained in detail in Chapters 3 and 4. A group of gratings to reflect all the desired spectral components are utilized and also another group of gratings to compensate the delays of desired components are used to incorporate them into a temporal pulse again. In this case, the grating's tuneable range becomes the main limitation in address reconfiguration. By using the MQC and MFH codes, each grating only needs to be tuneable within $1/(P + 1)$ of the total encoded bandwidth (TEB) where the TEB corresponds to the full length of each code sequence. Thus the required tuneable range for each grating is significantly reduced. Because of the lower weight of the new codes compared with Hadamard codes, the number of gratings required for implementing an encoder is much reduced. This is due to the fact that one grating is used to reflect one spectral component. However, the lower weight of the MQC and MFH codes also reduces the power of the effective photocurrent. Therefore, effects of other noise sources such as thermal and shot noises become stronger especially when the received power is low.

One code family with lower in-phase cross-correlation referred to as BIBD code is introduced where it has ideal in-phase cross-correlation when $m = 2$ too. On the other hand, another two codes (MCQ and MFH) with ideal in-phase cross-correlation have been studied. However, such codes cannot be applied in the original system using complementary coding scheme where in-phase cross-correlation between any two sequences $A$ and $B$ are required to be equal to that between sequences $\overline{A}$ (the complement of $A$) and $B$. Only Hadamard and $m$-sequence codes have such a property, as will be studied in detail in Chapter 4.

To apply BIBD codes in the SAC systems shown in Figure 2.29 [36], a pulse with specified spectral distribution is sent when the data bit is one and nothing is sent when

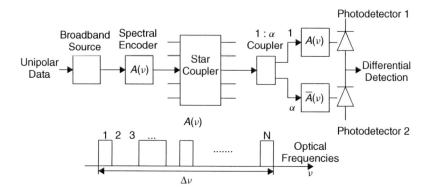

**Figure 2.29** Spectral-amplitude-coding OCDMA system using $(N, w, \lambda)$ codes (© 2001 IEEE. Reprinted with permission from Modified quadratic congruence codes for fiber Bragg-grating-based spectral-amplitude-coding optical CDMA systems, Z. Wei, H.M.H. Shalaby, and H. Ghafouri-Shiraz, *J. Lightw. Technol.*, **19** (9), 2001.)

the data bit is zero. At the receiver side, a 1:$a$ splitter is used at the beginning to divide the received signal into two parts. Then they are input respectively into two decoders with complementary decoding functions. Actually this function block is suitable for any codes with fixed in-phase cross-correlation. If a $(N, w, \lambda)$ code is used, the multiple-user interference (MUI) coming from $(K - 1)$ undesirable users at the first photodetector is equal to $(K - 1)\lambda$ and that at the second photodetector is equal to $\alpha(w - \lambda)(K - 1)$. When $\alpha = \lambda/(w - \lambda)$, these two MUI components are equal. Therefore, after balanced photodetection, the effects of MUI can be cancelled [20, 38]. It is easy to see that MQC and MFH codes can also be applied in this function block where $\lambda = 1$, as well as Hadamard and $m$-sequence codes.

When a $(N, w, \lambda)$ code is used in the system shown in Figure 2.29, the average signal-to-noise ratio (SNR) due to the intensity noise can be obtained and expressed as [20, 38]:

$$SNR = \frac{\Delta v (w - \lambda)}{BK\lambda(K + (w - 2\lambda)/\lambda)} \tag{2.56}$$

where $B$ is the noise equivalent electrical bandwidth of the receiver, $\Delta v$ is the encoded optical bandwidth in hertz, $K$ is the number of active users who are sending data bit one.

**Table 2.16** Signal-to-noise ratio of different spreading code families

| Code Family | Existence | Signal-to-Noise Ratio |
| --- | --- | --- |
| MFH | Prime Power $q$ | $\dfrac{\Delta v (q + 1)}{BK[((K - 1)/q) + q + K]}$ |
| MQC | Prime $P$ | $\dfrac{\Delta v (P + 1)}{BK[((K - 1)/P) + P + K]}$ |
| BIBD ($m = 2$) | Prime Power $q$ and $m \geq 2$ | $\dfrac{\Delta v}{BK[1 + (K - 1)(q^{m-1} - 1)/(q^{m-1} - 1)]}$ |

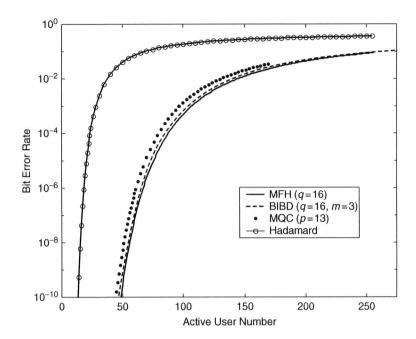

**Figure 2.30** BER performance of various spreading code families against the number of active users, $K$ (© 2002 IEEE. Reprinted with permission from Codes for optical spectral-amplitude-coding CDMA systems, Z. Wei and H. Ghafouri-Shiraz, *J. Lightw. Technol.*, **20** (8), 2002.)

Substituting the detailed values of code-weight and in-phase cross-correlation, the SNRs for different codes can be calculated as listed in Table 2.16. Assuming that the probability of PIIN has a Gaussian distribution profile, the corresponding bit-error rates (BER) using $P_e = \frac{1}{2}erfc(\sqrt{SNR/8})$ can also be calculated as shown in Figure 2.30. Note that complementary error function $erfc(x) = 1 - erf(x) = \int_x^\infty e^{-t^2}dt$. The following parameters are used in the analysis resulted in Figure 2.30: $\Delta v = 2.5\,$THz, i.e. equivalent to 20 nm linewidth, $B = 80\,$MHz, i.e. for the bit rate of 155 Mbps, and the operation wavelength is 1550 nm. The results in Figure 2.30 indicate that the MQC and MFH codes can suppress the effect of PIIN significantly and results in a much better performance than Hadamard codes. This suppression results from the higher in-phase auto-correlation to cross-correlation ratio. For Hadamard codes, this ratio is always equal to two. However, for the MQC and MFC codes it is equal to the code-weight, i.e. $(P + 1)$ and $(q + 1)$ respectively. The enhanced performance of MFH codes, as compared with the MQC codes, is due to the larger value of $q = 16$ compared with $P = 13$. When $m = 2$, the BER curve of BIBD codes will overlap with the MFH codes since the values of $q$ are both equal to 16. Also it can be easily concluded from Table 2.16 that the performance becomes better with the increase in $q$ value.

The system capacity against the number of simultaneous active users, $K$ is plotted in Figure 2.31 for different cases when Hadamard code with $N = 256$ and MFH codes with $q = 4$ and 16 are used. Given the system performance requirements, the network capacity is determined by the possible number of simultaneous users multiplied by the bit rate per

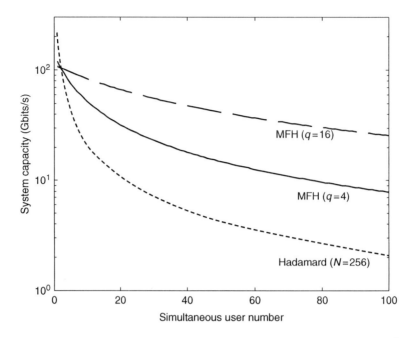

**Figure 2.31** Throughput against the number of active users, $K$ when BER $= 10^{-9}$ (© 2002 IEEE. Reprinted with permission from Codes for optical spectral-amplitude-coding CDMA systems, Z. Wei and H. Ghafouri-Shiraz, *J. Lightw. Technol.*, **20** (8), 2002.)

user, i.e. system throughput. In Figure 2.31, the desired BER is $10^{-9}$. It is shown that capacity of the system using MFH code is much larger than that using Hadamard codes. Moreover, when MFH codes are applied, the capacity will also increase with increase in $q$. With the employment of error correction techniques, as will be studied in Section 2.7, a much larger system capacity can be expected. Similar results can also be obtained when MQC and BIBD codes are employed in the SAC-OCDMA system.

## 2.6 Multidimensional Optical Codes

The signature codes are unique sequence codes for each subscriber, being used to distinguish multiple users' information from one another. According to the classification of spread spectrum techniques [44], the typical signature codes currently available are either one-dimensional (1D) or two-dimensional (2D).

The 1D signature codes are the direct spread (DS) sequence codes which employ only one channel for each sequence. For example, the 1D sequence codes can be represented as the mixture of ones and zeros in the time domain. By contrast, 2D signature codes are spread in both time and frequency (wavelength) domains such as fast frequency hopping (FFH) [44] and time-spreading sequence codes, which dynamically allocate multiple channels for each sequence. The channel index, by definition, from the fraction of the available bandwidth (e.g. 1) to the whole bandwidth (e.g. 160), substitutes the actual wavelength of the photons that largely enables the system administer to control frequency allocation more flexibly. For example, the channel index can be tangible in the following:

$$\text{Actual wavelength} = \text{Channel index} \times \text{Interval between two adjacent wavelengths}$$

$$+ \text{Minimum wavelength} \tag{2.57}$$

After the prudent selection of dazzling code measurements, there are four outstanding normalized standards: correlation, BER, cardinality and bandwidth efficiency. First of all, the correlation values of signature codes show the similarity between sequence codes, and the match filter can distinguish the information of each subscriber from noisy spread data, encoding by low cross-correlation signature codes. Second, the cardinality and bandwidth efficiency among different signature codes indicate the capability of accommodating simultaneous users and the effectiveness of unit length per single user. Finally, the BER constraint illustrates the overall system performance as the increasing number of subscribers within the capability of the signature codes.

The signature codes for each subscriber are deliberately designed to distinguish themselves from multiuser shared channel and noise (typically shot noise and thermal noise in optics). So the auto-correlation of signature codes should be as high as possible at the periodic time. However, the value between peaks should be eliminated to the lowest level which indicates that the code greatly differs from the time-shifted version of itself. Otherwise, the signature code cannot be implemented within asynchronous transceivers. As defined before, the correlation function is:

$$R_{XX}(\tau) = \sum_{i=0}^{n-1} x_i x_{i+\tau} \leq \lambda_a \tag{2.58}$$

where $x_i$ is the sequence code, $x_{i+\tau}$ is the $\tau$ time shift, and $\lambda_a$ is the largest value of auto-correlation function.

The cross-correlation value shows the similarity of the two sequence codes. Therefore, at any time of transmission, the cross-correlation should be as low as possible. Otherwise, the higher cross-correlation shows the increasing possibility of 'hits' and multiuser interference.

$$R_{XY}(\tau) = \sum_{i=0}^{n-1} x_i y_{i+\tau} \leq \lambda_c \tag{2.59}$$

where $x_i$ is one code sequence, $y_{i+\tau}$ is $\tau$ time-shifted version of another code sequence $y_i$ and $\lambda_c$ is the largest value of cross-correlation function.

The correlation function should satisfy a certain constraint at every dimension in multidimensional space. It is noted that the correlation function in the spreading or hopping domains can be interpreted differently. For example, in a 2D time-spreading wavelength-hopping coding scheme, the correlation function in the spreading domain (i.e. time) represents the intensity overlap of lightwaves causing relative intensity noise, whereas in the hopping domain (i.e. wavelength) the correlation function represents the waveforms mixing, resulting in beat noise. Accordingly, a lower cross-correlation value in the code family results in reduction or cancellation of such noise in the system.

The cardinality is the subscriber allocation capacity of 1D or 2D signature codes that support multiple users accessing the same channel simultaneously. Normally, the increasing length of signature codes can allow more subscribers to share the limited optical channels at the same time. However, not all the signature codes can accommodate the increasing subscribers with effective unit bandwidth. For 1D signature codes, the cardinality entirely depends on the algorithm of the code construction function. And the cardinality of 2D OOCs is the product of frequency-hopping coding (Chapter 3) and the cardinality of the spreading codes. The cardinality is the maximum allocation capability with massive errors. Therefore, the cardinality and bandwidth efficiency should be included with BER constraint in system or network performance discussions. There are around 160 different wavelengths available through the SMF-28e optical fibre [45]. The researcher aims at optimizing frequency efficiency, bandwidth over bit rate and unit bandwidth usage for each user. The comparative bandwidth efficiency between 1D and 2D signature codes highlights the sustainable code family in practice.

BER is the most important metric parameter of the performance analysis disclosing the occurrences of errors when the system allocates the increasing number of simultaneous users. With the consideration of the multiple-user interference (MUI), the higher constraint of BER is up to one error out of $10^9$ bits. The pulse separation and the reliable code-length under the correlation constraint are taken into account to accumulate the 'hits' within one channel. Most importantly, the highest constraint of BER indicates the largest number of sustainable subscribers that can access the same channels simultaneously with reliable system performance.

To increase the OCDMA system security and cardinality, various domains and parameters can be combined to create multidimensional codes such as time, space, polarization, phase and wavelength. For example, Sangin Kim *et al.* introduced three-dimensional (3D) space-wavelength-time-spreading codes for OCDMA communications application [46], followed by another set of 3D codes verified experimentally [47] in time-wavelength-polarization domains. There are a few 3D and other multidimensional coding techniques with various design algorithms to be found in the literature [48–52]. However, multidimensional coding brings much complexity to the system implementation and architecture, causing the focus on two-dimensional codes and encoding techniques. Here, in this section, popular 2D spreading codes are reviewed the details of which are found in research communities and other references more focused on optical spreading codes for OCDMA communications [1, 53, 54].

### 2.6.1 Two-Dimensional Optical Spreading Codes

The 2D optical orthogonal codes (OOC) are usually fast frequency hopping (FFH) codes spread over the time domain. Through comparative analysis between PC/OOC and OCFHC/OOC the optimal 2D-OOC is highlighted and studied in this section. The prime codes (PC) and one-coincidence frequency hop code (OCFHC) are taken as the fast frequency hopping sequences which indicate different carrier frequencies used in the different time-slots [55]. To construct a two-dimensional (2D) codeset, two one-dimensional (1D) codesets are selected to jointly generate a 2D codeset. As an example, the optical orthogonal code OOC($N$, 3, 1) (Section 2.3) and prime codesets are employed for time-spreading in the following.

### PC/OOC Code Family

The 2D optical orthogonal code family of PC/OOC, introduced in [56], is constructed by the combination of prime codes for a hopping scheme over optical orthogonal codes in the time-spreading domain. The time-shifted version of prime codes is generated with the prime number $P_h$, and the OOC($N_s$, 3, 1) is taken as the time-spreading codes. The PC/OOC allows the composite of time-spreading code with the row $j = 1$ of MPC, because of the ideally low correlation value of ($N_s$, 3, 1).

Secondly, the user-defined $P_h$ and the weight of OOC cannot be always equal. The segments of hopping code sequences are shown in Table 2.17 when $P_h = 5$ and OOC(7, 3, 1).

When $P_h = 5$ is larger than OOC code-weight $w = 3$, the sequence index of MPC should be modified and clipped as apparent from Table 2.18 for example. Finally the total cardinality of this 2D code sequences is:

$$\phi_h \times \phi_s = P_h^2 \times \left\lfloor \frac{N_s - 1}{w(w-1)} \right\rfloor = 25 \tag{2.60}$$

where $N_s, \phi_s$ and $\phi_h$ are OOC code-length, OOC cardinality and prime code cardinality respectively.

### OCFHC/OOC Code Family

The OCFHC/OOC is another family of 2D optical orthogonal codes, introduced in [57], which employs the one-coincidence frequency hop code (OCFHC) and the OOC ($N_s$, 3, 1)

**Table 2.17**  Two-dimensional PC/OOC code sequences

| $i$ | $j$ | Index $i$ | $j$ | Index $i$ | $j$ | Index $i$ | $j$ | Index $i$ | $j$ | Index $i$ |
|---|---|---|---|---|---|---|---|---|---|---|
| 0 |   | 11111 |   | 12345 |   | 13524 |   | 14253 |   | 15432 |
| 1 |   | 22222 |   | 23451 |   | 35241 |   | 42531 |   | 54321 |
| 2 | 1 | 33333 | 2 | 34512 | 3 | 52413 | 4 | 25314 | 5 | 43215 |
| 3 |   | 44444 |   | 45123 |   | 24135 |   | 53142 |   | 32154 |
| 4 |   | 55555 |   | 51234 |   | 41352 |   | 31425 |   | 21543 |

**Table 2.18**  Modified two-dimensional PC/OOC code sequences

| $i$ | $j$ | Index $i$ | $j$ | Index $i$ | $j$ | Index $i$ | $j$ | Index $i$ | $j$ | Index $i$ |
|---|---|---|---|---|---|---|---|---|---|---|
| 0 |   | 111 |   | 123 |   | 135 |   | 142 |   | 154 |
| 1 |   | 222 |   | 234 |   | 352 |   | 425 |   | 543 |
| 2 | 1 | 333 | 2 | 345 | 3 | 524 | 4 | 253 | 5 | 432 |
| 3 |   | 444 |   | 451 |   | 241 |   | 531 |   | 321 |
| 4 |   | 555 |   | 512 |   | 413 |   | 314 |   | 215 |

as the spreading codes. The code generation of OCFHC is the algorithm in which the non-zero primitive element of Galois field $GF(P^k)$ follows the mod-$P$ addition with all elements of $GF(P^k)$, where $P$ is the prime number for the hopping calculation and $k$ is an integer. If the Galois field $GF(P^k) = \{0, y_1, y_2, \ldots, y_{Pk-2}\}$, with non-zero primitive elements of $G(P^k) = \{0, x_1, x_2, \ldots, x_{Pk-2}\}$ we have:

$$F_{c_i} = \{f_{c_i}(0), f_{c_i}(1), \ldots, f_{c_i}(P^k - 2)\} \tag{2.61}$$

$$f_{c_i}(i) = x_j \oplus_P y_i + 1 \tag{2.62}$$

where the '+1' in frequency code $f_{c_i}(i)$ in Equation (2.63) is used to avoid the occurrence of channel index zero against the zeros in time-spreading sequences.

Accordingly, to obtain the best correlation value, $P^k$ should be greater than or equal to three to keep every channel appointed to each one in the spreading code once at least. To simplify the generation process, the prime number $P$ is assigned as two. Thus the formula can be re-expressed as:

$$f_{c_i}(i) = x_j \oplus_2 y_i = x_j \oplus y_i \tag{2.63}$$

where the symbol $\oplus$ represents logic exclusive-OR gate.

To better understand this 2D code family, here is an example when $P = 2, k = 3$ and $OOC(7, 3, 1)$.

$$GF(P^k) = GF(2^3) = \{0, y_1, y_2, \ldots, y_6\} = \{0, 2, 4, 3, 6, 7, 5\} \tag{2.64}$$

$$G(2^3) = \{0, x_1, x_2, \ldots, x_6\} = \{1, 2, 4, 3, 6, 7, 5\} \tag{2.65}$$

where $\{0, 2, 4, 3, 6, 7, 5\} \equiv \{000, 001, 010, 100, 011, 101, 111, 101\}$ and $\{1, 2, 4, 3, 6, 7, 5\} \equiv \{001, 010, 100, 011, 110, 111, 101\}$.

Now, when $i = 3$; $F_{c_3} = \{f_{c_3}(0), f_{c_3}(1), \ldots, f_{c_3}(6)\}$ and $f_{c_3}(3), -1 = xor(4, 3) = xor(100, 011) = (111) \equiv 7$.

An example of the OCFHC $(2^3)$ that generates seven groups within seven elements of sequences where the OOC code sequence of 1101000 from OOC (7, 3, 1) can be employed in each group is shown in Table 2.19. The three-code bundle has to be selected accordingly as in Table 2.19 for $w = 3$ in the deployed OOC.

The 2D OCFHC/OOC code sequences are illustrated in Table 2.20. As shown, seven groups of OCFHC are employed in the OOC channel (i.e. 1101000) generating 2D sequences. To increase the cardinality, the OCFHC group numbers can also be added as an additional group in Table 2.20, generating 56 2D spreading codes that can be spread in the time domain and hopped in the wavelength domain.

### 2.6.1.1 Evaluation and Analysis of Code Properties

The analysis of correlation function values, bit-error rate, codeset cardinality and bandwidth (spectral) efficiency are extensively taken into account for spreading-code performance and scalability. By the combination of these metrics, the composite and practical decisions should be made in the design stage of an OCDMA system and/or network architecture.

**Table 2.19** One-coincidence frequency hop codes

| i/j | Frequency Hopping Codes |
|---|---|
| | 0 1 2 3 4 5 6 0 1 |
| 0 | 2 3 5 4 7 8 6 2 3 |
| 1 | 1 4 6 3 8 7 5 1 4 |
| 2 | 4 1 7 2 5 6 6 4 1 |
| 3 | 6 7 1 6 3 4 2 6 7 |
| 4 | 3 2 8 1 6 5 7 3 2 |
| 5 | 8 5 3 6 1 2 4 8 5 |
| 6 | 7 6 4 5 2 1 3 7 6 |
| 7 | 5 8 2 7 4 3 1 5 8 |

**Table 2.20** OCFHC/OOC code sequences

| i | Group 1 | Group 2 | | Group 7 | Add. Group |
|---|---|---|---|---|---|
| 1 | 2 3 0 5 0 0 0 | 1 4 0 6 0 0 0 | | 5 8 0 2 0 0 0 | 1 1 0 1 0 0 0 |
| 2 | 3 5 0 4 0 0 0 | 4 6 0 3 0 0 0 | | 8 2 0 7 0 0 0 | 2 2 0 2 0 0 0 |
| 3 | 5 4 0 7 0 0 0 | 6 3 0 8 0 0 0 | | 2 7 0 4 0 0 0 | 3 3 0 3 0 0 0 |
| 4 | 4 7 0 8 0 0 0 | 3 8 0 7 0 0 0 | ......... | 7 4 0 3 0 0 0 | 4 4 0 4 0 0 0 |
| 5 | 7 8 0 6 0 0 0 | 8 7 0 5 0 0 0 | | 4 3 0 1 0 0 0 | 5 5 0 5 0 0 0 |
| 6 | 8 6 0 2 0 0 0 | 7 5 0 1 0 0 0 | | 3 1 0 5 0 0 0 | 6 6 0 6 0 0 0 |
| 7 | 6 2 0 3 0 0 0 | 5 1 0 4 0 0 0 | | 1 5 0 8 0 0 0 | 7 7 0 7 0 0 0 |

The 2D signature codes usually consist of (wavelength) hopping over the (time) spreading codes. The correlation process of 2D signature code can be considered as the decoding process of two sequence codes, where the correlator can collect the hits across different channels and then compare with the signature code in the memory bank as shown in Figure 2.32. If the position of ones and channel index are exactly the same at the time-slot, the output correlator will give a one hit, otherwise a zero hit.

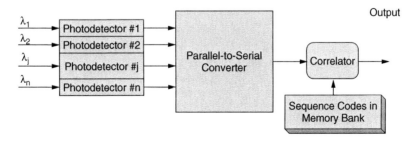

**Figure 2.32** Decoding process of 2D signature codes

The bit-error rate (BER) is the average possibility of error occurrences among different sequence codes and obviously increases when the number of simultaneous subscribers grows. The sustainable optical communications system performance is assumed error-free under $10^{-9}$.

The BER of 1D spreading codes can be calculated by the hits possibility between ones in the code sequences. According to the research by Jian-Guo Zhang et al. [33], the calculation is ideally comprised of the possibility of binary data occurrence, the threshold decoding range and the hits possibility between ones within different code sequences, expressed as:

$$\text{BER}_{1D} = \frac{1}{2} \sum_{i=0}^{w/2} \left[ (-1)^i \binom{w/2}{i} \left(1 - \frac{2f \times i}{w}\right)^{K-1} \right] \tag{2.66}$$

where $w$ is the code-weight depending on the number of ones, and $K$ is the number of subscribers. More importantly, $f = w^2/4L$ varies from different spreading code families depending on the maximum decoding slot distance (SD), representing the decoding distance between first and last successive ones in the sequences. $L$ denotes the code-length maintained under the correlation constraints. For example, $L$ equals $P^2$ in most prime code families and equals $2 \times SD + 2 = 2wP - 2$ in a modified prime code family [28].

The BER of 2D spreading codes can be expressed as the hits possibility of different code sequences occurring at the same time-slot and wavelength channel. According to Wing C. Wong et al. [58], the BER mainly depends on the possibility of threshold decoding, the product of hits possibility $f$ by non-hits possibility $1 - f$.

$$BER_{2D} = \frac{1}{2} \sum_{i=T_h}^{K-1} \left[ \binom{K-1}{i} f^i f^{K-1-i} \right] \tag{2.67}$$

where $T_h$ denotes the decoding threshold, normally as the code-weight or a bit lower than the code-weight and $K$ is the total number of simultaneous users. $f$ is the sum of the hits function to different groups depending on the code-weight $w$, code-length $N$, codeset cardinality $\phi_c$ and the hopping factor $P_h$.

For example in PC/OOC sequences the average hits from the first group to group $P_h^2$ can be expressed as:

$$f_0 = \frac{w^2(\phi_c \times P_h - 1) + (w - 1)^2}{2N_s(\phi_c \times P_h^2 - 1)} \tag{2.68}$$

And for the additional group is written as:

$$f_1 = \frac{w^2(\phi_c \times P_h - 1)}{2N_s(\phi_c \times P_h^2 - 1)} \tag{2.69}$$

Also, in OCFHC/OOC the average hits from the first group to group $P^k$ can be calculated as [55]:

$$f_0 = \frac{w^2\{C_1 + \frac{1}{2}[w^2(P^k - 1)C_1 - w] + C_2\}}{2N_s(\phi_c - 1)} \tag{2.70}$$

and for the additional group is written as:

$$f_1 = \frac{w^2\{C_1 - 1 + (P^k - 1)C_1 + C_2\}}{2N_s(\phi_c - 1)} \tag{2.71}$$

where $C_1 = \left\lfloor \frac{N_s - 1}{w(w-1)} \right\rfloor, C_2 = \left\lfloor \frac{P^k - 1}{w(w-1)} \right\rfloor$ and $k$ is the user-defined integer introduced in Section 2.6.1. Note that $\lfloor x \rfloor$ denotes the lower-round integer of $x$, i.e. $\lfloor 16.6 \rfloor = 16$.

Consequently, the hits possibility of each group is calculated and now the overall hits is $f = \frac{\phi_{c0}}{\phi_c}f_0 + \frac{\phi_{c1}}{\phi_c}f_1$ where $\phi_c$ is the total cardinality of additional group cardinality $\phi_{c_1}$ and normal cardinality of $\phi_{c_0}$.

As for the network scalability, the system capacity is the amount from the spreading codeset cardinality of 1D and/or 2D code families accommodated simultaneously in the optical channels under the particular BER constraint. The 1D codeset can be functionally segmented into prime code family and optical orthogonal code family, with the corresponding prime code $P$ and the code-length of $N_s$ as introduced in Sections 2.3 and 2.4. Table 2.21 compares the properties of various code families, including code-lengths, code-weights and cardinalities.

Also the cardinality of 2D spreading codes can be expressed as the product of the available hopping codes and the cardinality of spreading codes, as shown in Table 2.22.

In practice, the cardinality of all codesets should be extensively discussed under the sustainable BER constraints of $10^{-9}$. Therefore, the methods of BER and cardinality for

**Table 2.21** Properties of various 1D spreading codes

| Code Family | Code-Length | Code-Weight | Cardinality |
|---|---|---|---|
| PC | $P^2$ | $P$ | $P$ |
| MPC | $P^2$ | $P$ | $P^2$ |
| n-MPC | $P^2$ | $P + 1$ | $P^2$ |
| PMPC | $P^2$ | $P + 1$ | $P^2$ |
| GPMPC | $P^2$ | $P + 2$ | $P^2$ |
| T-MPC | $P^2$ | $P$ | $2 \times P^2$ |
| MQC | $P^2 + P$ | $P + 1$ | $P^2 - P$ |
| $(N_s, 3, 1)$ | $N_s$ | 3 | $\left\lfloor \dfrac{N_s - 1}{6} \right\rfloor$ |
| $(N_s, 4, 1)$ | $N_s$ | 4 | $\left\lfloor \dfrac{N_s - 1}{12} \right\rfloor$ |

**Table 2.22**  Properties of 2D spreading codes

| Code Family | Code-Length | Hopping Cardinality | Spreading Cardinality | Total Cardinality |
|---|---|---|---|---|
| PC/OOC | $N_s$ | $P^2$ | $\left\lfloor \dfrac{N_s - 1}{w(w - 1)} \right\rfloor$ | $P^2 \left\lfloor \dfrac{N_s - 1}{w(w - 1)} \right\rfloor$ |
| OCFHC/OOC | $N_s$ | $P^k$ | $\left\lfloor \dfrac{N_s - 1}{w(w - 1)} \right\rfloor$ | $P^k \left\lfloor \dfrac{N_s - 1}{w(w - 1)} \right\rfloor$ |

different codes are interactively combined to identify the system performance with the increasing number of subscribers.

## 2.7  Channel Encoding in OCDMA Systems

In OCDMA systems, the multiuser interference (MUI), which comes from those active users who are sending and receiving signals, could significantly degrade the overall performance of the communication link. Therefore, some research has been applied on channel coding techniques to improve the performance of OCDMA communications systems in the presence of MUI. Recently, a class of high performance error correcting codes known as Turbo codes has attracted a great deal of attention in optical communications mainly due to the high level of error correcting capability, reasonable encoding/decoding complexity and high coding gain [44]. In this section, error correcting codes as channel encoding techniques in OCDMA are introduced and studied, in order to enhance the overall system performance.

Due to the optical nature of signalling limitation, the spreading data in OCDMA should be unipolar-coded to minimize signal distortion, mainly caused by shot noise, thermal noise and photon dispersion. In the following, the Manchester code and the differential Manchester code as line codes, and the convolutional encoder and Viterbi decoder as forward error control code are practically implemented into an optical channel.

### 2.7.1  Manchester Codes

Manchester code and differential Manchester code [59] are return-to-zero (RZ) line codes and work as a special case of binary phase shift keying (BPSK) when modulating signals within square wave carriers as shown in Figure 2.33. The major drawback is to double the transmission bandwidth since the bit duration became half the normal duration. Nevertheless, RZ signalling are in favour of current ultra high-speed optical communication links (e.g. 40G and 100G) because the fewer bit duration than unipolar not-return-to-zero (NRZ) duration reduces the nonlinearity effect of fibre-optic.

At the same time, there are absolute advantages over other baseband transmission signalling including self-clocking or self-synchronizing function, simplified arbitrary binary codes and the longest repeating signals are only two same symbols. In optics, the Manchester code can represent one by a zero-to-one transition, otherwise zero by one-to-zero

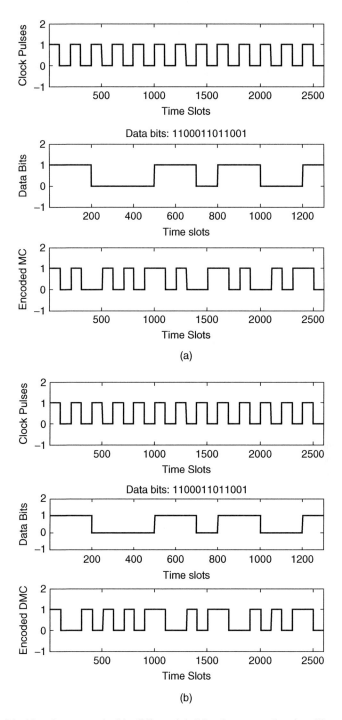

**Figure 2.33** (a) Manchester and (b) differential Manchester code signalling for data bits 1100011011001

transition. Compared with Manchester code, differential Manchester coding technique can perform less error-prone detection, depending on transmitting signal transitions, against shot noise and thermal noise in optical channel [60]. Accordingly, at the initialization, the differential Manchester code is defaulted using Manchester coding rule and then the one represents the first state following the previous and the second is complementary state. Otherwise zero indicates the first state against the previous and the rest is the reverse state.

## 2.7.2  Convolutional Codes

Claude E. Shannon showed that error-free communication can be achieved by using an infinitely long random code but this is quite impractical due to the enormous efforts required for decoding. However, he also introduced the principle of how to approach the capacity by splitting the incoming data into blocks containing as many bits as possible, which can be easily provided by the principle of code concatenation shown in Figure 2.34 [61]. As can be seen, the output of one encoder is fed into the input of next encoder. This process is carrying on up to the final encoder before the channel known as the inner encoder. As a result, the complexity of decoding at the receiver side is simply reduced by applying each of the 'component decoders' from the inner to the outer.

In order to randomize the input sequence an interleaver is used between component encoders. When we are dealing with concatenated code with two constituent codes, it may be placed between the outer and inner encoders to scramble the bits at the output of the outer encoder before it enters the inner encoder. Similarly, a de-interleaver with the same 'permutation pattern' is used at the receiver side between inner and outer decoders to scramble the bits at the output of inner decoder before it enters the outer decoder. Figure 2.35 illustrates the operation of interleaver and de-interleaver.

In Turbo codes, the convolutional codes are used as component or sometimes called 'constituent' encoders/decoders, separated by random interleavers. Therefore, in order to employ Turbo codes for the purpose of channel coding of the OCDMA systems, first of all it is necessary to have the knowledge of convolutional codes including:

- Design, construction, and fundamental parameters
- Encoding procedure and decoding algorithm
- Performance metrics and analysis in terms of bit-error rate

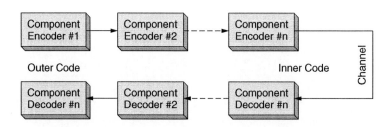

**Figure 2.34**  Code concatenation principle

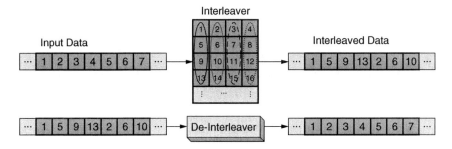

**Figure 2.35**  Interleaver and de-interleaver operation

In convolutional codes, the redundant bits are produced by inserting information bits into shift-registers, then adding the content of shift-registers and the information bits through a modulo-2 adder (e.g. XOR gate), and unlike block codes they may have infinite length [62].

In the case of Turbo codes, we are going to use the recursive systematic convolutional codes with the following characteristics:

- A systematic code is one in which the output sequence contains either a directly appeared input sequence or an easily recognizable sequence of the input bits, while the sequence of non-systematic code does not contain the data sequence.
- A recursive-systematic scheme can be easily obtained by introducing a feedback to the input of the encoder. Systematic codes allow quick-look-in (QLI) as compared with non-systematic codes, and they also require less hardware for encoding. Therefore, they are often preferable to the non-systematic codes. QLI is in a feedforward encoder where two parity sequences generated by one of the constituent recursive convolutional codes sum modulo-2 to the systematic message sequence [63, 64].

Convolutional codes are commonly specified by three parameters:

$k$: the number of parallel input information bits
$n$: the number of parallel output information bits
$m$: the maximum number of stages in any shift-register

The code rate $r$ is defined as a ratio of the parallel input information bits to the number of parallel output encoded bits, i.e. $r = k/n$. Moreover, the number of bits in the encoder memory bank, on which the output depends, is known as a constraint length and is denoted by $L$ [65].

### 2.7.2.1  Convolutional Encoding

Convolutional codes add redundancy in a quasi-continuous manner by adding the content of shift-registers with input bit stream through a modulo-2 adder. The encoder structure

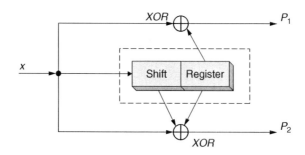

**Figure 2.36**  A convolutional code with rate 1/2

is easy to draw from its parameters and can be represented in different but equivalent ways [61]. Figure 2.36 depicts a rate $r = k/n = 1/2$ systematic convolutional encoder $L = 3$ with generator vectors $g_1 = (101)$ and $g_2 = (111)$.

Generator vectors are used to describe the connections between the shift-registers and the module-2 adders (the leftmost vector element or lowest-order coefficient represents the connection to the leftmost stage). The output of the upper and lower encoded channel sequences can then be expressed as multiplication of generator polynomials with input polynomial, i.e. $T^{(x)} = I^{(x)} \cdot g^{(x)}$.

The sequence generation can also be shown on a state diagram according to the encoder and decoder inputs/outputs. In order to draw a state diagram of the encoder, the state table should be derived. Table 2.23 and Figure 2.37 show the state table and state diagram of the designed encoder respectively.

Now, assuming that the input sequence is $x = (1011)$, from the state diagram the corresponding output would be $c = (11\ 01\ 00\ 10)$. The trellis diagram is basically an alternate way of viewing the state diagram but it is generally preferred because it represents events in a linear time sequencing manner and is widely used as a decoding tool. The trellis diagram is drawn by connecting all the possible states as time progresses [66]. Figure 2.38 shows the trellis diagram and the output path for the input sequence $x = (1011)$.

Note that the upward branch is the output path assigned to input zero, and conversely the downward branch is assigned to the input information bit one. In order to show the state transitions and their corresponding input/output values a legend usually accompanies the trellis diagram [67].

**Table 2.23**  State table of the encoder

| Present state | Input | Output | Next state |
|---|---|---|---|
| 00 | 0 | 00 | 00 |
|    | 1 | 11 | 10 |
| 01 | 0 | 11 | 00 |
|    | 1 | 00 | 10 |
| 10 | 0 | 01 | 01 |
|    | 1 | 10 | 11 |
| 11 | 0 | 10 | 01 |

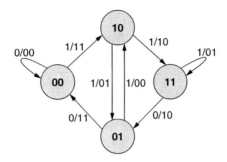

**Figure 2.37** State diagram of designed encoder

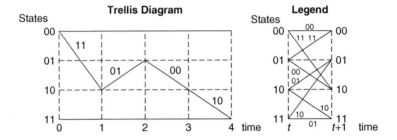

**Figure 2.38** Trellis diagram of the encoder

At the end of the transmission, the encoder should end in a defined state (usually the all-zero state) using tail bits. Although tail bits cause a loss in spectral efficiency of the system because they are extra bits that do not carry information, they significantly reduce the probability of error for the latest transmitted data.

### 2.7.2.2 Convolutional Decoding

The analysis of decoding convolutional code is performed by using a powerful technique known as the Viterbi algorithm generated in 1967 by Professor Andrew J. Viterbi [68]. The Viterbi algorithm is an iterative process which computes maximum-likelihood estimation on the transmitted code sequence $y$ from the received sequence $r$ such that it maximizes the probability that sequence $r$ is received, conditioned on the estimated code sequence [67].

In order to find the maximum-likelihood selection, we should compute a metric for every possible path through the trellis diagram. The Hamming distance can be used as a metric if the demodulator provides only hard decisions [40]. Recall that the Hamming distance $d_H(x, y)$ between two codewords is the number of different bits, where $x, y$ are code vectors and $x_n, y_n \in \{0, 1\}$ [61]:

$$d_H(x, y) = \sum_n |x_n - y_n| \qquad (2.72)$$

The encoded sequence $c = (11\ 01\ 00\ 10)$ is obtained as a result of input $x = (1011)$. Assuming that sequence $r = (10\ 01\ 00\ 10)$ is the received one with an error on the second bit. For a clear understanding, the complete convolutional code system of this example is depicted in Figure 2.39.

Now, start decoding of the received sequence $r = (10\ 01\ 00\ 10)$ using the trellis diagram and the Hamming distance between the received and codeword sequences.

As can be seen in Figure 2.40, at time 1, the Hamming distance between the first branch of the received sequence and each of the corresponding two possible trial sequence branches is 1. By iterating this process for the second received sequence, the distances are 2, 1, 2 and 1. Note that the underlined numbers at each node are the cumulative distances between the received sequence and one of the paths from the base of the trellis.

Following the same process toward the third level node, as seen in Figure 2.41, it happens for the first time that there are two unique paths leading to each node and as a result there are two distinct Hamming distances for each node. In this case, for each node we choose the sequence having the shorter distance; for instance for a node at the top-right corner of Figure 2.41 at time 3, we choose distance 2 between 2 and 3. When there are equal distances, we arbitrarily choose a particular branch into such a node.

The situation at the fourth node is also shown in Figure 2.41. Now, according to Viterbi algorithm we should compare the two paths of each node and only the path with the shorter distance is retained. Therefore, among four possible paths, a path with the smallest Hamming distance is the finally selected. The Hamming distances and corresponding output sequences of the final stage are represented in Table 2.24 where the selected path distance and corresponding output sequence are found in the forth sequence number.

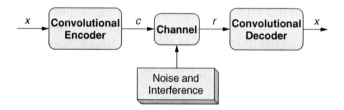

**Figure 2.39**   Generic convolutional code system

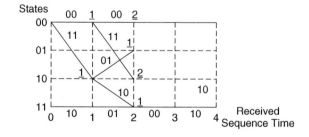

**Figure 2.40**   Decoding of first and second branches

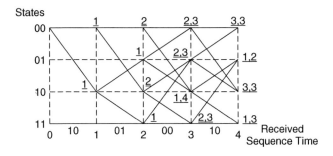

**Figure 2.41**  Last stage of decoding the received sequence

**Table 2.24**  Final stage output sequences

| Sequence Number | Sequences | Hamming Distance from Received Sequence |
|---|---|---|
| 1 | 11 11 01 11 | 3 |
| 2 | 11 01 00 01 | 1 |
| 3 | 11 10 10 00 | 3 |
| 4 | 11 01 00 10 | 1 |

It is apparent from Table 2.24 that the sequence numbers of two and four have the minimum Hamming distance among the other trial sequences. Since the fourth sequence is much more similar to the received sequence, it is chosen as a correct received sequence. However, the decision between the second and fourth sequences would also depend on channel characteristics.

The error on the second bit of the output sequence has been corrected simply by measuring the closeness or similarity of two binary sequences (Hamming distance) and using the smaller distance to find the most likely output path in the trellis diagram. According to trellis diagram, the decoded sequence is obtained as $\acute{x} = (1011)$.

In this example, we have assumed that only the hard decision is provided in the demodulator. There will be better performance (up to 2 dB) when soft decision-making is used. The details of decoding based on the soft decision algorithms for this convolutional code will be briefly discussed in Section 2.7.4.7.

### 2.7.3  Turbo Codes

In this section, the main three types of Turbo code structures namely parallel, serial and hybrid concatenated convolutional codes are studied. Note that due to the fixed length interleavers, Turbo codes are a class of block codes though they are a concatenation of convolutional codes [62].

### 2.7.3.1 Parallel Concatenation Convolutional Codes (PCCC) Encoder/Decoder

The conventional Turbo code introduced by C. Berrou [69] comprises parallel concatenation of two recursive systematic convolutional codes (PCCC) separated by a random interleaver. The recursive systematic codes (RSC) used in PCCC Turbo coding have short constraint length to optimize computation complexity, and the total latency of the system as shown in Figure 2.42.

Although low-rate encoders offer very strong protection to the transmitted message, they are not bandwidth efficient [70]. To improve the bandwidth efficiency, high-rate systems are normally considered, e.g. 1/3. Figure 2.43 depicts the structure of a PCCC with rate $r = 1/3$.

The outputs of the encoders consist of systematic bit $y_k^{1,s}$ which directly appears in the output sequence, and two parity bits $y_k^{1,p}$ and $y_k^{2,p}$ from the two RSC encoders. These rate-one separate outputs from each encoder are known as code fragments. Accordingly a Turbo code with rate $r = 1/3$ can be interpreted as the parallel concatenation of three

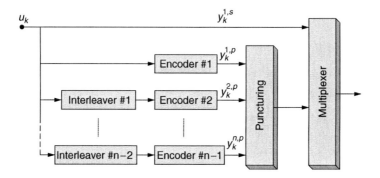

**Figure 2.42**   PCCC Turbo code encoder with rate $1/n$

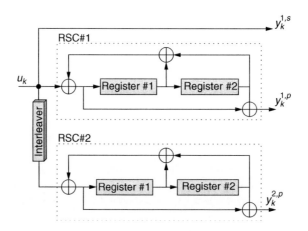

**Figure 2.43**   PCCC encoder with rate 1/3

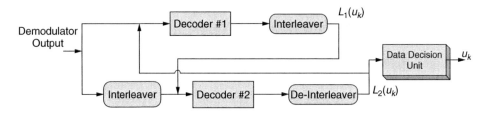

**Figure 2.44**   PCCC decoder with rate 1/3

code fragments. Representing the generator vector of each fragment in octal, the encoder in Figure 2.43 could be expressed as (1, 5/7, 5/7) code.

The first RSC encoder in Figure 2.43 directly encodes the input sequence while the second RSC encoder encodes the permuted version of the input sequence generated from the pseudo-random interleaver in another branch. As a result of this process, a large constraint length and randomized sequence is generated which helps in approaching the idea of Shannon.

In the receiver side, the Turbo decoding cannot utilize the conventional Viterbi decoding algorithm due to the code-length and random sequences. Turbo decoding is typically performed in an iterative manner where iteration consists of one decoding phase for each code sequence. The decoder structure for the encoder in Figure 2.43 is shown in Figure 2.44.

At every stage, the code sequence is decoded separately and the output of the decoder, the extrinsic information $L(u_k)$, is used as a priori information for the subsequent decoder during the second stage. This process continues until some criterion of the reliability of the decision is fulfilled.

### Puncturing

Puncturing is the act of adaptation of a code rate against the source rate and available bandwidth. For the encoder in Figure 2.43, it is assumed that the coded sequences $y_k^{1,s} = (111)$ as systematic bits, $y_k^{1,p} = (101)$ as parity bit stream of first RSC encoder and $y_k^{2,p} = (100)$ as parity bit stream of second RSC encoder, are a result of the three-bit trial input sequence. The parity check bits can be equally punctured to obtain PCCC Turbo code with rate $r = 1/3$ as illustrated in Figure 2.45.

The advantage of the puncturing technique is that it requires no changes in the decoder rate. Nevertheless, by puncturing we also miss some of the parity check bits and therefore it degrades the system performance. So whether the puncturing technique should be applied depends on the channel characterization [70].

### 2.7.3.2   Serial Concatenated Convolutional Code (SCCC) Encoder/Decoder

Serial concatenated convolutional code (SCCC) with rate $r = 1/3$ consists of two cascaded constituent codes of rate 1/2 and 2/3 separated by an interleaver as shown in Figure 2.46 [71].

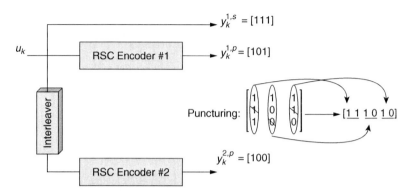

**Figure 2.45**   Puncturing of a PCCC Turbo code with rate 1/3

**Figure 2.46**   SCCC encoder with rate 1/3

The encoding process in SCCC is slightly different from the PCCC scheme. The input sequence is first encoded by the outer encoder, then the inner encoder uses the permuted version of the outer encoder output, and then the overall rate of the encoder is equal to the multiplication of the rate of all the constituent codes.

At the bit-error rate of $10^{-6}$, SCCC with rate 1/2 and $N = 256$ is almost 1.5 dB more power efficient than the PCCC with the same rate and Interleaver size [72]. However, they yield algebraically increasing decoding complexity [73]. The block diagram of SCCC decoder is shown in Figure 2.47.

The outer decoder in Figure 2.47 makes no use of direct input where the output of the outer decoder is solely obtained from the extrinsic information of the inner decoder. Thus, the key problem is how to obtain extrinsic information from the outer decoder and

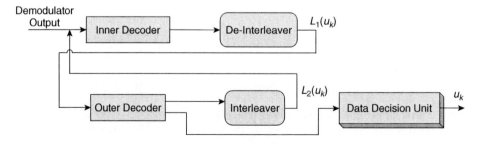

**Figure 2.47**   SCCC decoder with rate 1/3

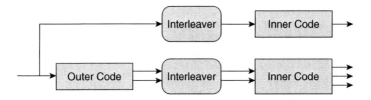

**Figure 2.48** HCCC encoder with rate 1/4

insert it into the inner decoder as a priori information. Due to the decoding complexities, the PCCC will be utilized for the Turbo coding of OCDMA system in Section 2.8.

A hybrid concatenated convolutional code (HCCC) is a combined version of PCCC and SCCC schemes. An example is shown in Figure 2.48.

The low coding rate and delay of its components result in a significant latency in comparison to PCCC and SCCC schemes. Therefore, this scheme is only implementable in extremely high data rates where the resultant delay is tolerable [71].

## 2.7.4 Turbo Decoding Algorithms

The decoding of Turbo codes is more complex than convolutional codes, because the transmitted sequence is randomized, and it also has a larger constraint length. In order to design an efficient Turbo decoding in terms of power efficiency, the optimal and sub-optimal Turbo decoding algorithms should be investigated and analysed in terms of bit-error probability against the signal-to-noise ratio. Since Turbo codes are employed for channel coding of the communications systems, currently both encoding and decoding operations are performed in the electrical domain. Here, the performance of the decoding algorithms is considered for BPSK modulation and the additive white Gaussian noise (AWGN) channel. As discussed in Section 2.7.2, the maximum-likelihood (ML) hard decision decoding is used for convolutional codes, where a path in the trellis diagram with minimum Hamming distance from the received sequence is chosen to be the ML estimate on the transmitted codeword. In this case, the decoding is based on the decision as to whether the survivor or the competing path is an ML solution, whereas no information about the 'decision reliability' is provided. Therefore, in order to provide some sort of information about the reliability of the decision, some side information known as 'soft information', which is a measure of confidence for the decision together with the hard decision, should be provided. This type of decision is known as a 'soft decision' [69].

The soft decision is applied into Viterbi decoder to form a decoder which incorporates reliability values and known as 'soft output Viterbi algorithm (SOVA)' decoder. Similar to conventional Viterbi algorithm which Hamming distance is calculated for the ML survivor path, the SOVA calculates the reliability value only for the ML survivor path in order to find the most probable information sequence that was transmitted [71]. However, if we could calculate the most probable information bit that has been transmitted instead of a sequence, a measure of confidence for decision would inevitably approach the 'optimal decoding'. The algorithm for this purpose is known as the maximum a posteriori (MAP) algorithm [65] which considers all possible paths of the trellis diagram in order to minimize the bit-error probability.

### 2.7.4.1  Iterative Decoding

Although in Turbo codes it is not possible to obtain the maximum likelihood solution due to the presence of the interleaver, approaching an optimal solution is quite feasible by means of an 'iterative process'. In Turbo codes, an iterative process is used to improve performance where each component code is decoded separately by using the most recent decoded information from the other code's component. The iterative process can be applied to any type of decoding technique. Hence, an iterative Turbo decoder could be composed of concatenated SOVA or MAP decoders [65, 67, 71, 74].

### 2.7.4.2  Maximum A-Posteriori (MAP) Algorithm

Now, the thorough analysis of the MAP algorithm is discussed, followed by the modified versions of the MAP algorithm for reducing the computations.

In the MAP algorithm, all possible paths in the trellis diagram are considered to minimize the bit-error probability. Similar to the other decoding algorithms that we have already discussed, first of all we should derive a metric for MAP decoding algorithm as in the following sections.

#### Log-Likelihood Ratio

A log-likelihood ratio (LLR) is defined as the base-10 logarithm of the ratio of the probabilities for information bit value $u$ (0 or 1) at time $t$ representing ($+1$ or $-1$) respectively [69, 75]:

$$LLR(u_t) = \log(\frac{P(u_t = +1)}{P(u_t = -1)}) \qquad (2.73)$$

The sum of all probabilities of an event is equal to one:

$$P(u_t = +1) = 1 - P(u_t = -1) \qquad (2.74)$$

This Equation (2.74) can be rewritten in terms of probability of one event as:

$$LLR(u_t) = \log(\frac{P(u_t = +1)}{1 - P(u_t = +1)}) \qquad (2.75)$$

Basically, the sign of LLR is used for the hard decision while the magnitude is a measure for soft decision [40].

### 2.7.4.3  MAP Algorithm for Turbo Decoding

In a Turbo coded system with rate 1/4, the decoding process based on MAP algorithm can be interpreted as a iterative improvement of LLR of a-posteriori probabilities (APP) of each information bit value $u_t = +1$ and $u_t = -1$ given by [69]:

$$LLR(u_t) = \log \frac{P(u_t = +1|y_1^N)}{P(u_t = -1|y_1^N)} \qquad (2.76)$$

where $y_1^N$ is the decoder output sequence and $P(u_t = \pm 1)$ is the APP of bit value (+1 or −1). Each constituent Turbo decoder makes full use of the reliability (i.e. soft) information generated by the previous MAP decoder (at the first stage of iteration) as a priori information known as extrinsic information [75]. Extrinsic information is simply the processed output of each constituent decoder which is used as the a priori input to the second stage of iteration. The decoder carries on using soft information until the final iteration when some stopping criterion is met, based on the output at each iteration. Then, the final data bit sequence $\hat{u}_t$ is decided to be the final output of the decoder. The stopping criterion for the number of iterations chiefly depends on the signal-to-noise ratio (SNR). At higher SNR the results converge faster, thus a lower number of iterations is required as compared to the low SNR values.

The reliability of information data bits can be computed through the Bahl–Cocke–Jelinek–Raviv (BCJR) algorithm which is introduced in the following.

### 2.7.4.4  BCJR Algorithm

According to Bayes' theorem for conditional probability:

$$P(B|A) = \frac{P(A,B)}{P(A)} \tag{2.77}$$

Accordingly, we have here,

$$P(u_t = +1|y_1^N) = \frac{P(y_1^N, u_t = +1)}{P(y_1^N)} \tag{2.78}$$

$$P(u_t = -1|y_1^N) = \frac{P(y_1^N, u_t = -1)}{P(y_1^N)} \tag{2.79}$$

Thus, Equation (2.77) could be reformulated to:

$$LLR(u_t) = \log \frac{P(y_1^N, u_t = +1)}{P(y_1^N, u_t = -1)} \tag{2.80}$$

The joint probability in the numerator and denominator of Equation (2.81) is the same as replacing the information bit with starting state $\hat{s}$ and ending state $s$ in the trellis diagram illustrated in Figure 2.49.

As can be seen, $P(y_1^N, u_t = +1)$ can be interpreted as $P(\hat{s}, s, y_1^N)$ for $u_t = +1$ to express the probability that the upper path in the trellis diagram is selected. Hence Equation (2.81) can be derived as [66]:

$$LLR(u_t) = \log \frac{P(y_1^N, u_t = +1)}{P(y_1^N, u_t = -1)} = \frac{\sum_{u_t = +1} P(\hat{s}, s, y_1^N)}{\sum_{u_t = -1} P(\hat{s}, s, y_1^N)} \tag{2.81}$$

In order to identify $\hat{s}$ and $s$, the $N$ bits of received sequence $y_1^N$ can be split into three manageable pieces:

$$P(\hat{s}, s, y_1^N) = P(\hat{s}, s, y_1^{k-1}, y_k, y_{k+1}^N) \tag{2.82}$$

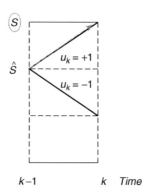

**Figure 2.49**   Starting and ending states in the trellis

where, by using Bayes' theorem:

$$P(A, B|C) = P(A|C) \cdot P(B|A, C) \tag{2.83}$$

Equation (2.83) can be expressed in terms of past $y_1^{k-1}$, present $y_k$ and future $y_{k+1}^N$ parts of the sequence as:

$$P(\hat{s}, s, y_1^{k-1}, y_k, y_{k+1}^N) = P(y_{k+1}^N|\hat{s}, s, y_1^{k-1}, y_k) \cdot P(\hat{s}, s, y_1^{k-1}, y_k) \tag{2.84}$$

According to Bayes' rule and Equation (2.84), assuming that the future sequence is only dependent on present state $s$:

$$P(\hat{s}, s, y_1^{k-1}, y_k, y_{k+1}^N) = P(y_{k+1}^N|s) \cdot P(s, y_k|\hat{s}, y_1^{k-1}) \cdot P(\hat{s}, y_1^{k-1}) \tag{2.85}$$

In the MAP decoding algorithm, the 'forward, backward and transition metrics' are regarded as three components of Equation (2.86) as described by:

- Forward metric: $\alpha_k(\hat{s}) = P(\hat{s}, y_1^{k-1})$
- Backward metric: $\beta_{k+1}(s) = P(y_{k+1}^N|s)$
- Transition metric: $\gamma_k(\hat{s}, s) = P(s, y_k|\hat{s}, y_1^{k-1})$

The forward and backward metrics could be obtained by using the following forward and backward recursions [76]:

$$\alpha_k(s) = \sum_{\hat{s}} \alpha_{k-1}(\hat{s}) \cdot \gamma_k(\hat{s}, s) \tag{2.86}$$

$$\beta_{k-1}(\hat{s}) = \sum_{s} \beta_k(s) \cdot \gamma_k(\hat{s}, s) \tag{2.87}$$

Thus, the log-likelihood ratio of the information bit in Equation (2.82) can be derived in terms of its metrics as:

$$LLR(u_t) = \log \frac{\sum_{(\hat{s}, s), u=+1} \hat{\alpha}_k(\hat{s}) \gamma_k(\hat{s}, s) \hat{\beta}_{k+1}(s)}{\sum_{(\hat{s}, s), u=-1} \hat{\alpha}_k(\hat{s}) \gamma_k(\hat{s}, s) \hat{\beta}_{k+1}(s)} \tag{2.88}$$

where $\hat{\alpha}_k(\hat{s})$ and $\hat{\beta}_{k+1}(s)$ are normalized forward and backward metrics given by [66]:

$$\hat{\alpha}_k(\hat{s}) = \frac{\alpha_k(\hat{s})}{\sum_{\hat{s}} \alpha_k(\hat{s})} \qquad (2.89)$$

$$\hat{\beta}_{k+1}(s) = \frac{\beta_{k+1}(s)}{\sum_{\hat{s}} \sum_{s} \hat{\alpha}_k(s) \cdot \gamma_{k-1}(s, \hat{s})} \qquad (2.90)$$

### 2.7.4.5 Log-MAP Algorithm

Since the MAP algorithm is very complex due to the amount of computation needed, it could be simplified by performing the operations in the logarithmic domain. The forward, backward and transition metrics could be calculated in the logarithmic domain using the Jacobian logarithm [77]. Using this approach, the multiplication in Equations (2.87) and (2.88) is replaced by addition to save computation [78]:

$$\hat{\alpha}_k(s) = \max(\hat{\alpha}_{k-1}(\hat{s}) + \hat{\gamma}_k(\hat{s}, s)) \qquad (2.91)$$

$$\hat{\beta}_{k-1}(\hat{s}) = \max(\hat{\beta}_k(s) + \hat{\gamma}_k(\hat{s}, s)) \qquad (2.92)$$

where $\hat{\alpha}_k(s)$, $\hat{\beta}_k(s)$ and $\hat{\gamma}_k(\hat{s}, s)$ are logarithmic values of forward, backward and transition metrics. In order to preserve the original MAP algorithm, all maximizations can be augmented over two values with the correction function of:

$$\max(x_1 + x_2) = \max(x_1, x_2) + \log(1 + e^{-|x_1 - x_2|}) \qquad (2.93)$$

Thus, the log-likelihood of the information bit in Equation (2.82) can be derived in terms of the new metrics as follows [78]:

$$LLR(u_t) = \max_{(\hat{s},s),u=+1} [\hat{\alpha}_{k-1}(\hat{s}) + \hat{\beta}_k(s) + \hat{\gamma}_k(\hat{s}, s)]$$

$$- \max_{(\hat{s},s),u=-1} [\hat{\alpha}_{k-1}(\hat{s}) + \hat{\beta}_k(s) + \hat{\gamma}_k(\hat{s}, s)] \qquad (2.94)$$

It is also noted that the significant saving on computations arises since the division operation is replaced by subtraction in the logarithmic domain.

### 2.7.4.6 Max-Log-MAP Algorithm

In the log-MAP algorithm, the logarithmic term corrects the approximation of the 'selection and comparison' operation. In the max-log-MAP algorithm the approximation is made by omitting the logarithmic term of Equation (2.94). Thus, the approximation is only based on the selection and comparison operations [75]:

$$\max(x_1 + x_2 + \cdots + x_n) = \max(x_1, x_2, \ldots, x_n) \qquad (2.95)$$

Using this approximation, the max-log-MAP algorithm results in sub-optimal performance unlike with the log-MAP and the original MAP. However, this algorithm is also considered

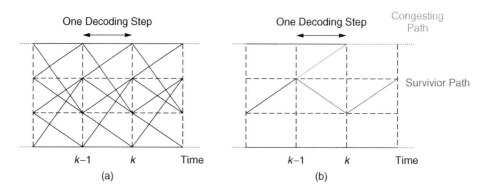

**Figure 2.50**   (a) MAP versus (b) SOVA decoding algorithms

for comparison with other optimal MAP algorithms and can be employed to the OCDMA system due to the following benefits:

- The log-likelihood of the information bit could be obtained from Equation (2.95), while the logarithmic term in Equation (2.94) is omitted, thus the max-log-MAP is comparatively easier to implement.
- Similar to the original MAP, the max-log-MAP algorithm attempts to minimize bit-error probability by approximating the metrics of the log-likelihood of the codeword. Thus, by using max-log-MAP, a significant improvement could be obtained in the performance of the system in comparison with the algorithms where estimation is based on computing the sequence error probability such as the following SOVA algorithm.

### 2.7.4.7   Soft-Output Viterbi Algorithm

As mentioned briefly in Section 2.7.2.2, a metric for soft decision could be incorporated in the conventional (hard) decision Viterbi algorithm. In the soft-output Viterbi algorithm, only two paths in the trellis diagram are considered, although the competing path may not be the best path. As shown in Figure 2.50, only one competing path at every decoding step is computed at each decoding iteration.

To compare with the MAP algorithm where bit-error probability is minimized, the SOVA minimizes the sequence error probability. Therefore, the MAP algorithm is significantly more accurate than SOVA at the expense of complexity.

If the aim is to employ optimal decoding algorithms for channel coding of the OCDMA systems, the analysis of algorithms is usually carried out based on bit-error probability rather than sequence error probability [67].

## 2.8   Turbo-Coded Optical CDMA

The addresses in an OCDMA system require the code sequence to be composed of unipolar (0,1) with good correlation properties. Good correlation properties refers to the fact that:

- The large auto-correlation peak value for each code sequence enables effective detection of the desired signal.
- The small out-of-phase auto-correlation values, known as auto-correlation side lobes, acquire synchronization between transmitter and desired receiver.
- The ideal and/or small cross-correlation value between code sequences reduces the interference from the other users, which is known as multiuser interference (MUI).

A codeset $\varphi(N, w, \lambda_a, \lambda_c)$ is a family of unipolar $(0, 1)$ sequences of length $N$ and weight $w$ with auto- and cross-correlation values of $\lambda_a$ and $\lambda_c$, respectively. The auto- and cross-correlation functions of sequences for any $X, Y \in \varphi$ and integer $l$ are described as:

$$C_{X,X}(l) = \sum_{n=0}^{N-1} X_n \cdot X_{n+l} = \begin{cases} \lambda_a & l = 0 \\ < \lambda_a & 0 < l \leq N - 1 \end{cases} \tag{2.96}$$

$$C_{X,Y}(l) = \sum_{n=0}^{N-1} X_n \cdot Y_{n+l} \leq \lambda_c \quad 0 \leq l \leq N - 1 \tag{2.97}$$

Here for the analysis, a weight-one coincidence optical orthogonal code (OOC) family where the out-of-phase auto-correlation and maximum cross-correlation values are bounded by one [4] is chosen as:

$$C_{C^k,C^l}(l) = \sum_{n=0}^{N-1} C_n^k \cdot C_{n-i}^l = \begin{cases} w & k = l \text{ and } j = 0 \\ \leq 1 & k = l \text{ and } 1 \leq j \leq N - 1 \\ \leq 1 & k \neq l \text{ and } 1 \leq j \leq N - 1 \end{cases} \tag{2.98}$$

As discussed in Section 2.3, the number of available users, $K$, for strict optimal OOC is bounded by:

$$K = |\varphi| \leq \left\lfloor \frac{(N-1)}{w(w-1)} \right\rfloor \tag{2.99}$$

Simply, the number of users is the maximum admissible number of available address codes for length $N$ and weight $w$.

The most interesting modulation schemes that are popular for incoherent OCDMA applications are amplitude shift keying (ASK), also called on-off keying (OOK), and pulse position modulation (PPM).

In the OOK scheme, bit one is used to represent the transmission of the code sequence while binary bit zero is used to represent the non-existence of a transmitted code sequence. For example, a three-bit symbol of 011 is OOK and PPM modulated as shown in Figure 2.51.

The $M$-ary PPM signalling is an interesting modulation scheme due to the significant enhancements in spectral efficiency of the overall OCDMA system. In this scheme, each transmitted symbol is represented by a laser pulse on one of $M$ disjoint time-slots. The simplest case of $M$-ary PPM is when $M = 2$, known as binary-PPM (BPPM). In BPPM, a time frame is only divided into two time-slots. To represent zero, the code sequence is transmitted within the first time-slot while bit one is represented by transmitting the code

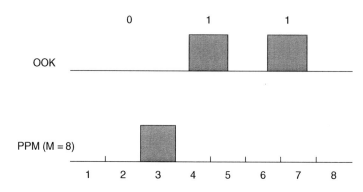

**Figure 2.51**   OOK versus PPM signalling for symbol 011

sequence within the second time-slot. Then each time-slot is further spread into $N$ chips at the OCDMA encoder as shown in Figure 2.52.

It is observed from Figure 2.51 that transmitting symbol 011 in OOK signalling requires two pulses, whereas in PPM signalling only one pulse represents the entire symbol. While OOK scheme is relatively simple to implement [79], it has poor power efficiency as compared to the PPM signalling format. Since the signal power is modulated in the OOK scheme, easy eavesdropping techniques are feasible by simple power detection even with no need for knowledge of the spreading codes in OCDMA [80, 81]. On the other hand, the bit duration, and as a result the chip duration, must be reduced significantly to support high bit rate applications such as video streaming. To do so, ultrashort pulses are required which are challenging in terms of generation, propagation and detection [82].

Since the major drawback of OCDMA systems is their spectral efficiency, we can increase the number of bits per symbol by increasing the number of time-slots. In *M-ary* PPM-OCDMA system, a periodic time frame is established and the transmitted symbols of length $\log_2^M$ bits are placed in one of $M$ disjoint time-slots, each with duration $T_s$. Figure 2.53 illustrates each time-slot which is further spread into $N$ chips of duration $T_c$ after the OCDMA encoder where $N$ is the OOC code-length. The code-length $N$ is also known as the 'spreading factor'.

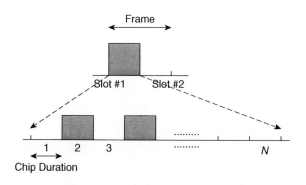

**Figure 2.52**   BPPM-OCDMA signalling format

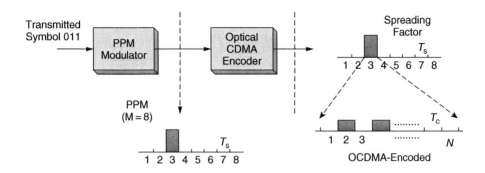

**Figure 2.53** PPM signalling for M = 8

## 2.8.1 Turbo-Coded Optical CDMA Transceivers

The block diagram of a Turbo-coded OCDMA transmitter is shown in Figure 2.54 where each block's responsibility will be introduced and discussed accordingly.

### Turbo Encoder

The information sequence of each user is fed into the Turbo encoder (PCCC, for example, see Section 2.7.3.1). Then the encoded output is fed into the multiplexer for parallel-to-serial conversion. Since Turbo codes are applied as error-correcting codes in this system, both encoding and decoding are performed in the electrical domain.

### Modulator

Before OCDMA encoding, the coded bit stream is first modulated by an *M-ary* PPM scheme, for example to further improve the system performance. As can be seen in Figure 2.53, one frame with duration $T = M \cdot T_s$ consists of $M$ time-slots where $M$ is a number of possible transmitted symbols. For example, for a symbol that is three bits long, $M = 8$ (i.e. $\log_2^8$) time-slots are needed to represent the transmitted symbols. Meanwhile, the laser is pulsed on the first chip position of the selected slot [83].

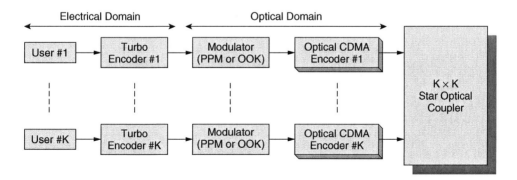

**Figure 2.54** Turbo-coded OCDMA transceiver block diagram

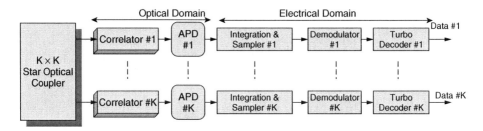

**Figure 2.55**  Turbo-coded OCDMA receiver block diagram

### *Optical CDMA Encoder*

The intended address code is formed by passing the output pulse of the modulator into the OCDMA encoder. For example, for a time-spreading scheme, each time-slot with width $T_s$ is further spread into $N$ chips with duration $T_c$. The corresponding position of these pulses is determined depending on the address codes. Recalling that $w$ is the OOC's code weight, the $w$ optical tapped-delay lines allocate the pulses according to the corresponding OOC sequences of the destination user. Different OCDMA encoder structures in OCDMA schemes, such as temporal and spectral, will be discussed in detail in Chapters 3, 4 and 5. The resultant optical signals from several users are combined and sent into the OCDMA network infrastructure independently.

Now, the block diagram of a Turbo-coded OCDMA receiver is shown in Figure 2.55, where the functionality of each block is introduced in the following.

### *Optical Correlator*

The received signal at the desired receiver includes not only the desired information but all other users' information and the background noise. Thus the intended information needs to be extracted by an optical correlator which can be a set of optical tapped-delay lines again, in the case of a time-spreading scheme, matched to the pulse positions and code sequences assigned at the transmitter [84]. Hence, each receiver could correctly decode its own signals with the same code sequence. If the received signal is correlated with the correct signature sequence, it hits the auto-correlation peak value, otherwise the output yields the cross-correlation value resulting from other users' interfering signals, known as multiuser interference (MUI). Consequently, since the MUI has a considerable negative effect on the system performance, the design and choice of optical spreading codes in itself is a science.

### *Avalanche Photodiode (APD)*

At the photodetector, the output of the correlator which is the desired signal at the presence of the MUI, is converted into an electrical signal. In this example, an avalanche photodiode (APD) is chosen for photodetection whereas other types such as PIN photodiodes are also applicable.

### *Integration and Sampler*

The electrical signal converted from the photodetector is directly integrated over chip duration $T_c$. Then, in order to estimate the photon-counts over each time-slot, the sampler extracts the $M = 8$ sampling signals at time $t = j \cdot T_s$ where $j = (0, 1, \ldots, 7)$, i.e. at $t = 0, T_s, 2T_s, \ldots, 7T_s$. Thus, the number of photon-counts over each time-slot is obtained at the output of the sampler to make the detection decision [85].

### Demodulation

The number of photon-counts collected from each time-slot is sent to the PPM demodulator to recover the original transmitted symbols. The highest number of photon-counts over each time-slot in the presence of MUI is chosen to be the transmitted symbol, similar to a maximum-likelihood decision.

### Turbo Decoder

The transmitted symbols in the presence of MUI are fed into the Turbo decoder to estimate and correct the transmitted bit stream errors. Then, the decoded information bits are generated.

## 2.8.2  Analysis of Uncoded OCDMA

The upper-bounds on bit-error probability of an uncoded PPM-OCDMA system are derived in this section. The following assumptions are made for the analysis of the system [75, 84]:

- The optical communication channel is intensity modulated.
- The optical characteristics of all users are identical.
- Synchronization between the transmitter and the intended receiver properly exists.
- Code sequences are weight-one coincidence OOCs.
- The signal photon rate is constant for optical tapped-delay line of all transmitters.
- The information of the first user is intended to be extracted among all $K$ users.

The output of the photodetector in the $i^{th}$ time-slot can be modelled as a Poisson random distribution:

$$Y_i = D_i + \sum_{k=2}^{K} I_i^k + N_i \quad i \in \{1, 2, \ldots, M\} \tag{2.100}$$

where $Y_i$ is the average photon-count of the $i^{th}$ time-slot at the output of the photodetector composed of Poisson photon-count components: the desired user signal $D_i$ (e.g. for user #1), the interference from the other users $I_i^k$ and the additive optical noise component $N_i$. In some optical fibre networks, the additive optical noise component is neglected for all time-slots [84]. Nevertheless, the actual background light is a fundamental factor of optical communication systems, so this noise parameter in the analysis of the system should be considered. Now, the expression for the lower-bound on probability of a correct slot decision is derived for any $i, j \in \{1, 2, \ldots, M\}$, and $i \neq j$:

- When $Y_i > Y_j$, then the symbol $i$ is chosen
- When $Y_i = Y_j$, assuming data symbols are equally likely, either symbol $i$ or $j$ is chosen randomly.

The probability of a correct decision for the first time-slot is lower-bounded by:

$$P_c \geq \sum_{i=1}^{M} P_r\{Y_i > Y_1, Y_i > Y_2, \ldots, Y_i > Y_{i-1}, Y_i > Y_{i+1}, \ldots, Y_i > Y_M | b_1 = i\} \cdot P_r\{b_1 = 1\}$$

(2.101)

where $b_t = i$ denotes the symbol $i$ at time $t$. To find the photon-counts of the first time-slot, i.e. $b_1 = 1$, Equation (2.102) can be rewritten as [75],

$$P_c \geq P_r\{Y_i > Y_1, Y_i > Y_2, \ldots, Y_i > Y_{i-1}, Y_i > Y_{i+1}, \ldots, Y_i > Y_M | b_1 = 1\}$$

$$\geq \sum_{l_1=0}^{K-1} \sum_{l_2=0}^{K-1-l_1} \cdots \sum_{l_M=0}^{K-1-l_{M-1}} P_{c_1} \cdot P_r\{k = 1\}$$

(2.102)

where $P_r\{k = 1\}$ can be calculated using the multinomial distribution of $k$ given by:

$$P_r\{k = 1\} = \frac{1}{M^{K-1}} \cdot \frac{(K-1)!}{l_1! \cdot l_2! \cdot \ldots \cdot l_M!}$$

(2.103)

From the Poisson distribution of the avalanche photodiode output (i.e. photodetector) $P_{c_1}$ is expressed as [75],

$$P_{c_1} = \sum_{k=1}^{\infty} POS(k, K_s + K_b + \lambda_s T_c l_1) \cdot \prod_{j=2}^{M} \left\{ \sum_{i=0}^{k-1} POS(i, K_b + \lambda_s T_c l_j) \right\}$$

(2.104)

where $POS(x, y) = y^x \cdot e^{-y}/x!$ denotes the Poisson mass function and $K_s = w\lambda_s T_c$ and $K_b = w\lambda_b T_c$ are the average photon-counts per symbol due to the desired signal and noise respectively. $w$ represents the code-weight and $\lambda_s$ is the photon absorption rate of the APD photodetector equal to $\eta P/h\nu$, where $\eta$ is the detector's quantum efficiency, $h$ is Planck's constant ($6.624 \times 10^{-34}$), $\nu$ is the optical frequency and $P$ is the average optical power. Also, $\lambda_b$ is the photon absorption rate of the detector due to the background noise or dark current. Accordingly, the probability of error can be obtained from the probability of a correct decision as:

$$P_e = 1 - P_c$$

(2.105)

Finally, the upper-bounded (subtracted from the lower-bounded BER) on bit-error probability of an uncoded PPM-OCDMA system can be derived as:

$$P_E \leq \frac{1}{2}\left(\frac{M}{M-1}\right) P_e$$

(2.106)

## 2.8.3 Analysis of Turbo-Coded OCDMA

In this section, the upper-bounded bit-error probability of Turbo-coded PPM-OCDMA system is derived. In addition to the assumptions that have been made for uncoded system, in this case we also assume that all users utilize parallel concatenated convolutional codes with rate 1/3 as Turbo encoder/decoder described in Section 2.7.3.1.

### 2.8.3.1   Bounds on Bit-Error Probability

As introduced in Section 2.7.3, the three bits at the encoder output can be expressed as a systematic bit $y_k^{1,s}$ (uncoded output) and two parity bits $y_k^{1,p}$ and $y_k^{2,p}$. These rate-one separate outputs from each encoder are referred to the code fragments. Each code fragment is denoted by $g(x,i,d)$ which is the number of paths of length $x$, input Hamming weight $i$, and output Hamming weight $d$. In a Turbo encoder with rate 1/3 introduced in Figure 2.43, the conditional probability of producing code fragments of Hamming weights $d_1$ and $d_2$ is given by [65, 75]:

$$P(d_1,d_2|i) = \frac{g_1(N,i,d_1) \cdot g_2(N,i,d_2)}{\binom{N}{i}} \tag{2.107}$$

where $N$ is the number of information bits and $\binom{N}{i}$ is the total number of codewords of Hamming weight $i$. From Equation (2.108), the number of codewords generated by information sequence of weight $i$ and two parity bit streams with weights of $d_1$ and $d_2$ are derived as:

$$T(i,d) = \sum_{d_1=0}^{N} \sum_{d_2=0}^{N} P(d_1,d_2|i) \tag{2.108}$$

Since the desired path and all-zero path at the trellis diagram differ in $i+d$ bit positions, $P_{c_1}$ should be replaced by $\acute{P}_{c_1}$, where the Hamming weight of the codeword is considered as $i+d$ in Equation (2.106):

$$\acute{P}_{c_1} = \sum_{k=1}^{\infty} POS(k, K_s + K_b + \lambda_s T_c l_1(i+d)) \cdot \prod_{j=2}^{M} \left\{ \sum_{z=0}^{k-1} POS(z, K_b + \lambda_s T_c l_j(i+d)) \right\} \tag{2.109}$$

Now from Equations (2.106), (2.109) and (2.110) the bit-error probability of (1, 5/7, 5/7) Turbo-coded PPM-OCDMA system is upper-bounded as:

$$P_E = \sum_{i=0}^{N} \sum_{d=0}^{2N} \frac{i}{N} T(i,d) \cdot \acute{P}_e \tag{2.110}$$

where $\acute{P}_e$ is the new probability of error calculated with the new $\acute{P}_{c_1}$ introduced in Equation (2.106).

### 2.8.3.2   Input-Output Weight Coefficients

In order to find an upper-bound on the bit-error probability of the overall system, e.g. Equation (2.111), the weight distribution of the Turbo code which is the number of codewords generated by an information sequence of weight $i$ and two parity bit streams of weights $d_1$ and $d_2$ should be determined. In doing so, the algorithm for weight distribution computation of the Turbo code is shown in a flow chart in Figure 2.56 [75, 84].

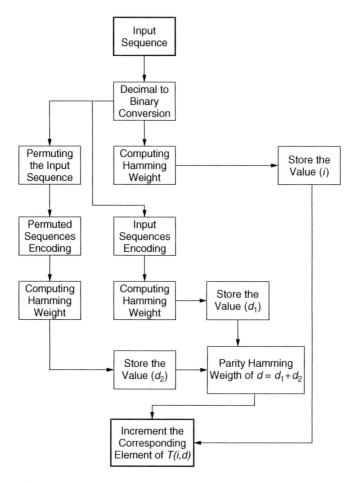

**Figure 2.56** Computing I/O weight coefficients of Turbo code

By following these steps, the weight distribution matrix will be obtained:

1. The maximum number of codewords for the interleaver size $N$ is $2^N$. For each of length-$N$ codeword, the Hamming weight $i$ is $0 \leq i \leq N$. This value is computed and stored.
2. Similarly, the Hamming weights of parity bit streams $d_1$ and $d_2$ are computed where the total Hamming weight of parity output $d = d_1 + d_2$.
3. Since $0 \leq i \leq N$ and $0 \leq d \leq 2N$, a $(N + 1) \times (2N + 1)$ matrix is defined where each element represents the number of codewords with input Hamming weight $i$ and parity output Hamming weight $d$:

$$T(i,d) = \begin{bmatrix} T_{(0,0)} & \cdots & T_{(0,N)} \\ \vdots & \ddots & \vdots \\ T_{(N,0)} & \cdots & T_{(N,2N)} \end{bmatrix} \tag{2.111}$$

The initial value for all elements is set to zero. For each input codeword and corresponding parity output, the Hamming weight is evaluated. Then the corresponding element of the matrix is either incremented or remains zero. This procedure should be carried out for all combinations of inputs of length $N$ to obtain a matrix which represents the number of codewords for input Hamming weight $i (0 \leq i \leq N)$ and output Hamming weight $d (0 \leq d \leq 2N)$.

The problem here is that $2^N$ is an enormous number and this procedure may take a very long time to run. Thus, the algorithm is implemented for the weight distribution computation of the upper-bound on the entire system. However, to better understand the procedure, here is an example of how to compute $T(3,d)$ elements of the weight distribution matrix for a small interleaver with size $N = 5$.

First of all, the elements of the weight distribution matrix $T(i,d)$ are initially set to zero:

$$T(i,d) = \begin{bmatrix} 0 & 0 & 0 & 0 & 0 & 0 & 0 & 0 & 0 & 0 & 0 \\ 0 & 0 & 0 & 0 & 0 & 0 & 0 & 0 & 0 & 0 & 0 \\ 0 & 0 & 0 & 0 & 0 & 0 & 0 & 0 & 0 & 0 & 0 \\ 0 & 0 & 0 & 0 & 0 & 0 & 0 & 0 & 0 & 0 & 0 \\ 0 & 0 & 0 & 0 & 0 & 0 & 0 & 0 & 0 & 0 & 0 \\ 0 & 0 & 0 & 0 & 0 & 0 & 0 & 0 & 0 & 0 & 0 \end{bmatrix}_{6 \times 11} \qquad (2.112)$$

where $0 \leq i \leq 5$ and $0 \leq d \leq 10$.

Now, considering the possible input sequences for $i = 3$, the corresponding parity bit streams are represented in Table 2.25. There are $\binom{5}{3} = \frac{5!}{2!3!} = 10$ codewords for $i = 3$.

The last column of the table represents the elements of the $4^{th}$ row in the weight distribution matrix. As can be observed from Table 2.25, there are four codewords with the input Hamming weight $i = 3$ and $d = 6$ as the maximum number of repeated codewords with the same code weight, and there is one codeword with the input Hamming weight $i = 3$ and $d = 3$; there are also two codewords with the input Hamming weight $i = 3$

**Table 2.25** Parity bit streams for input sequences with Hamming weight $i = 3$

| $i = 3$ No. | Input Bit Stream | Parity #1 | $d_1$ | Parity #2 | $d_2$ | $d$ | $T_{(3,d)}$ |
|---|---|---|---|---|---|---|---|
| 1 | 00111 | 00100 | 1 | 01001 | 2 | 3 | $T(3,3)$ |
| 2 | 01011 | 01110 | 3 | 10010 | 2 | 5 | $T(3,5)$ |
| 3 | 10011 | 11010 | 3 | 01011 | 3 | 6 | $T(3,6)$ |
| 4 | 01101 | 01011 | 3 | 11101 | 4 | 7 | $T(3,7)$ |
| 5 | 11001 | 10101 | 3 | 11111 | 5 | 8 | $T(3,8)$ |
| 6 | 01110 | 01001 | 2 | 10111 | 4 | 6 | $T(3,6)$ |
| 7 | 11100 | 10010 | 2 | 11010 | 3 | 5 | $T(3,5)$ |
| 8 | 11010 | 10111 | 4 | 10101 | 3 | 7 | $T(3,7)$ |
| 9 | 10110 | 11101 | 4 | 10010 | 2 | 6 | $T(3,6)$ |
| 10 | 10101 | 11111 | 5 | 00100 | 1 | 6 | $T(3,6)$ |

and $d = 5$, and so on. Then the corresponding zero element in matrix $T(i,d)$ should be set to the number of codewords seen in Table 2.25, while there is no codeword for $i = 3$ and $d = (0, 1\ 2, 4, 9, 10)$, thus the corresponding elements remains zero:

$$T(i,d) = \begin{bmatrix} 0 & 0 & 0 & 0 & 0 & 0 & 0 & 0 & 0 & 0 & 0 \\ 0 & 0 & 0 & 0 & 0 & 0 & 0 & 0 & 0 & 0 & 0 \\ 0 & 0 & 0 & 0 & 0 & 0 & 0 & 0 & 0 & 0 & 0 \\ 0 & 0 & 0 & 1 & 0 & 2 & 4 & 2 & 1 & 0 & 0 \\ 0 & 0 & 0 & 0 & 0 & 0 & 0 & 0 & 0 & 0 & 0 \\ 0 & 0 & 0 & 0 & 0 & 0 & 0 & 0 & 0 & 0 & 0 \end{bmatrix} \qquad (2.113)$$

## 2.9  Summary

This chapter is dedicated to reviewing well-known optical spreading codes including bipolar $m$-sequence, Gold codes, Hadamard–Walsh codes, and unipolar optical orthogonal codes, prime code families and codes with ideal cross-correlation. For codes with ideal in-phase cross-correlation, three algebraically constructed codeset families of balanced incomplete block design (BIBD), modified quadratic congruence (MQC) and modified frequency hopping (MFH) are introduced and analysed.

The codes with ideal in-phase cross-correlation are very important in the application of spectral-amplitude-coding OCDMA systems due to the fact that multiuser interference and/or beat noise in the wavelength domain significantly reduce. It has been shown that the overall system performance can be improved significantly when codes with ideal in-phase cross-correlation are used instead of the Hadamard code. These code families can also be utilized in synchronous OCDMA systems to cancel the multiple access interference as well as to increase the system capacity by supporting a greater number of subscribers.

On the other hand, to increase the OCDMA system security and cardinality, various domains and parameters are combined to create multidimensional codes which are encoded in time, space, polarization, phase and wavelength domains. However, multidimensional coding brings a lot of complexity to the system implementation and architecture, and so the focus is mainly on two-dimensional codes and encoding techniques in practice. Here in this chapter, typical two-dimensional spreading codes are studied and investigated.

Since various kinds of noise and impairments in the optical communications link cause variation of the bit values resulting in bit-errors, employing forward error correction techniques attracts a lot of attention in the (optical) communications signal processing research community. Popular convolutional codes and well-established Turbo codes are studied and introduced in this chapter along with numerous encoding and decoding methods and structures that can be utilized in the error-correction channel coding scheme. For example, Turbo codes are utilized for an OCDMA system with optical orthogonal codes as spreading code sequences as well as $M$-$ary$ pulse position modulation. Accordingly, the bit-error bounds on two uncoded and Turbo-coded systems are also derived and analysed.

Additionally, different methods of encoding and decoding Turbo codes – such as optimal decoding algorithm (i.e. original MAP algorithm), max-log-MAP and log-MAP as approximation algorithms – are studied. In the following chapters, the application of spreading codes in different OCDMA systems such as temporal or spectral will be studied with a detailed analysis and practical encoding/decoding structures.

# References

1. Yin, H. and Richardson, D.J. (2007) *Optical code division multiple access communication networks: theory and applications*. Tsinghua University Press, Beijing, China and Springer Verlag GmbH, Berlin, Germany.
2. Dixon, R.C. (1976) *Spread spectrum system*. Wiley-Interscience Publication, USA.
3. Buehrer, R.M. (2006) *Code division multiple access (CDMA)*. Morgan & Claypool Publishers, Colorado, USA.
4. Chung, F.R.K., Salehi, J.A. and Wei, V.K. (1989) Optical orthogonal codes: design, analysis and application. *IEEE Trans. on Info. Theory*, **35** (3), 595–605.
5. Chung, H. and Kumar, P. (1990) Optical orthogonal codes – new bounds and an optimal construction. *IEEE Trans. on Info. Theory*, **36** (4), 886–873.
6. Maric, S.V. (1993) New family of algebraically designed optical orthogonal codes for use in CDMA fiber optic networks. *Electronics Letters*, **29** (6), 538–539.
7. Liang, W. *et al*. (2008) A new family of 2D variable-weight optical orthogonal codes for OCDMA systems supporting multiple QoS and analysis of its performance. *Photonic Network Communications*, **16** (1), 53–60.
8. Kwong, W.C. and Yang, G.C. (2004) Multiple-length multiple-wavelength optical orthogonal codes for optical CDMA systems supporting multirate multimedia services. *J. Selected Areas Comm.*, **22** (9), 1640–1647.
9. Huang, J. *et al*. (2005) Multilevel optical CDMA network coding with embedded orthogonal polarizations to reduce phase noises. In: *ICICS*, Bangkok, Thailand.
10. Tarhuni, N. *et al*. (2005) Multiclass optical orthogonal codes for multiservice optical CDMA networks. *J. Lightw. Technol.*, **24** (2), 694–704.
11. Gu, F. and Wu, J. (2005) Construction of two-dimensional wavelength/time optical orthogonal codes using difference family. *J. Lightw. Technol.*, **23** (11), 3642–3652.
12. Kwon, H.M. (1994) Optical orthogonal code-division multiple-access system – part i: APD noise and thermal noise. *IEEE Trans on Comm.*, **24** (7), 2470–2479.
13. Mcyntyre, R.J. (1972) The distribution of gains in uniformly multiplying avalanche photodiodes: Theory. *IEEE Trans. Electron Devices*, **ED-19** (6), 703–713.
14. Abshire, J.B. (1984) Performance of OOK and low-order PPM modulations in optical communications when using APD-based receivers. *IEEE J. on Comm.*, **COM-32** (10), 1140–1143.
15. Yang, G.-C. (1994) Performance analysis for synchronization and system on CDMA optical fiber networks. *IEICE Trans. on Comm.*, **E77B** (10), 1238–1248.
16. Keshavarzian, A. and Salehi, J.A. (2005) Multiple-shift code acquisition of optical orthogonal codes in optical CDMA systems. *IEEE Trans on Comm.*, **53** (4), 687–697.
17. Griner, U.N. and Arnon, S. (2004) A novel bipolar wavelength-time coding scheme for optical CDMA systems. *IEEE Photonics Tech. Letters*, **16** (1), 332–334.
18. Hamarsheh, M.M.N., Shalaby, H.M.H. and Abdullah, M.K. (2005) Design and analysis of dynamic code division multiple access communication system based on tunable optical filter. *J. Lightw. Technol.*, **23** (12), 3959–3965.
19. Jau, L.L. and Lee, Y.H. (2004) Optical code-division multiplexing systems using Manchester coded Walsh codes. *IEE Optoelectronics*, **151** (2), 81–86.
20. Wei, Z. and Ghafouri-Shiraz, H. (2002) Proposal of a novel code for spectral amplitude coding optical CDMA systems. *IEEE Photonics Tech. Letters*, **14** (3), 414–416.
21. Weng, C.S. and Wu, J. (2001) Perfect difference codes for synchronous fiber-optic CDMA communication systems. *J. Lightw. Technol.*, **19** (2), 186–194.
22. Yang, G.C. and Kwong, W.C. (1995) Performance analysis of optical CDMA with prime codes. *Electronics Letters*, **31** (7), 569–570.
23. Kwong, W.C., Perrier, P.A. and Prucnal, P.R. (1991) Performance comparison of asynchronous and synchronous code-division multiple-access techniques for fiber-optic local area networks. *IEEE Trans. on Comm.*, **39** (11), 1625–1634.
24. Liu, M.Y. and Tsao, H.W. (2000) Cochannel interference cancellation via employing a reference correlator for synchronous optical CDMA system. *J. Microw. & Opt. Tech. Let.*, **25** (6), 390–392.
25. Zhang, J.G. and Kwong, W.C. (1997) Design of optical code-division multiple-access networks with modified prime codes. *Electronics Letters*, **33** (3), 229–230.

26. Liu, F. and Ghafouri-Shiraz, H. (2005) Analysis of PPM-CDMA and OPPM-CDMA communication systems with new optical code. *SPIE Proc.*, Shanghai, China, vol. 6021.

27. Liu, F., Karbassian, M.M. and Ghafouri-Shiraz, H. (2007) Novel family of prime codes for synchronous optical CDMA. *J. Optical and Quantum Electronics*, **39** (1), 79–90.

28. Zhang, J.G., Sharma, A.B. and Kwong, W.C. (2000) Cross-correlation and system performance of modified prime codes for all-optical CDMA applications. *J. Opt. A: Pure Appl. Opt.*, **2** (5), L25–L29.

29. Liu, M.Y. and Tsao, H.W. (2001) Reduction of multiple access interference for optical CDMA systems. *J. Microw. & Opt. Tech. Let.*, **30** (1), 1–3.

30. Karbassian, M.M. and Kueppers, F. (2010) Synchronous optical CDMA networks capacity increase using transposed modified prime codes. *J. Lightw. Technol.*, **28**(17), 2603–2610.

31. Lin, C.H. *et al.* (2005) Spectral amplitude-coding optical CDMA system using Mach–Zehnder interferometers. *J. Lightw. Technol.*, **23** (4), 1543–1555.

32. Murugesan, K. (2004) Performance analysis of low-weight modified prime sequence codes for synchronous optical CDMA networks. *J. Optical Communications*, **25** (2), 68–74.

33. Zhang, J.G., Kwong, W.C. and Sharma, A.B. (2000) Effective design of optical fiber code-division multiple access networks using the modified prime codes and optical processing. In: *IEEE WCC-ICCT*, Beijing, China.

34. Kavehrad, M. and Zaccarin, D. (1995) Optical code division-multiplexed systems based on spectral encoding of noncoherent sources. *J. Lightw. Technol.*, **13** (3), 534–545.

35. Smith, E.D.J., Blaikie, R.J. and Taylor, D.P. (1998) Performance enhancement of spectral-amplitude-coding optical CDMA using pulse position modulation. *IEEE Trans. on Comm.*, **46** (9), 1176–1185.

36. Zhou, X. *et al.* (2000) Code for spectral amplitude coding optical CDMA systems. *Electronics Letters*, **36** (8), 728–729.

37. Kostic, Z. and Titlebaum, E.L. (1994) The design and performance analysis for several new classes of codes for optical synchronous CDMA and for arbitrary-medium time-hopping synchronous CDMA communication systems. *IEEE Trans on Comm.*, **42** (8), 2608–2617.

38. Wei, Z. and H. Ghafouri-Shiraz (2002) Codes for spectral-amplitude-coding optical CDMA systems. *J. Lightw. Technol.*, **20** (8), 1284–1291.

39. Einarsson, G. (1980) Address assignment for a time-frequency-coded spread-spectrum system. *J. of Bell Syst. Tech.*, **59** (7), 1241–1255.

40. Michelson, A.M. and Levesque, A.H. (1985) *Error-control techniques for digital communication*. Wiley-interscience publication.

41. Anderson, I. (1990) *Combinatorial designs*. Ellis Horwood Limited, New York, USA.

42. Salehi, J.A. and Brackett, C.A. (1989) Code division multiple-access technique in optical fiber networks – part II: system performance analysis. *IEEE Trans. on Comm.*, **37** (8), 834–842.

43. Wei, Z., Shalaby, H.M.H. and Ghafouri-Shiraz, H. (2001) Modified quadratic congruence codes for fiber Bragg-grating-based spectral-amplitude-coding optical CDMA systems. *J. Lightw. Technol.*, **19** (9), 1274–1281.

44. Viterbi, A.J. (1995) *CDMA, principles of spreading spectrum communication*. Addison Wesley, Boston, USA.

45. Corning.com, Corning SMF-28e optical fiber product information, In: Corning, Inc.

46. Kim, S., Yu, K. and Park, N. (2000) A new family of space/wavelength/time spread three-dimensional optical code for OCDMA networks. *J. Lightw. Technol.*, **18** (4), 502–511.

47. McGeehan, J.E. *et al.* (2005) Experimental demonstration of OCDMA transmission using a three-dimensional (time–wavelength–polarization) codeset. *J. Lightw. Technol.*, **23** (10), 3282–3289.

48. Yeh, B.C., Lin, C.H. and Wu, J. (2009) Noncoherent spectral/time/spatial optical CDMA system using 3-D perfect difference codes. *J. Lightw. Technol.*, **27** (6), 744–759.

49. Lin, C.L. and Wu, J. (1999) Large capacity ATM switching fabric using three-dimensional optical CDMA technology. In: *IEEE Symposium on Computers and Communications*.

50. Singh, L. and Singh, M.L. (2009) A new family of three-dimensional codes for optical CDMA systems with differential detection. *Optical Fiber Technology*, **15** (5-6), 470–476.

51. Kumar, M.R., Pathak, S.S. and Chakrabarti, N.B. (2009) Design and analysis of three-dimensional OCDMA code families. *Opt. Switching and Networking*, **6** (4), 243–249.

52. Kumar, M.R., Pathak, S.S. and Chakrabarti, N.B. (2009) Design and performance analysis of code families for multidimensional optical CDMA. *IET Communications*, **3** (8), 1311–1320.

53. Prucnal, P.R. (2005) *Optical code division multiple access: fundamentals and Applications*. CRC Taylor & Francis Group, Florida, USA.

54. Yang, G.C. and Kwong, W.C. (2002) *Prime codes with applications to CDMA: optical and wireless networks*. Artech House.

55. Wan, S.P. and Hu, Y. (2001) Two-dimensional optical CDMA differential system with prime/OOC codes. *IEEE Photonics Tech. Letters*, **13** (12), 1373–1375.

56. Tancevski, L. and Andonovic, I. (1994) Wavelength hopping/time spreading code division multiple access systems. *Electronics Letters*, **30** (17), 1388–1390.

57. Shparlinski, I. (1999) *Finite fields: theory and computation*. Springer, Dordrecht, The Netherlands.

58. Kwong, W.C. *et al.* (2005) Multiple-wavelength optical orthogonal codes under prime-sequence permutations for optical CDMA. *IEEE Trans on Comm.*, **53** (1), 117–123.

59. Stallings, W. (2004) *Data and computer communications*. Prentice Hall, 7th Edition, New Jersey, USA.

60. Al-Sammak, A.J. (2002) Encoder circuit for inverse differential Manchester code operating at any frequency. *Electronics Letters*, **38** (12), 567–568.

61. Molisch, A.F. (2005) *communications*. John Wiley & Sons, Chichester, England.

62. Huffman, W. and Pless, V. (2003) *Fundamentals of error-correcting codes*. Cambridge University Press, Cambridge, UK.

63. Massey, P.C. and Costello Jr., D.J. (2001) Turbo codes with recursive nonsystematic quick-look-in constituent codes. In: *IEEE Proc. International Symposium on Information Theory*, Washington DC, USA, pp. 141 (doi: 10.1109/ISIT.2001.936004).

64. Shamir G.I. and Kai Xie, "Universal Source controlled channel decoding with nonsystematic quick-look-in turbo codes", *IEEE Transactions on Communications*, **57**, (4), 960–971.

65. Valenti, M. (1998) Turbo codes and iterative processing. In: *Proc. IEEE New Zealand Wireless Communications Symposium*, New Zealand.

66. Schlegel, C. and Perez, L. (1997) *Trellis coding*. IEEE Press, New Jersey, USA.

67. Dasgupta, U. and Narayanan, K.R. (2001) Parallel decoding of turbo codes using soft output T-algorithms. *IEEE Comm. Letters*, **5** (8), 352–354.

68. Viterbi, A.J. (1967) Error bounds for convolutional codes and an asymptotically optimum decoding algorithm. *IEEE Transactions on Information Theory*, **13** (2), 260–269.

69. Berrou, C., Glavieux, A. and Thitimajshima, P. (1993) Near Shannon limit error-correcting coding and decoding: turbo-codes. In Proc. IEEE International Conference on Communication (ICC).

70. Marti, S. and Ahmad, M.O. (2008) A bandwidth efficient Turbo coding scheme for VDSL systems. *J. of Circuits, Systems, and Signal Processing*, **27** (5), 563–597.

71. Achiba, R., Mortazavi, M. and Fizell, W. (2000) Turbo code performance and design trade-offs. In: *Proc. IEEE MILCOM*, Los Angeles, USA.

72. Divsalar, D. and Pollara, F. (1997) Serial and hybrid concatenated codes with applications. In: *Proc. Intl. Symp. Turbo Codes and Appls*, Brest, France.

73. Forney, G.D. (1966) *Concatenated codes*. MIT Press, Massachusetts, USA.

74. Le Bidan, R. *et al.* (2008) Article ID 658042. Reed-Solomon turbo product codes for optical communications: from code optimization to decoder design. *EURASIP Journal on Wireless Communications and Networking*.

75. Kim, J.Y. and Poor, H.V. (2001) Turbo-coded optical direct-detection CDMA system with PPM modulation. *J. Lightw. Technol.*, **19** (3), 312–322.

76. Bahl, L. *et al.* (1974) Optimal decoding of linear codes for minimizing symbol error rate. *IEEE Transactions on Information Theory*, **20** (2), 284–287.

77. Robertson, P., Villebrun, E. and Hoeher, P. (1995) A comparison of optimal and sub-optimal MAP decoding algorithms operating in the Log domain. In: *Proc. ICC*, Seattle, USA.

78. Talakoub, S. and Shahrrava, B. (2004) A linear Log-MAP algorithm for Turbo decoding over AWGN channels. In: *Proc. Electro/Information Technology*, Milwaukee, WI, USA.

79. Farhadi, G. and Jamali, S.H. (2006) Performance analysis of fiber-optic BPPM CDMA systems with single parity-check product codes. *IEEE Trans on Comm.*, **54** (9), 1643–1653.

80. Jiang, Z., Leaird, D.E. and Weiner, A.M. (2006) Experimental investigation of security issues in OCDMA. In: *OFC*, Anaheim, CA, USA.

81. Shake, T.H. (2005) Security performance of optical CDMA against eavesdropping. *J. Lightw. Technol.*, **23** (2), 655–670.

82. Arbab, V.R. *et al*. (2007) Increasing the bit rate in OCDMA systems using pulse position modulation techniques. *Optics Express*, **15** (19), 12252–12257.

83. Bazan, T.M., Harle, D. and Andonovic, I. (2006) Mitigation of beat noise in time–wavelength optical code-division multiple-access systems. *J. Lightw. Technol.*, **24** (11), 4215–4222.

84. Ohtsuki, T. and Kahn, J.M. (2000) BER performance of Turbo-coded PPM-CDMA systems on optical fiber. *J. Lightw. Technol.*, **18** (12), 1776–1784.

85. Karbassian, M.M. and H. Ghafouri-Shiraz (2007) Fresh prime codes evaluation for synchronous PPM and OPPM signaling for optical CDMA networks. *J. Lightw. Technol.*, **25** (6), 1422–1430.

# 3

# Optical CDMA Review

## 3.1 Introduction

Interest in OCDMA has been steadily growing during recent decades and this trend is accelerating due to the optical fibre penetration in the first-mile and the establishment of passive optical network (PON) technology as a pragmatic solution for residential access. In OCDMA, an optical code represents a user address and it signs each transmitted data bit. We define optical coding as the process by which a code is inscribed into, and extracted from, an optical signal. Although a prerequisite for OCDMA is the optical coding as thoroughly introduced in Chapter 2, it has a wide range of novel and promising applications, such as access protocol and label switching. Most previous reviews of OCDMA have focused on physical layer (PHY) implementations. This chapter is an overview of OCDMA technologies and presents an overview of networking applications.

## 3.2 Optical Coding Principles

Optical code-division multiplexing (OCDM) is a process by which each communication channel is distinguished by a specific optical code rather than a wavelength, as in WDM, or a time-slot, as in TDM. An encoding operation optically transforms each data bit before transmission. The encoding and decoding operations alone constitute optical coding. OCDMA is the use of optical network technology to arbitrate channel access among multiple network nodes in a distributed fashion. Encoding involves multiplying the data bit by a code sequence either in the time domain or in the wavelength domain. A combination of both can also be used which is referred to as two-dimensional (2D) coding [1–4], which was discussed in Section 2.6. Time-domain coding that manipulates the phase of the optical signal requires phase-accurate coherent sources. Alternatively, positive encoding manipulates the power of the optical signal but not its phase, and typically uses incoherent sources. In wavelength-domain coding, transmitted bits consist of

*Optical CDMA Networks: Principles, Analysis and Applications*, First Edition. Hooshang Ghafouri-Shiraz and M. Massoud Karbassian.
© 2012 John Wiley & Sons, Ltd. Published 2012 by John Wiley & Sons, Ltd.

a unique subset of wavelengths forming the code sequences. 2D-coding combines both wavelength selection and time-spreading. A data bit is encoded as consecutive chips of different wavelengths, the unique wavelength sequence constituting the code. Regardless of the coding domain, the coding operation broadens the spectrum of the data signal, hence the designation of spread spectrum as discussed in Chapter 1. It is worth mentioning that encoding can also be performed in the space-domain [5], whereby the code determines the positions of chips within a dense fibre array or a multicore fibre.

An OCDMA local area network (LAN) is based on a broadcast medium as illustrated in Figure 3.1. Multiple en/decoders in the central office are communicating with end-users in LAN as shown in Figure 3.1. The encoded signals are coupled and transmitted over the channel. Receivers are receiving the sum of the encoded signals, but the intended receiver picks up only the signal which was encoded with corresponding assigned codes. Unwanted signals appear as noise to the decoder and are called multiple access interference (MAI). MAI is the principal source of noise in OCDMA and is the limiting factor to system performance. In a well-designed OCDMA-LAN where MAI is overcome, users can successfully communicate asynchronously regardless of network traffic as shown in Figure 3.1.

The decoding process is based on the correlation function. The signal-to-signal correlation implies the similarities between two signals. Correlating the same codes (i.e. referred to as auto-correlation) gives the maximum value of correlation function, while correlating two different codes (i.e. cross-correlation) indicates the two codes differences. For example, when cross-correlation is zero the two signals are orthogonal. Accordingly,

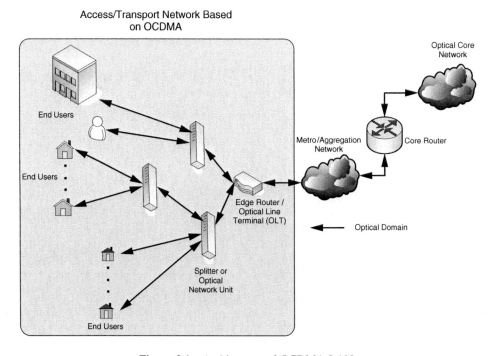

**Figure 3.1**   Architecture of OCDMA-LAN

high auto-correlation and low cross-correlation values are highly desirable in CDMA communication systems. Decoding an encoded signal by the same code represents a logical auto-correlation of a single code. Otherwise, the operation represents a cross-correlation between two different codes. Code design aims at generating codes with high auto-correlation and low cross-correlation properties. More particular requirements on code design arise from the use of specific transmission media or components. Various code families were introduced in Chapter 2, addressing such requirements [2, 3, 6–15]. An important feature of a code is its code-length which plays an important role in the design and security of the system.

A larger code-length improves correlation properties among code sequences, thus raising the system performance in terms of lower bit-error rate (BER) and less MAI, but this means lower throughput as well [16–18].

Employing incoherent versus coherent light sources is often used to classify OCDMA systems because that choice has important cost and performance implications [19]. Note that the choice of coherent sources does not always imply that chips are phase rather than power modulated. In addition, whether the encoding occurs in a fibre or in a planar lightwave circuit or in an out-of-fibre external device, it has an important impact on system design [20].

Throughout this book, we distinguish multiplexing techniques (xDM) from multiple accesses (xDMA) to make the entire scheme more understandable. The former focus on transport and the latter denotes distributed access methods. For example, in an OCDMA-over-WDM network, all nodes require an OCDMA transceiver for access purposes, while each wavelength is used as a medium for a number of OCDMA channels for transport purpose.

## 3.3   OCDMA Networking: Users Are Codes

OCDMA is based on the principle that codes are mapped to the identities or addresses of users following a code–user relation. Accordingly, OCDMA was initially proposed to implement ultrafast broadcast LAN.

### 3.3.1   From LAN to PON

LAN and PON contender techniques include TDMA, WDMA and subcarrier multiple-access (SCMA) [21, 22]. In TDMA, a time-slot is allocated to each user statically or dynamically. TDMA is already deployed in two forms: ATM-PON and EPON, which will be introduced in Chapter 9, while in WDMA, each user is assigned a specific wavelength like in WDM-PON.

Both TDMA and WDMA benefit from the maturity of electrical multiplexing and optical transmission gained in backbone networks. TDMA-WDM has been proposed as a viable extension to TDMA that achieves dynamic bandwidth allocation (DBA) on multiple wavelengths [22]. In SCMA, microwave channels are multiplexed electrically, and the composite signal modulates the optical carrier. SCMA is commonly used in hybrid fibre-coax (HFC) networks to carry broadcast community access television (CATV) channels and SCMA-WDM has notable applications in radio-over-fibre networks [23].

The combination of three potential advantages makes OCDMA attractive from a networking perspective. First, OCDMA offers a larger channel count than the spectral division of WDMA. Second, asynchronous transmission simplifies access control to the medium as compared to TDMA. Finally, multiclass multirate services can be implemented by using variable code-lengths code-weights sequences as discussed in Chapter 10 [1, 24, 25].

The motivation for OCDMA-LAN is reinforced by the expectation of bursty LAN traffic patterns. In terms of channel count, dense WDM (DWDM) may surpass OCDMA by offering enough channels for LAN applications in addition to simpler implementations. Access environments, however, require even larger channel counts than LAN and stand to benefit from the simplicity of asynchronous transmission and quality-of-service (QoS) differentiation, while it is doable as there has been a change of focus in PON access networks to OCDMA research.

### 3.3.2  OCDMA for Access Networks

OCDMA has recently become viewed as a candidate technology for future PON access networks which will be introduced in Chapter 9 [22, 26–32].

A well-designed OCDMA access network eliminates channel contention. In other words, provided that interference is controlled, an upstream or downstream connection can be established asynchronously with no collisions or blocking. The networking implications are substantial. The following is a brief description of the main assets that OCDMA exhibits from a networking perspective [18, 33]:

- Ideally, no channel control mechanism is required to avoid collisions or allocate bandwidth. In addition, optical network units (ONU) are not required to report the instantaneous bandwidth requirements to the optical line terminal (OLT), thereby reducing round-trip walk time and delay.
- OCDMA supports a larger number of users than TDMA or WDMA, especially multidimensional-OCDMA systems [1–3, 34, 35] where codes exploit time, wavelength and/or space/polarization. It is also possible for an even larger number of codes to be assigned in OCDMA if an access protocol is used to avoid collisions.
- Like WDMA, OCDMA offers a virtual point-to-point topology over the physical tree architecture. However, to do so, WDMA requires an in-field WDM multiplexer or individual wavelength filters at the ONU. OCDMA, however, requires the cheaper power splitter and/or correlator but incurs the larger power losses associated.
- A new TDMA or WDMA user reduces free bandwidth irreversibly due to the bandwidth guaranteed under ONU service-level agreements (SLA), thus requiring changes to bandwidth allocations. For example, for every new user added in TDMA, the OLT may be required to run the admission control process. In OCDMA, a new user does not reduce the other users' bandwidth or time allocation.
- Unlike WDMA and TDMA-WDM, OCDMA can accommodate a large number of low bit rate users on the same optical medium. Moreover, using multirate OCDMA techniques, low and high bit rate channels can be provided on the same link. Such aspects correspond to access traffic patterns and are highly desirable as they eliminate electronic grooming.

- Finally, ATM-PON design was conducted on the assumption that downstream traffic largely exceeds upstream traffic, while OCDMA offers the capability of supporting high-speed symmetric traffic for a larger number of network units than WDMA [29].

Moreover, serious propositions must include the gradual migration paths from WDMA to OCDMA as is the case currently from TDMA to WDMA [36]. Such migration paths offer partial implementations that postpone some of the research elements required for full OCDMA-PON. An OCDMA-PON proposal is also analysed in Chapter 9.

## 3.4 Optical CDMA Techniques

A variety of approaches to OCDMA have been suggested [11, 19, 37–39]. They share a common strategy of distinguishing data channels not by wavelength or time-slot, but by distinctive spectral or temporal code as a signature impressed onto the bits of each channel. Carefully designed receivers isolate channels by code-specific detection. There is no global optimum topology for fibre-optic LAN interconnection yet. Thus each topology has its own advantages and disadvantages, which may become significant or insignificant depending on the specific applications [40, 41].

In the intensity-modulated on-off keying (OOK) system, each user information source modulates the laser diode directly [42] or indirectly using an external modulator [43]. Figure 3.2 illustrates the basic intensity-modulated (OOK) OCDMA network. The optical signal is encoded optically in an encoder that maps each bit into very high rate optical sequences (i.e. the product of code-length and data rate). The encoded lightwaves from all active users are broadcast in the network by a star coupler. The star coupler can be a passive or active device. The optical decoder or matched filter at the receiving node is matched to the transmitting node giving a high correlation peak that is detected by the photodetector. Other users using the same network at the same time but with different codes give rise to MAI.

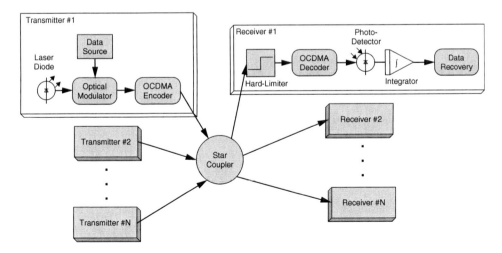

**Figure 3.2**   Intensity-modulated OCDMA star network

This MAI can be high enough to make the LAN useless if the employed codes in the network do not satisfy specific cross-correlation properties. Other factors affecting the performance of the network are shot noise and thermal noise at the receiver. These receiver noise sources are considered and analysed in following chapters of this book. It should be noted that the received signal may also be affected by polarization mode dispersion (PMD), chromatic dispersion, Kerr nonlinearities, nonlinear phase noise, stimulated Raman scattering, amplified-spontaneous emission (ASE), linear filtering effects, inter-symbol interference and linear crosstalk effects from the optical channel [39].

## 3.4.1   Coherent vs. Incoherent OCDMA

Generally, OCDMA systems can be classified as incoherent or coherent schemes. Incoherent schemes are based on the intensity-modulation/direct-detection (IM-DD) scheme that incorporates non-coherent direct-detection of superimposed optical power of all users. The operation of direct-detection makes the procedure simple and the receiver is cost effective. The photodetector detects the power of the optical signal but not the instantaneous phase variations of the optical signal. Thus, only incoherent signal processing techniques can be used to process the signature sequences composed of only ones and zeros restricting the type of codes that can be used in incoherent OCDMA systems [39]. In coherent OCDMA, the phase information of the optical carrier is crucial for the de-spreading process. Due to the nature of optical fibre transmission and its nonlinear effects the complexity of the coherent OCDMA receiver makes this approach more difficult to realize. However, the performance of the coherent scheme is superior to the incoherent one since the receivers are more sensitive to signal-to-noise ratio (SNR), which makes the overall system performance better [19, 44–46].

Depending on how an optical signal is encoded, OCDMA can be classified as (i) spectral OCDMA or (ii) temporal OCDMA. Each of these classifications will be discussed briefly in the following sections and in more details in Chapters 4 and 5 respectively. Temporal OCDMA performs the coding in time domain by using very short optical pulses, e.g. 10 ps at data rate of 1 Gbps and code-length of 100, using optical tapped-delay lines (OTDL) to compose the coded optical signal [39]. Spectral OCDMA, on the other hand, encodes the phase or intensity of the spectral content of a broadband optical signal by using phase or amplitude masks [39]. Wavelength-hopping can also be considered as a temporal-spectral coding where the coding is done in both dimensions [39].

## 3.4.2   Synchronous vs. Asynchronous OCDMA

Synchronous OCDMA (S-OCDMA) dramatically improves efficiency in the trade-offs between code-length, MAI and address space. Since, in S-OCDMA, the receiver examines the correlator output only at one instant in the chip-interval, codesets for S-OCDMA are described by the triple $(N, w, \lambda)$ where $N$ is the code-length, $w$ is the code-weight and $\lambda$ is the maximum cross-correlation value. In general, an OOC set of $C_a$ with $(N, w, \lambda_a, \lambda_c)$, with cardinality of $|C_a|$ designed for asynchronous OCDMA (A-OCDMA) can be used as a codeset of $C_s$ with $(N, w, \max(\lambda_a, \lambda_c))$ and cardinality $|C_s| = n.|C_a|$ for S-OCDMA,

**Table 3.1**  OOC with $F = 32$ and $w = 4$

| No. of Chips between Subsequent Ones | Optical Orthogonal Codes (OOC) |
|---|---|
| 9,3,15,5 | 10000000010010000000000000010000 |
| 4,7,19,2 | 10001000000100000000000000000010 |

since each of the $n$ time-shifts of each code sequence of $C_a$ can be used as a unique code sequence in $C_s$ with the same correlation properties.

In contrast, prime code families can also be utilized in A-OCDMA; however, a smaller number of subscribers is then accommodated due to the lack of time-shifting feature used in the synchronous one. In the OOC, which are normally applied to A-OCDMA, we have to set the weight (i.e. the number of ones) and the code sequence independently and keep the number of spreading codes small to obtain a good correlation value. Table 3.1 shows the OOCs with frame-length $F = 32$ and weight $w = 4$ [6]. The total number of OOC sequences is given by the integer part of $(F - 1)/(w^2 - w)$, hence there are only two codes to satisfy the cross-correlation value of *one*. To assure this condition, the distance of any two ones should be different in all codes, as shown in Table 3.1. Therefore, to increase the number of spreading codes in OOC, either the frame-length or the weight has to increase. In practice, in order to make 25 code sequences with a weight of 7, the code-length needs to be 1051. Therefore, OOC's frame-length needs to be 30 times larger than the group-padded modified prime code (GPMPC), introduced in Chapter 2, for example $((w^2 - w) \times 25 + 1)/(P^2 + 2P) = 30$ where $P = 7$ and $w = 7$) that accordingly decreases the bit rate dramatically. When the OOC weight decreases, its correlation properties degrade.

The effect of channel interference (i.e. MAI) is inherent in temporal OCDMA: when the number of simultaneous active users increases, the optical pulses from the intended user and the interfering users overlap and BER tends to have an error-floor. Therefore, it is necessary to reduce the probability of overlapped pulses from interfering users to mitigate the effect of co-channel interference. The probability of overlapping pulses has been reduced by changing the modulation scheme OOK to *M-ary* pulse position modulation (PPM) due to the variable pulse positions of pulse occurrence in PPM which is discussed in detail in Chapter 5, although the error-floor cannot be totally cancelled [47–49].

### 3.4.3  Wavelength-Hopping Coding

A fast wavelength-hopping OCDMA system can be implemented by using a fibre Bragg grating (FBG) [50–53]. Multiple Bragg gratings are used to generate the CDMA hopping frequencies (i.e. wavelengths). Due to the linear first-in-first-reflected nature of multiple Bragg gratings, the time-frequency hopping pattern is determined by the order of the grating frequencies in the fibre. The order of the grating frequencies in the decoder is the reverse of that in the encoder to achieve the matched filtering operation. Figure 3.3 shows the encoder and decoder in a star-coupled network; the wavelengths are circulated

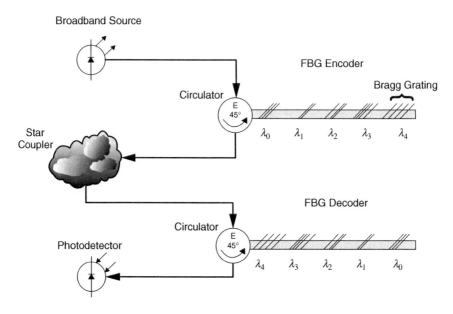

**Figure 3.3**   Principle of FBG encoder and decoder

and reflected through the fibre Bragg gratings according to the corresponding spreading code. If the central wavelength of the incoming lightwave equals the Bragg wavelength, it will be reflected by the FBG, or it will be transmitted. With a proper assigned CDMA coding pattern the reflected light field from the FBG will be spectrally encoded onto an address code. To reduce the effect of the MAI, codes with minimum cross-correlation properties are required [11]. These codes fall into the category of one-coincidence sequences and are characterized by the following three properties:

- All of the sequences are of the same length.
- In each sequence, each frequency is used at most once.
- The maximum number of hits between any pair of sequences for any time shift equals *one*.

### 3.4.4   Spectral Phase Coding (SPC)

Figure 3.4(a) shows an encoder and decoder of the spectral phase encoding system. The information source modulates ultrashort laser pulses. The generated short pulses are Fourier transformed and the spectral components are multiplied by the code corresponding to a phase shift of 0 or $\pi$ [54, 55]. A Fourier transform can be implemented by the grating as well as a lens pair as shown in Figure 3.4(b).

As a result of phase encoding, the original optical ultrashort pulse is transformed into a low intensity signal with longer duration. The liquid crystal modulator (LCM) can be utilized to set the spectral phase to maximum-sequence phase [56]. The LCM has a fully programmable linear array and individual pixels can be controlled by applying drive levels resulting in phase shifts of 0 or $\pi$. The dispersed pulse is partitioned into $N_c$ frequency

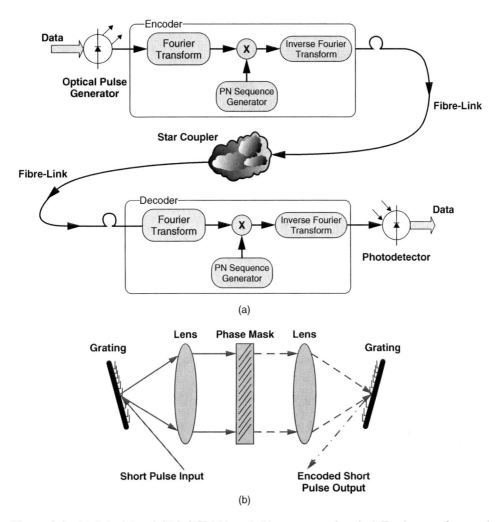

**Figure 3.4**  (a) Principle of SPC-OCDMA and (b) structure of optical Fourier transform and spectral phase coding

chips by the aid of a phase mask that can be an LCM. Each chip is assigned a phase shift depending on the users' address code sequences. The detailed analysis and discussions of such a system are in Chapter 4.

### 3.4.5  *Spectral-Amplitude-Coding (SAC)*

In SAC-OCDMA format, frequency components of the signal from a broadband optical source are encoded by selectively blocking or transmitting them in accordance with a signature code [19, 57]. Compared to SPC-OCDMA, SAC-OCDMA is less expensive due to incoherent optical source. For the access environment, where cost is one of the most decisive factors, the SAC-OCDMA seems therefore to be a promising candidate.

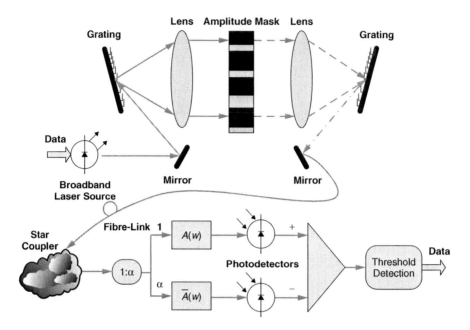

**Figure 3.5**   Principle of the SAC-OCDMA scheme

Figure 3.5 shows the principle of the structure of an SAC-OCDMA system. The receiver filters the incoming signal through the same direct decoder filter $A(w)$ at the transmitter as well as its complementary decoder $\overline{A}(w)$. The outputs from these decoders are detected by two photodetectors connected in a balanced structure. For an interfering signal, depending on the assigned signature code, a part of its spectral components will match the direct decoder, and the other part will match the complementary decoder. Since the output of the balanced receiver represents the difference between the two photodetector outputs, the interfering channels will be cancelled, whereas the matched channel is demodulated, i.e. MAI is cancelled in this SAC-OCDMA system.

Several signature codesets have been proposed for an SAC-OCDMA, including $m$-sequence [52], Hadamard [18], and modified-quadratic congruence (MQC) codes [58]. As introduced in Chapter 2, each of these signature codesets can be represented by $(N, w, \lambda)$ where $N$, $w$ and $\lambda$ denote the code-length, code-weight and in-phase cross-correlation value respectively. In the $m$-sequence codeset $w = (N + 1)/2$ and $\lambda = (N + 1)/4$; also the weight and in-phase cross-correlation of the Hadamard codeset are $N/2$ and $N/4$ respectively. In MQC code, $\lambda = 1$, and for a prime number $P$ we have a code-length of $N = P^2 + P$ and weight of $W = P + 1$. Let $C_d = \{C_d(0), C_d(1), \ldots, C_d(N - 1)\}$ and $C_k = \{C_k(0), C_k(1), \ldots, C_k(N - 1)\}$ be two $(0,1)$ signature codes, and the correlation properties are given by:

$$C_{c_d c_k} = \sum_{i=0}^{N-1} C_d(i).C_k(i) = \begin{cases} W & d = k \\ \lambda & d \neq k \end{cases} \quad (3.1)$$

The correlation between $\overline{C}_d$ (a complementary of $C_d$) and $C_k$ is then:

$$C_{\overline{c}_d c_k} = \sum_{i=0}^{N-1} \overline{C}_d(i).C_k(i) = W - C_{c_d c_k} = \begin{cases} 0 & d = k \\ W - \lambda & d \neq k \end{cases} \qquad (3.2)$$

To completely cancel MAI, it is necessary to have an optical coupler with ratio $1 : \alpha = \lambda / (w - \lambda)$ [19] at the input of the two detectors, as shown in Figure 3.4. [19]. The cancellation of the interfering signal (i.e. $d \neq k$) by the balanced receiver thus can be seen as:

$$C_{c_d c_k} - \alpha C_{\overline{c}_d c_k} = 0 \qquad (3.3)$$

A broadband optical source is used in frequency-domain encoding so the optical beating interference (OBI), or beat noise, is the major performance degrading factor. OBI occurs when a photodetector simultaneously receives two or more optical signals at nearly the same wavelength. One of the solutions to the beat noise in coherent SAC-OCDMA is to employ an optical spreading codeset with the lowest possible weight and the longest possible length for a given bit rate [11].

A lower code-weight causes a lower SNR which means that the received optical power is low because it is further reduced by a factor of $\alpha$ in one of the branches as shown in Figure 3.5, which makes OBI negligibly small. In the case of high optical power, a long low-weight code causes a higher SNR (i.e. better system performance) due to low in-phase cross-correlation which finally results in lower OBI [30].

Another major degrading factor in spectral-coded schemes, SPC and SAC, is phase-induced intensity noise (PIIN) that is highly proportional to the electric current generated by the photodetectors as [59]:

$$\delta_{PIIN} = I^2.B.\tau_c \qquad (3.4)$$

where $I$ is the photocurrent, $B$ is the receiver's noise-equivalent electrical bandwidth and $\tau_c$ is the coherence time of source. Chapter 4 gives a detailed analysis of such a system.

### 3.4.6 Time-Spreading Coding

The temporal OCDMA signal can be generated by the splitting and combining of ultrashort optical pulses. A high-peak optical pulse is encoded into a low intensity pulse train using parallel OTDLs at the transmitter in a star-coupled architecture. The decoding is performed by intensity correlation at the receiver using matched parallel OTDLs [15, 49, 60].

Incorrectly positioned pulses in the pulse train will form a background interference signal. The research into these incoherent OCDMA schemes led to the invention of a few major code groups such as optical orthogonal codes (OOC) and prime code families. In order to reduce crosstalk (i.e. MAI), these codes are all designed to have long code-length and low code-weight to reduce temporal overlap between pulses from different users at the intensity correlator output. On the other hand, there should be a trade-off, since the long code-length can cause spectral inefficiency. Even with very carefully designed codes, the co-channel interference due to non-orthogonal sequences carries a severe performance penalty. The bit-error rate (BER) is usually quite high and the number of allowable active users becomes very limited [6, 39, 48, 49, 61, 62].

Due to the extremely fast growth in bandwidth demand in recent years, it is now necessary to make full use of the entire bandwidth capacity available in optical fibres. However, at the time that OCDMA systems using delay-line networks were proposed, it was believed that the terabit communication capacity in fibre-optics would never be fully utilized. Even multigigabit networks were highly respected at that time. Although optics has been used for carrying signals, all the switching and multiplexing operations were performed in the electronic domain. Optoelectronic (OE) and electro-optic (EO) conversions occur at the terminal equipment. The OE–EO conversion was regarded as the bottleneck to high-speed multiplexing. Therefore, due to recent progress in photonics technologies, ultrafast switching, multiplexing and signal processing in optical domain are feasible in transparent optical networks, and accordingly temporal OCDMA has drawn a lot of attention recently to make use of the redundant bandwidth in optical fibres to alleviate the electronic processing overhead at the networks interface. This scheme will be analysed in detail in Chapter 5.

## 3.5   Free-Space and Atmospheric Optical CDMA

Free-space optical (FSO) communication has been attracting growing attention in recent years as a solution to the last-mile access bottleneck [63].

Service providers are very interested in FSO because it utilizes free space (air) as a communications channel which is inherently high capacity, cheap and unregulated in the optical spectrum (yet!). These benefits motivated the research community to test FSO's feasibility. Nykolak *et al.* [64] reported a transmission rate of up to 2.5 Gbps over a 4.4 km FSO link. Narrow optical beams and precise filtering made the FSO link secure and mitigated interference remarkably.

Figure 3.6(a) shows the free-space physical placement of optical links. The main challenges in FSO links are the atmospheric instabilities such as scattering and scintillation as well as pointing line-of-sight problems arising from obstacles and beam wander [65]. To date, FSO communication links have mostly been deployed as point-to-point links to extend the network reach as an economically revenue-generating feasible alternative to physically extending the network reach through cable or optical fibre. Highly efficient optical phase arrays [66] have no need for mechanical steering and can use a single wireless optical transmitter to communicate simultaneously with several receivers.

Wireless mesh networks have also attracted significant interest as a future broadband technology offering merits of broadband access to end-users [67]. This mesh network is scalable to the number of users and traffic volume and, at the same time, network reliability increases proportionally to its size [68]. A wireless optical-mesh network with capabilities in terms of capacity, dynamic reconfiguration and path protection is obtained through the integration of FSO links into wireless mesh networks. The current RF technology capacity is unmatched by the capacity level of FSO links. Using optical phase arrays brings reconfigurable and flexible transceiver architectures with dynamically self-organizing features.

Free-space optical links with huge optical bandwidth have inherent advantages over radio and cable solutions. However, atmospheric loss blocks reliable data communication

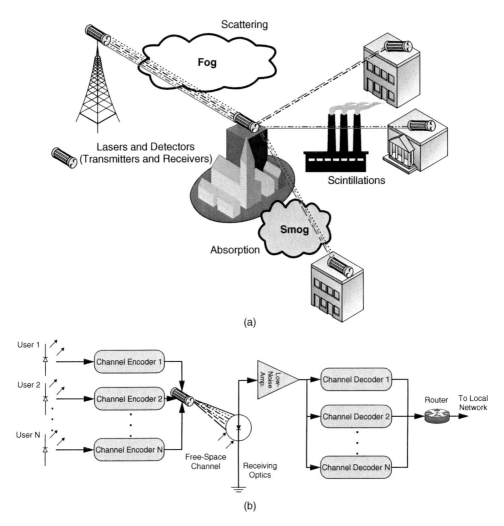

**Figure 3.6** Free-space optical links: (a) placement of transceivers on buildings and contributing noises and (b) basic free-space OCDMA subsystems

except in clear weather. Fortunately, it has been experimentally proven that building links with a practically low probability of outage are feasible in almost all actual climatic conditions [69]. Atmospheric links are also problematic in that they suffer from air turbulence that causes slow fading in the received power. Some standard diversity techniques can mitigate this. Spatial multiple-input multiple-output (MIMO) diversity techniques are very common, being the most efficient solution. Wireless OCDMA as well as radio CDMA suffers from the near–far effect, so the power control in the uplink transmissions becomes a challenging task. Aminzadeh-Gohari *et al.* [70] analysed a power control system and its effectiveness for optical orthogonal codes (OOC) infrared networks in various cases. The power control algorithm can also be used for quality of service (QoS) provisioning in terms of reliability (i.e. BER).

Ohtsuki [71] proposed atmospheric OCDMA for short-range wireless communications. The overall performance of the system was analysed, using a correlator receiver, in terms of BER for pulse position modulation (PPM) based on a chip synchronous and slot asynchronous scheme. The results revealed that the atmospheric OCDMA can realize high-speed communications when the variance due to scintillation is small, while error-correction codes are required for large values of variance. Accordingly, Jazayerifar et al. [72] investigated the issue by considering various detection structures, namely, correlator and chip-level receivers using optical amplifiers, receiver diversity techniques and some kind of internal coding scheme as shown in Figure 3.6(b). In this work they took into account all sources of noise in a semi-classical photon-counting approach leading to an analytically exact evaluation of performance under practical conditions. Their results indicate that such systems can be implemented using practical parameters.

Synchronous and asynchronous FSO OCDMA have also been investigated in the concept of atmospheric link in [68] while considering effects of scintillation, receiver noise and multiuser interference. It has been mentioned that synchronous OCDMA using Walsh–Hadamard codes reaches rates that are not achievable in an asynchronous system under strong fading. Since the free-space optical channel is much noisier than fibre-optic, the spectrally encoded schemes of OCDMA are more considerable to compare with temporally encoded schemes.

## 3.6  Summary

An overview of coding fundamentals in the optical domain has been briefly introduced and discussed here. The details of these systems are explored comprehensively in the following chapters. OCDMA's potential as an access protocol for optical networking has been discussed. We have also reviewed the most common encoding techniques in optical spread spectrum communications in time, frequency and spectra domains by considering their merits and drawbacks in fibre-optic and free-space communication links. In the following chapters, various transceiver architectures with respect to advanced coherent and incoherent modulation and multiple access interference (MAI) cancellation techniques will be analysed and discussed with the applications of optical access networks in mind.

## References

1. Liang, W. et al. (2008) A new family of 2D variable-weight optical orthogonal codes for OCDMA systems supporting multiple QoS and analysis of its performance. *Photonic Network Communications*, **16** (1), 53–60.
2. Griner, U.N. and Arnon, S. (2004) A novel bipolar wavelength–time coding scheme for optical CDMA systems. *IEEE Photonics Tech. Letters*, **16** (1), 332–334.
3. Gu, F. and Wu, J. (2005) Construction of two-dimensional wavelength/time optical orthogonal codes using difference family. *J. Lightw. Technol.*, **23** (11), 3642–3652.
4. Teixeira, A.L.J. et al. (2001) All-optical time–wavelength code router for optical CDMA networks. In: *LEOS, The 14th Annual Meeting of the IEEE*, San Diego, USA.
5. Phoel, W.G. and Honig, M.L. (1999) MMSE space-domain interference suppression for multi-rate DS-CDMA. In: *Vehicular Technology Conf*, Houston, TX, USA.
6. Chung, F.R.K., Salehi, J.A. and Wei, V.K. (1989) Optical orthogonal codes: design, analysis and application. *IEEE Trans. on Info. Theory*, **35** (3), 595–605.

7. Hamarsheh, M.M.N., Shalaby, H.M.H. and Abdullah, M.K. (2005) Design and analysis of dynamic code division multiple access communication system based on tunable optical filter. *J. Lightw. Technol.*, **23** (12), 3959–3965.

8. Jau, L.L. and Lee, Y.H. (2004) Optical code-division multiplexing systems using Manchester coded Walsh codes. *IEE Optoelectronics*, **151** (2), 81–86.

9. Liu, F. (2006) Estimation of new-modified prime code in synchronous incoherent CDMA network MPhil Dissertation at School of EECE, University of Birmingham.

10. Maric, S.V. (1993) New family of algebraically designed optical orthogonal codes for use in CDMA fiber optic networks. *Electronics Letters*, **29** (6), 538–539.

11. Wei, Z. and Ghafouri-Shiraz, H. (2002) Proposal of a novel code for spectral amplitude coding optical CDMA systems. *IEEE Photonics Tech. Letters*, **14** (3), 414–416.

12. Weng, C.S. and Wu, J. (2001) Perfect difference codes for synchronous fiber-optic CDMA communication systems. *J. Lightw. Technol.*, **19** (2), 186–194.

13. Yang, G.C. and Kwong, W.C. (1995) Performance analysis of optical CDMA with prime codes. *Electronics Letters*, **31** (7), 569–570.

14. Kwong, W.C., Perrier, P.A. and Prucnal, P.R. (1991) Performance comparison of asynchronous and synchronous code-division multiple-access techniques for fiber-optic local area networks. *IEEE Trans. on Comm.*, **39** (11), 1625–1634.

15. Liu, M.Y. and Tsao, H.W. (2000) Co-channel interference cancellation via employing a reference correlator for synchronous optical CDMA system. *J. Microw. & Opt. Tech. Let.*, **25** (6), 390–392.

16. Viterbi, A.J. (1995) *CDMA, principles of spreading spectrum communication*. Addison Wesley, Boston, USA.

17. Prasad, R. (1996) *CDMA for wireless personal communications*. Artech House, Boston, USA.

18. Prucnal, P.R. (2005) *Optical code division multiple access: fundamentals and Applications*. CRC Taylor & Francis Group, Florida, USA.

19. Kavehrad, M. and Zaccarin, D. (1995) Optical code division-multiplexed systems based on spectral encoding of noncoherent sources. *J. Lightw. Technol.*, **13** (3), 534–545.

20. Agraval, G.P. (1992) *Fiber-optic communication systems*. John Wiley & Sons Inc, USA.

21. Gibson, J.D. (1993) *Principles of digital & analog communications*. Maxwell MacMillan, Canada.

22. Killat, U. (1996) *Access to B-ISDN via PON-ATM communication in practice*. Wiley Teubner Communications, Chichester, England.

23. Mestdagh, D.J.G. (1995) *Fundamentals of multi-access optical fiber networks*. Artech House Inc, Boston, USA.

24. Kwong, W.C. and Yang, G.C. (2004) Multiple-length multiple-wavelength optical orthogonal codes for optical CDMA systems supporting multirate multimedia services. *J. on Selected Areas in Comm.*, **22** (9), 1640–1647.

25. Lin, J.Y., Jhou, J.S. and Wen, J.H. (2007) Variable-length code construction for incoherent optical CDMA systems. *Optical Fiber Technology*, **12** (2), 180–190.

26. Ohara, K. (2003) Traffic analysis of Ethernet-PON in FTTH trial service. In: *OFC*.

27. Ahn, B. and Park, Y. (2002) A symmetric-structure CDMA-PON system and its implementation. *IEEE Photonics Tech. Letters*, **14** (9), 1381–1383.

28. Gupta, G.C. *et al.* (2007) A simple one-system solution COF-PON for metro/access networks. *J. Lightw. Technol.*, **25** (1), 193–200.

29. Kitayama, K., Wang, X. and Wada, N. (2006) OCDMA over WDM PON – solution path to gigabit symmetric FTTH. *J. Lightw. Technol.*, **24** (4), 1654–1662.

30. Yamamoto, F. and Sugie, T. (2000) Reduction of optical beat interference in passive optical networks using CDMA technique. *IEEE Photonics Tech. Letters*, **12** (12), 1710–1712.

31. Zhang, C., Qui, K. and Xu, B. (2007) Passive optical networks based on optical CDMA: design and system analysis. *Chinese Science Bulletin*, **52** (1), 118–126.

32. Kramer, G. (2005) *Ethernet passive optical network*. McGraw-Hill, New York, USA.

33. Stok, A. and Sargent, E.H. (2002) The role of optical CDMA in access networks. *IEEE Comm. Mag.*, **40** (9), 83–87.

34. Yeh, B.C., Lin, C.H. and Wu, J. (2009) Noncoherent spectral/time/spatial optical CDMA system using 3-D perfect difference codes. *J. Lightw. Technol.*, **27** (6), 744–759.

35. McGeehan, J.E. *et al.* (2005) Experimental demonstration of OCDMA transmission using a three-dimensional (time–wavelength–polarization) codeset. *J. Lightw. Technol.*, **23** (10), 3282–3289.

36. Fouli, K. and Maier, M. (2007) OCDMA and optical coding: principles, applications and challenges. *IEEE Comm. Mag.*, **45** (8), 27–34.
37. Yang, C.C. (2008) The application of spectral-amplitude-coding optical CDMA in passive optical networks. *Optical Fiber Technology*, **14** (2), 134–142.
38. Cooper, A.B. *et al*. (2007) High spectral efficiency phase diversity coherent optical CDMA with low MAI. In: *Lasers and Electro-Optics (CLEO)*, Baltimore, USA.
39. Azizoghlu, M., Salehi, J.A. and Li, Y. (1992) Optical CDMA via temporal codes. *IEEE Trans on Comm.*, **40** (8), 1162–1170.
40. Gumaste, A. and Zheng, S. (2006) Light-frames: A pragmatic solution to optical packet transport – extending the ethernet from LAN to optical networks. *J. Lightw. Technol.*, **24** (10), 3598–3615.
41. Chapman, D.A., Davies, P.A. and Monk, J. (2002) Code-division multiple-access in an optical fiber LAN with amplified bus topology: the SLIM bus. *IEEE Trans on Comm.*, **50** (9), 1405–1408.
42. Shalaby, H.M.H. (2002) Complexities, error probabilities and capacities of optical OOK-CDMA communication systems. *IEEE Trans on Comm.*, **50** (12), 2009–2017.
43. Abshire, J.B. (1984) Performance of OOK and low-order PPM modulations in optical communications when using APD-based receivers. *IEEE J. on Comm.*, **COM-32** (10), 1140–1143.
44. Liu, X. *et al*. (2004) Tolerance in-band coherent crosstalk of differential phase-shift-keyed signal with balanced detection and FEC. *IEEE Photonics Tech. Letters*, **16** (4), 1209–1211.
45. Foschini, G.J. and Vannucci, G. (1988) Noncoherent detection of coherent lightwave signals corrupted by phase noise. *IEEE Trans on Comm.*, **36** (3), 306–314.
46. Koshi, T., Kikuchi, K. and Kikuchi, H. (1988) *Coherent optical fiber communications*. KTK Scientific Publisher, Japan.
47. Shalaby, H.M.H. (1995) Performance analysis of optical synchronous CDMA communication systems with PPM signaling. *IEEE Trans. on Comm.*, **43** (2/3/4), 624–634.
48. Lee, T.S., Shalaby, H.M.H. and Ghafouri-Shiraz, H. (2001) Interference reduction in synchronous fiber optical PPM-CDMA systems *J. Microw. & Opt. Tech. Let.*, **30** (3), 202–205.
49. Shalaby, H.M.H. (1998) Co-channel interference reduction in optical PPM-CDMA systems. *IEEE Trans. on Comm.*, **46** (6), 799–805.
50. Wang, X. *et al*. (2005) 10-user, truly-asynchronous OCDMA experiment with 511-chip SSFBG en/decoder and SC-based optical thresholder. In: *OFC*, Anaheim, CA, USA.
51. Yang, C.C. (2006) Optical CDMA-based passive optical network using arrayed-waveguide-grating. In: *IEEE ICC, Circuits and Systems*, Istanbul, Turkey.
52. Huang, J. *et al*. (2006) Hybrid WDM and optical CDMA implementation with M-sequence coded waveguide grating over fiber-to-the-home network. In: *IEEE ICC, Circuits and Systems*, Istanbul, Turkey.
53. Tsang, W.T. *et al*. (1993) Control of lasing wavelength in distributed feedback lasers by angling the active stripe with respect to the grating. *IEEE Photonics Tech. Letters*, **5** (9), 978–980.
54. Heritage, J.P., Salehi, J.A. and Weiner, A.M. (1990) Coherent ultrashort light pulse code-division multiple access communication systems. *J. Lightw. Technol.*, **8** (3), 478–491.
55. Chang, C.C., Sardesai, H.P. and Weiner, A.M. (1998) Code-division multiple-access encoding and decoding of femtosecond optical pulses over a 2.5-km fiber link. *IEEE Photonics Tech. Letters*, **10** (1), 171–173.
56. Weiner, A.M. (1995) Femtosecond optical pulse shaping and processing. *Progress in Quantum Electronics*, **3** (9), p. 161.
57. Smith, E.D.J., Blaikie, R.J. and Taylor, D.P. (1998) Performance enhancement of spectral-amplitude-coding optical CDMA using pulse position modulation. *IEEE Trans. on Comm.*, **46** (9), 1176–1185.
58. Wei, Z. and Ghafouri-Shiraz, H. (2002) Codes for spectral-amplitude-coding optical CDMA systems. *J. Lightw. Technol.*, **20** (8), 1284–1291.
59. Wei, Z., Ghafouri-Shiraz, H. and Shalaby, H.M.H. (2001) Performance analysis of optical spectral-amplitude-coding CDMA systems using super-fluorescent fiber source. *IEEE Photonics Tech. Letters*, **13** (8), 887–889.
60. Lin, C.L. and Wu, J. (2000) Channel interference reduction using random Manchester codes for both synchronous and asynchronous fiber-optic CDMA systems. *J. Lightw. Technol.*, **18** (1), 26–33.
61. Salehi, J.A. (1989) Code division multiple-access techniques in optical fiber networks – part I: fundamental principles. *IEEE Trans. on Comm.*, **37** (8), 824–833.
62. Salehi, J.A. and Brackett, C.A. (1989) Code division multiple-access technique in optical fiber networks – part II: system performance analysis. *IEEE Trans. on Comm.*, **37** (8), 834–842.

63. Willebrand, H. and Clark, G. (2001) Free space optics: a viable last-mile alternative. In: *Proc. SPIE Wireless, Mobile Communications*, Beijing, China.
64. Nykolak, G. *et al*. (1999) 2.5 Gbit/s free-space optical link over 4.4 km. *Electronics Letters*, **35** (4), 578–579.
65. Arnon, S. (2003) Effects of atmospheric turbulence and building sway on optical wireless-communication systems. *Optics Letters*, **28** (2), 129–131.
66. Dorschner, T. *et al*. (1996) An optical phased array for lasers. In: *Proc. IEEE Int. Symp. Phased Array Systems and Technology*, Boston, USA.
67. Schrick, B. and Riezenman, M. (2002) Wireless broadband in a box. *IEEE Spectrum Mag.*, June 38–43.
68. Hamzeh, B. and Kavehrad, M. (2004) OCDMA-coded free-space optical links for wireless optical-mesh networks. *IEEE Trans on Comm.*, **52** (12), 2165–2174.
69. Salehi, J.A. (2007) Emerging OCDMA communication systems and data networks. *J. Optical Networking*, **6** (9), 1138–1178.
70. Aminzadeh-Gohari, A. and Pakravan, M.R. (2006) Analysis of power control for indoor wireless infrared CDMA communication. In: *Proc. IPCCC*, Phoenix, AZ, USA.
71. Ohtsuki, T. (2003) Performance analysis of atmospheric optical PPM CDMA systems. *J. Lightw. Technol.*, **21** (2), 406–411.
72. Jazayerifar, M. and Salehi, J.A. (2006) Atmospheric optical CDMA communication systems via optical orthogonal codes. *IEEE Trans on Comm.*, **54** (9), 1614–1623.

# 4

# Spectrally Encoded OCDMA Networks

## 4.1 Introduction

In Chapter 3, we introduced different kinds of OCDMA schemes alongside various transceiver structures composed of lenses, masks and gratings for spectral-encoded systems. However, as pointed out earlier, for this system the basic transceiver structure is too complex which makes it too expensive to build.

An important requirement for transceiver design in a practical OCDMA system is the address reconfiguration ability. The transmitter structure which was introduced in Section 3.4.4 is composed of gratings and a mask. In such a transmitter, we need to mechanically change the amplitude mask to generate a different sequence. In doing so, every user needs a set of the same masks, which makes the transmitter tremendously expensive.

To overcome these disadvantages, many encoder/decoder structures have been proposed based on the present optical devices. For example, arrayed waveguide gratings (AWG) have been employed to construct a spectral-amplitude-coding (SAC) OCDMA system in [1]. This chapter is devoted to a discussion of various system implementation methods using different optical devices.

Fibre Bragg gratings (FBGs) have also been used to implement the encoder/decoder in OCDMA [2]. A coherent spectral phase-encoder composed of FBGs has been proposed in [3]. In a similar approach, we have designed both the transmitter and the receiver structures for a SAC system based on the SAC-appropriate modified quadratic congruence (MQC) and modified frequencyhopping (MFH) codes, introduced in detail in Chapter 2.

The review of the AWG-based system implementation, acoustically tuneable optical filters (ATOFs) [4] and a SAC system implemented by ATOF will be discussed in this chapter as well as the transceiver structure based on FBG technology. Tuneable range requirement for address reconfiguration and factors affecting bit rate limitation are also investigated. Regarding spectral encoding schemes and code features introduced in

*Optical CDMA Networks: Principles, Analysis and Applications*, First Edition. Hooshang Ghafouri-Shiraz and M. Massoud Karbassian.

Chapters 2 and 3, it should be noted that the MQC and MFH codes are the most suitable and most recent encoding schemes for spectral coding OCDMA, and thus they have been analysed and considered throughout this chapter.

## 4.2  Spectral-Amplitude-Coding Schemes

### 4.2.1  Arrayed Waveguide Gratings (AWGs)

AWGs are made by using planar waveguide technology. They can realize the following functions which are also shown in Figure 4.1:

- The AWG can take a multichannel (i.e. multiwavelength) input appearing on a single input waveguide (i.e. port) and separate the channels into different output ports.
- The AWG can combine many inputs of different wavelengths from different input ports into the same output port.
- The AWG can operate bidirectionally.

Generally AWGs are used as the multiplexers/demultiplexers in wavelength-division multiplexing (WDM) optical communications systems. Typical AWGs can handle a wavelength spacing of 0.8 nm (i.e. 100 GHz). Devices with 64 inputs and 64 outputs have been demonstrated and up to 16-channel versions are available commercially. The advertised insertion loss of current commercial devices is less than 3 dB for each channel. This compares favourably with other devices performing similar functions.

An SAC system implemented by AWG technology is shown in Figure 4.2. In this system, AWGs are used as a spectral separator, through which the broadband pulses from the source are divided into many spectral components. The desired components are coupled into another AWG while undesired components are eliminated (reflected by mirrors or absorbed by black masks). At the receiver, three AWGs are connected according to the address code. Where Hadamard code is applied, half of the spectral components (separated by the first AWG) are connected to one AWG and the rest are connected to the other one. Undesired pulses from other users will result in equal amounts of multiple user interference (MUI) in both photodetectors. Therefore, desired data can be retrieved after the balanced detection. In this structure, address reconfiguration of the transmitter may be realized by mechanically changing the masks' distribution. Another

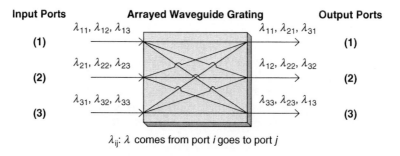

$\lambda_{ij}$: $\lambda$ comes from port $i$ goes to port $j$

**Figure 4.1**  AWG function diagram

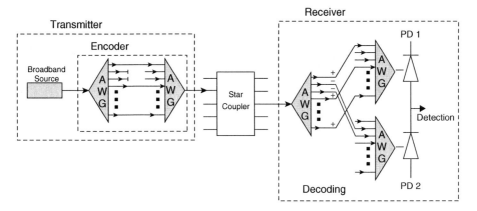

**Figure 4.2**   AWG-based SAC-OCDMA system

possible reconfiguration method is to insert a liquid crystal mask (LCM) between the two AWGs in the encoder. In this case, the address sequence can be reconfigured flexibly by LCM rearrangement which is controlled by electrical signals.

## 4.2.2   Acoustically Tuneable Optical Filters (ATOF)

The basic concept of ATOF is the interaction of an acoustic wave (ATOF) and a light-wave. In an LiNbO$_3$ photonic integrated circuit an acoustic waveguide (with a width of about 100 µm) is integrated around an optical single-mode waveguide. The propagation constants of the optical polarization mode (TE and TM) differ because of birefringence. By applying of an acoustic wave with the same wavelength as the beat wavelength of the two polarization modes an interaction between these two modes can occur and light will be converted from TE to TM, for example. The birefringence is wavelength dependent. By applying a single electrical frequency to such a device, only one optical wavelength conversion can be observed. If we assume a constant acoustic power between the transducer (transmitter of the acoustic wave) and the absorber, the filter characteristic is like a Sinc function (according to the Fourier transformation of a rectangle). The power of the acoustic wave, and therefore the voltage of the electrical driver signal, determine the degree of conversion. For 100% conversion, voltages of some 100 mV will be enough [4] (available filter for the first telecommunication window: 800–900 nm). Increasing the voltage over the 100% conversion point gives a back conversion into the incoming polarization mode.

In ATOF, if the incoming polarization mode is TE and it is also filtered in TE mode at the output, then the ATOF will act like a band-pass filter. Whereas, if the incoming TE polarized lightwave is filtered in TM mode at the output, then ATOF will act like a band-stop filter.

The linear characteristic of the acoustic wave transmitter allows the use of several drive frequencies (more than 100) at the same time. The limit is given by the maximum applied voltage at the transducer. By the choice of suitable drive frequencies and voltages, almost all filter functions can be realized. The spectral resolution is determined by the interaction length of the acoustic wave on the ATOF (a few *cm* of length will give a resolution of a few tens of nm in the first telecommunication window). Moreover, with variation of the voltage of the various drive frequencies a spectral flattening of the Gaussian power spectrum is realized. Therefore an ATOF has some important properties for use as a spectral encoder, including the following:

- The filter structure can be determined by the choice of the spectrum of the electrical drive signal.
- Very long sequences are possible by the use of about 100 frequencies at the same time.
- Only low drive power (in the range of mW) are required.
- Spectral flattening of the superluminescent diode (SLD) spectrum is simple to make by variation of the drive voltages of several lines.
- Band-pass and stop-band filters can be realized with different polarization filters (or a polarization dividing splitter gives the inverse function at two outputs).
- Because of using only one waveguide without any free-space optics the insertion loss of the element can be minimized.
- The size of the filter is in the *cm* range.
- the filter function can be changed in just a few μs.

Also the disadvantages of ATOF can be noted as the necessity of a permanent drive signal and the temperature dependence of the birefringence.

Figure 4.3 shows the system implementation by ATOFs. In this structure, ATOF is used to encode/decode the data stream. The acoustic waves corresponding to the desired address code are activated by the output of the electrical sequence-generator. At the transmitter, only one output of the ATOF in TM or TE mode is applied, while at the receiver side, the two inverse outputs in the TM and TE modes are utilized for the balanced detection to cancel the effect of multiuser interference (MUI). Since polarization filters are required at the input of each ATOF, 3 dB loss is inevitably inserted. In this structure, address reconfiguration can be easily realized by changing the output of the electrical sequence-generator. This reconfiguration method is very flexible since there is no relationship between the address codes and internal devices' connections (as in the case of AWG implementation). Therefore we can use any possible code in this structure without changing the hardware connections.

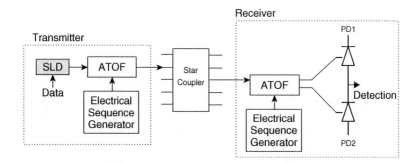

**Figure 4.3**   ATOF-based SAC-OCDMA system

## 4.2.3   Fibre Bragg Gratings

Fibre Bragg grating (FBG) fabrication technology is now well established, and grating devices are finding many applications in optical communication networks and sensors. Of particular interest for optical communications is the correction of chromatic dispersion in existing fibre links by chirped FGBs, allowing an increase in the transmission data rate and distance.

Basically, a grating is a periodic structure or perturbation in a material. This variation in the material has the property of reflecting or transmitting light in a certain direction depending on the wavelength. An FBG is constructed within an optical fibre. Therefore it has advantages such as low cost, low loss (around 0.1 dB at 1.55 µm wavelength), ease of coupling with other fibres, polarization insensitivity, low temperature coefficient and simple packaging. A fibre grating is a narrowband reflection filter that is fabricated through a photo-imprinting process [5]. The technique is based on the observation that germanium-doped silica fibre exhibits high photosensitivity. This means that one can induce a change in the refractive index of the core by exposing it to 244 nm ultraviolet radiation. Optical bandwidths of 100 GHz and less have been demonstrated in such a photo-induced grating [6]. When a multiwavelength signal encounters the grating, those wavelengths phase-matched to the Bragg reflection condition are not transmitted and reflected back.

The imprinted grating can be represented as a uniform sinusoidal modulation of the refractive index along the core [7]:

$$n(z) = n_{core} + \delta n \cdot \left[ 1 + \cos\left(\frac{2\pi \cdot z}{\Lambda}\right) \right] \tag{4.1}$$

where $z$ is the direction along the fibre axis, $\Lambda$ is the period of interference pattern (and hence the period of the grating), $n_{core}$ is the unexposed core refractive index and $\delta_n$ is the photo-induced change in the index. The maximum reflectivity $R$ of the grating occurs when the Bragg condition holds, that is at a reflection wavelength $\lambda_{Bragg}$ where [7]:

$$\lambda_{Bragg} = 2\Lambda n_{eff} \tag{4.2}$$

and $n_{eff}$ is the mode effective index of the core. At this wavelength, the peak reflectivity $R_{max}$ for the grating of length $L$ and coupling coefficient $\kappa$ can be expressed as [7]:

$$R_{max} = \tan h^2(\kappa L) \tag{4.3}$$

The full bandwidth $\Delta\lambda$ over which the maximum reflectivity holds is written as [7]:

$$\Delta\lambda = \frac{\lambda_{Bragg}^2}{\pi n_{eff} L} [(\kappa L)^2 + \pi^2]^{1/2} \tag{4.4}$$

Furthermore, an approximation for the full-width half-maximum (FWHM) bandwidth can be expressed as [7]:

$$\Delta\lambda_{FWHM} \approx \lambda_{bragg} \cdot s \cdot \left[ \left(\frac{\delta n}{2n_{core}}\right)^2 + \left(\frac{\Lambda}{L}\right)^2 \right]^{1/2} \tag{4.5}$$

where $0.5 \leq s \leq 1$ in that $s \approx 1$ for strong gratings with near 100 per cent reflectivity, and $s \approx 0.5$ for weak gratings. For a uniform sinusoidal modulation of the index throughout the core, the coupling coefficient $\kappa$ is given by [7, 8]:

$$\kappa = \frac{\pi \cdot \delta n \cdot \eta}{\lambda_{Bragg}} \tag{4.6}$$

with $\eta$ being the fraction of optical power contained in the fibre core. Under the assumption that the grating is uniform in the core, $\eta$ can be approximated by [9]:

$$\eta \approx 1 - V^{-2} \tag{4.7}$$

where $V$ is the normalized frequency of the fibre [9] and it is often used in the context of step-index fibres. It is defined as:

$$V = \frac{2\pi}{\lambda} \cdot a \cdot NA = \frac{2\pi}{\lambda} \cdot a \cdot \sqrt{n_{core}^2 - n_{cladding}^2} \tag{4.8}$$

where $\lambda$ is the vacuum wavelength, $a$ is the radius of the fire core, and NA is the numerical aperture.

The normalized frequency $V$ is related to the following various essential properties of an optical fibre:

- For $V \leq 2.405$ an optical fibre supports only one mode and is referred to as a single-mode fibre.
- An optical fibre with $V > 2.405$ is referred to as a multimode fibre. For large values of $V$ the number of supported modes $M$ of a step-index fibre can be approximated as [8]:

$$M \approx \frac{4}{\pi^2} V^2 \tag{4.9}$$

which is usually even more approximated by $M \approx V^2/2$.
- The confinement factor in a single-mode fibre (SMF) (i.e. the ratio of light power in the core to that of the total power) is a function of $V$. Its values for $V = 1$ and $V = 2$ are 7% and 70%, respectively. Hence, in most fibre transmission systems SMFs are designed to operate in the range $2 < V < 2.4$.
- A low $V$ value makes a fibre sensitive to micro-bend losses and to absorption losses in the cladding. However, a higher $V$ value may increase scattering losses in the core or at the core–cladding interface.

For certain types of photonic crystal fibres, an effective $V$ number can be defined, where $n_{cladding}$ is replaced with an effective cladding index. The same equations as for the step-index fibres can be used for calculating quantities such as the single-mode cut-off, mode radius and splice losses. A more precise evaluation is needed for non-uniform or non-sinusoidal index variations.

When a broadband pulse is input into a group of FBGs, as the spectral components corresponding to $A(\omega)$ are reflected back, the output at the other end of the grating group contains all the complementary components corresponding to $\overline{A}(\omega)$ as shown in Figure 4.4. The reflected light can be considered as a spectrally encoded pulse after delay compensation, while the transmitted one is its complementary.

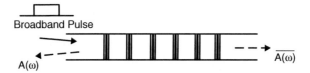

**Figure 4.4** Graphic representation of the FBG group (© 2001 IEEE. Reprinted with permission from Modified quadratic congruence codes for fiber Bragg-grating-based spectral-amplitude-coding optical CDMA systems, Z. Wei, H.M.H. Shalaby and H. Ghafouri-Shiraz, *J. Lightw. Technol.*, **19** (9), 2001.)

Figures 4.5 and 4.6 show the transmitter and receiver structures based on the FBGs and modified quadratic congruence (MQC) or modified frequency hopping (MFH) coding scheme, introduced in Chapter 2, respectively, for SAC-OCDMA where $P$ is a prime number. In the receiver shown in Figure 4.6, the output from the top of the first FBG group is used directly as the complementary-code-correlated output $R(\omega)\overline{A_k}(\omega)$, where $R(\omega)$ is the received signal and $\overline{A_k}(\omega)$ is the complement code of the receiver address sequence.

When a data bit one is sent, an optical pulse from a broadband thermal source launches into the encoder; whereas no optical pulse is sent if the data bit is zero. The optical pulse passes through the first fibre-grating groups and correspondent spectral components are reflected. For the reconfiguration of the destined address code, the gratings in the encoder are all tuneable, which means that the central wavelength of the reflected spectral component of each grating can be changed. The second group of FBGs in the transmitter is used to compensate the round-trip delay of different spectral components caused by the first one so that all the reflected components have the same time delay and can be incorporated into a pulse again. In address reconfiguration we need not only to tune the first group of gratings in the transmitter, but also to tune the second group correspondingly. At the receiver, each grating is fixed according to the receiver's address sequence. MUI can

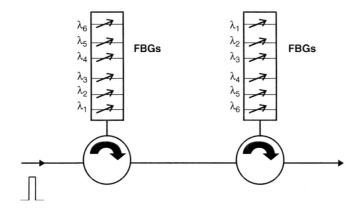

**Figure 4.5** Transmitter structure (© 2001 IEEE. Reprinted with permission from Modified quadratic congruence codes for fiber Bragg-grating-based spectral-amplitude-coding optical CDMA systems, Z. Wei, H.M.H. Shalaby and H. Ghafouri-Shiraz, *J. Lightw. Technol.*, **19** (9), 2001.)

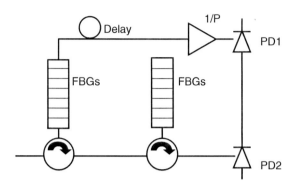

**Figure 4.6**   Receiver structure (© 2001 IEEE. Reprinted with permission from Modified quadratic congruence codes for fiber Bragg-grating-based spectral-amplitude-coding optical CDMA systems, Z. Wei, H.M.H. Shalaby and H. Ghafouri-Shiraz, *J. Lightw. Technol.*, **19** (9), 2001.)

be cancelled by balanced detection when codes with fixed in-phase cross-correlation are applied.

It should be noted that it is unnecessary to divide the source spectrum precisely in such a system because each transmitted pulse is generated by the reflection of a fibre grating. Data can be recovered accurately at the receiver so long as most of the power in the original spectrally encoded pulse can be reflected by the gratings at the desired receiver. This can be easily realized by using a grating with a length in the cm range.

Combined with MQC and MFH codes, these transmitter and receiver structures have following advantages:

- The weights of MQC and MFH codes are much lower than Hadamard code. Since one grating is used to reflect a spectral component, corresponding to a pulse in a code sequence, the number of gratings we need in the implementation of the transmitter and receiver are significantly reduced. This makes it very easy to build a transmitter and receiver pair. If Hadamard code is implemented in the above structures, the number of required gratings will be too large. This will make the system more complex and expensive.
- Elements in each sequence of the new codes can be divided into subsequences, and each subsequence contains just a single one. This property makes it much easier to realize address reconfiguration by grating tuning. In this case, the grating's tuneable range is the main limitation for address reconfiguration. By using the new constructed codes explained in Section 4.3.3, each grating only needs to be tuneable within $1/(P+1)$ of the total encoded bandwidth (TEB), where the TEB corresponds to the full length of each code sequence. Thus the required tuneable range for each grating is significantly reduced.
- At the receiver side, we can utilize all the received optical power efficiently since outputs at both ends of grating group are used for complementary decoding procedures.

## 4.3 System Considerations

### 4.3.1 Tuneable Range

Grating tuning can be implemented by fibre stretching using piezoelectric devices or by temperature adjustment. When MQC and MFH codes are used due to their property, the required tuneable spectral range (changing range by tuning) for each grating is equal to $1/(P + 1)$ of the total utilized bandwidth, see Chapter 2. Therefore, we can greatly reduce the required tuneable range for each grating by using MQC and MFH codes. A grating's tuneable linewidth range $\Delta\lambda_s$ can be expressed as $\Delta\lambda_s = 0.8\lambda\Delta L/L$, where $L$ is the grating length and $\Delta L$ is its change by stretching [11]. Let $\Delta L/L = 0.005$ and operating wavelength $\lambda = 1550$ nm, the tuneable linewidth range is given by $\Delta\lambda_s = 6.2$ nm. Since there are $(P + 1)$ gratings in each encoder, we can utilize 37 nm linewidth if $P = 5$. On the other hand, when the total linewidth is 30 nm, the required value of $\Delta L/L$ can be reduced to 0.004. Therefore, the application of MQC and MFH codes makes it more convenient to realize the encoder reconfiguration.

### 4.3.2 Bit Rate Limitation

In the FBG-based encoding/decoding structure, the length of the FBG group no longer limits the bit rate, as was the case in [11]. This is because the delay of each spectral component is the same after delay compensation. Therefore the bit rate is mainly limited by two factors: (i) the maximal modulation rate of the broadband thermal source and (ii) the bandwidths of the photodetectors used in the balanced detection. It is easy to see that the first factor can be significantly enlarged when an external modulator is applied.

### 4.3.3 Low-Cost Structure by Using Couplers

In the transmitter and receiver structures shown in Figures 4.5 and 4.6, circulators are used along with FBGs to provide a high separation between the input port and the output port. When requirements are loose, low-cost couplers can be used, taking place of the circulators [12]. We have shown the transmitter and receiver structures using couplers in Figures 4.7 and 4.8, respectively. One major disadvantage of these structures lies in the high loss due to the couplers. For example, in Figure 4.7, the input pulse is first coupled into the upper arm with a grating group. Then the reflected spectral components are coupled back by the same $2 \times 2$ coupler and input into another coupler for a second time.

Spectral-amplitude-coding (SAC) OCDMA systems have the potential to completely eliminate the multiple user interference (MUI) when code sequences with fixed in-phase cross-correlation (such as $m$-sequence or Hadamard code) are employed. However as the broadband thermal sources (BTS) are used in such systems, phase-induced intensity noise (PIIN) due to the spontaneous emission severely affects the system performance. It has been shown that the effect of PIIN is proportional to the square of photocurrent, and performance cannot be improved by a simple increase of the received optical power [13].

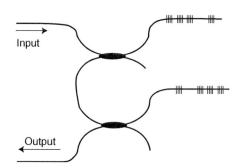

**Figure 4.7**   Encoder structure ($P = 5$)

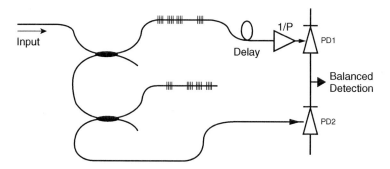

**Figure 4.8**   Decoder structure using coupler

To suppress this effect, a new code defined as $\left( \frac{q^{m+1}-1}{q-1}, \frac{q^{m}-1}{q-1}, \frac{q^{m-1}-1}{q-1} \right)$ has been introduced in [14], using the theorem of block designs whose construction methods were reviewed in Chapter 2. It has been found that the intensity noise can be effectively suppressed by using these codes and hence a higher signal-to-noise ratio (SNR) results. This higher SNR is in fact due to the higher ratio of auto-correlation peak to the fixed in-phase cross-correlation as compared with that of $m$-sequence or Hadamard codes. In Chapter 2, SNR expressions were listed and the performance results from Gaussian approximation plotted regarding MQC and MFH to show the improvement brought by these new codes. In this chapter, partially we focus on the detailed analysis of the SAC-OCDMA system considering MQC code. Similar performance improvement can also be obtained for MFH code.

Based on these codes and the transmitter/receiver pair described in Section 4.3.3, we have analysed the SAC-OCDMA systems by Gaussian approximation taking into account intensity, shot and thermal noise. Due to the use of the MQC code, a significantly improved performance has been obtained compared with a former system using Hadamard code. Gaussian approximation is accurate only when the simultaneous user number is large and hence the central limit theorem is valid to be applied. But when there are only a few simultaneous users, the system performance will depend on the exact probability distribution of PIIN. It has been pointed out that when both the intensity and shot noise are considered; the count of the arrived photons obeys negative binomial (NB) distribution [15]. If there is only one active user, this negative-binomial-distributed photoelectron counting will result in a lower bit-error rate (BER) than Gaussian approximation because of its lower probability tails [16]. This result has also been verified experimentally by Tasshi Dennis [17]. In this chapter we have also analysed the system by NB distribution and compared the results with that from Gaussian distribution. It has been shown that Gaussian approximation can give as accurate an estimate as the NB distribution, especially when the number of users is large.

## 4.3.4 Noise Sources at the Receivers

When a PIN diode is used in an SAC system, the noise sources at the receiver side include shot noise, thermal noise and dark current noise. Dark current noise refers to the noise generated when there's no light input. This noise is due to the imperfection of the receiver PIN photodiode (PIN-PD) and can be alleviated with the development of its manufacturing technology. Shot noise, also referred to as quantum noise, comes from the quantum fluctuations of the incoming light, which is generally modelled as a Poisson process. Thermal noise is generated after the photon-to-electron conversion, which is caused by the thermal interaction of the electrons. Its statistics obey Gaussian distribution. The effect of shot noise can be given by $\langle i_{sh}^2 \rangle = 2eIB$, where $e$ is the electron's charge, $I$ is the average photocurrent, $B$ is the noise-equivalent electrical bandwidth of the receiver; and the effect of thermal noise can be written as $\langle i_{th}^2 \rangle = 4K_bT_nB/R_L$, where $K_b$ is the Boltzmann's constant, $T_n$ is the receiver's noise temperature, and $R_L$ is the receiver's load resistor.

When incoherent light fields are mixed and incident upon a photodetector, the phase noise of the fields causes an intensity noise term in the photodetector output, labelled as phase-induced intensity noise (PIIN). For the detection of light from a thermal source, assuming spatial coherence at the detector, the variance of the photodetector current can be expressed as [10]:

$$\langle i^2 \rangle = 2eIB + I^2(1 + P_D^2)B\tau_c + 4K_bT_nB/R_L \qquad (4.10)$$

where $\tau_c$ is the source coherence time, and $P_D$ is the degree of polarization of the source. Here, the first item results from shot noise, the second item denotes the effect of PIIN and the third item represents the effect of thermal noise. The details of the definition of degree of polarization $P_D$ can be found in [15]. If $G(v)$ denotes the single sideband power spectral density of the thermal source, its coherence time $\tau_c$ can be expressed as:

$$\tau_c = \frac{\int_0^\infty G^2(v)d_v}{\left[\int_0^\infty G(v)d_v\right]^2} \qquad (4.11)$$

Intensity noise can be closely approximated by a gamma variable. More precisely, the underlying photoelectron count statistics, including both shot and intensity noise are known to be negative binomial (NB). Both the gamma and NB distributions are asymptotically Gaussian as the ratio of optical bandwidth to the maximum electrical bandwidth tends to infinity [15].

## 4.4   Gaussian Approach Analysis

### 4.4.1   Optimal Threshold

Noise in the communication systems is generally assumed as a random variable obeying Gaussian distribution whose probability density function (PDF) can be expressed as:

$$p(x) = \frac{1}{\sqrt{2\pi\sigma^2}} e^{-(x-\mu)^2/2\sigma^2} \tag{4.12}$$

where $\mu$ is the mean and $\sigma^2$ is the variance of the variable $x$. In digital communication system, there are only two signal patterns: 0 and 1. When signal pattern is 0, the received signal can be considered as a Gaussian variable with mean $\mu_0$ and standard deviation $\sigma_0$. When signal pattern is 1, we consider the received signal as a Gaussian variable with mean $\mu_1$ and standard deviation $\sigma_1$ as shown in Figure 4.9. For any given decision threshold, the BER in the case of a 0 signal pattern (denoted as $P_e(0)$) corresponds to the area under the PDF curve $(\mu_0, \sigma_0)$ to the right of the threshold line; while the BER for a 1 signal pattern (denoted as $P_e(1)$) corresponds to the area under the PDF curve $(\mu_1, \sigma_1)$ to the left of the threshold line. If the probability of sending 1 and 0 are $P(1)$ and $P(0)$ *respectively*, the total BER can be written as:

$$P_e = P(0)P_e(0) + P(1)P_e(1) \tag{4.13}$$

Assuming that probabilities of sending 0 and 1 are equal, we have:

$$P_e = \frac{1}{2}[P_e(0) + P_e(1)] \tag{4.14}$$

In this case, the optimal threshold lies at the point where $P_e(0) = P_e(1)$ and the total BER corresponds to the shadowed area in Figure 4.9. A linear approximation of the optimal threshold can be expressed as:

$$Th_{sub-optimal} = \frac{\mu_1\sigma_0 - \mu_0\sigma_1}{\sigma_1 + \sigma_0} \tag{4.15}$$

### 4.4.2   Signal-to-Noise Ratio and Bit-Error Rate

Signal-to-noise ratio (SNR) is an important parameter in an analogue communication system. Since all digital signals are finally transmitted as analogue waveforms, there's a relation between the SNR of the received analogue waveforms and the BER after decision. Here we define SNR by the following equation [10]:

$$SNR = \frac{[\mu_1 - \mu_0]^2}{[\sigma_1^2 + \sigma_0^2]/2} \tag{4.16}$$

when $\sigma_1^2 = \sigma_0^2 = \sigma^2$, this becomes:

$$SNR = \frac{[\mu_1 - \mu_0]^2}{\sigma^2} \tag{4.17}$$

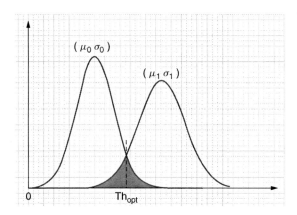

**Figure 4.9**  Optimal threshold in Gaussian approximation

Assuming that the statistics of the received signals obey Gaussian distribution, thus $Th_{opt} = (\mu_1 + \mu_0)/2$ and the BER can be calculated as [10, 18]:

$$P_e(0) = P_e(1) = \frac{1}{\sqrt{2\pi}\sigma} \int_{Th_{optimal}}^{\infty} e^{-(x-\mu_0)^2/2\sigma^2} dx = \frac{1}{\sqrt{2\pi}} \int_{\frac{Th_{optimal}-\mu_0}{\sigma}}^{\infty} e^{-t^2/2} dt \quad (4.18)$$

By defining $Q(x) = \frac{1}{\sqrt{2\pi}} \int_x^{\infty} e^{-t^2/2} dt$ and $erfc(x) = \frac{1}{\sqrt{\pi}} \int_x^{\infty} e^{-t^2} dt$, the final BER can be expressed as:

$$P_e = [P_e(0) + P_e(1)]/2 = Q\left(\frac{Th_{optimal} - \mu_0}{\sigma}\right) = Q\left(\frac{\mu_1 - \mu_0}{2\sigma}\right) \quad (4.19)$$

or as [10, 19, 20]:

$$P_e = \frac{1}{2} erfc\left(\frac{\mu_1 - \mu_0}{\sqrt{8}\sigma}\right) = \frac{1}{2} erfc\left(\sqrt{\frac{SNR}{8}}\right) \quad (4.20)$$

### 4.4.3  Gaussian Performance Analysis

The following assumptions should be considered for mathematical simplicity of the system analysis:

- Each light source is ideally unpolarized ($P_D = 0$) and its spectrum is flat over the bandwidth $[v_0 - \Delta v/2, v_0 + \Delta v/2]$, where $v_0$ is the central optical frequency and $\Delta v$ is the optical source bandwidth in Hertz.
- Each power spectral component has identical spectral width.
- Each user has equal power at the receiver.
- Each bit-stream from each user is synchronized.

Based on the developed MQC and MFH codes, and FBG technology, a transmitter and receiver pair has been proposed in Section 4.3.3 and its advantages have been discussed.

In this section, we will analyse that system by Gaussian approximation taking into account intensity noise as well as the shot noise and thermal noise at the optical receiver. The effect of the receiver's dark current is neglected.

According to Equation (4.10), the variance of photocurrent due to the detection of an ideally unpolarized thermal light (where $P_D = 0$) can be expressed as:

$$\langle i^2 \rangle = 2eIB + I^2 B \tau_c + 4KT_n B/R_L \tag{4.21}$$

If $c_k(i)$ and $\bar{c}_k(i)$ denote the $i$th element of the $k$th MQC code sequence and its complement respectively, code properties can be written as:

$$\sum_{i=1}^{N} c_k(i)c_l(i) = \begin{cases} P+1, & k = l \\ 1, & k \neq l \end{cases} \tag{4.22}$$

Hence,

$$\sum_{i=1}^{N} c_k(i)\bar{c}_l(i) = \begin{cases} 0, & k = l \\ P, & k \neq l \end{cases} \tag{4.23}$$

Based on the assumptions, we can analyse the system performance using Gaussian approximation now. The power spectral density (PSD) of the received optical signals can be written as [10]:

$$r(v) = \frac{P_{sr}}{\Delta v} \sum_{k=1}^{K} d_k \sum_{i=1}^{N} c_k(i) \left\{ u\left[ v - v_0 - \frac{\Delta v}{2N}(-N + 2i - 2) \right] \right.$$

$$\left. - u\left[ v - v_0 - \frac{\Delta v}{2N}(-N + 2i) \right] \right\} \tag{4.24}$$

where is $P_{sr}$ the received signal power at the receiver (if the loss due to transmission and star coupler is $\xi$ and source pulse power is $P_o$ then the received signal power is $\xi P_o$), $K$ is the number of active users less than or equal to $P^2$ where $P$ is a prime number; $N$ is the MQC code-length given by $N = P^2 + P$, $d_k$ is the data bit of the $k$th user that is 1 or 0, and $u(t)$ is the unit step function expressed as:

$$u(t) = \begin{cases} 1, & t \geq 0 \\ 0, & t < 0 \end{cases} \tag{4.25}$$

Therefore the PSDs at the photodetectors PD$_1$ and PD$_2$ in Figure 4.6 of the $l$th receiver during one bit period can be written as [10]:

$$G_1(v) = \frac{1}{P} \frac{P_{sr}}{\Delta v} \sum_{k=1}^{K} d_k \sum_{i=1}^{N} c_k(i)\bar{c}_l(i) \left\{ u\left[ v - v_0 - \frac{\Delta v}{2N}(-N + 2i - 2) \right] \right.$$

$$\left. - u\left[ v - v_0 - \frac{\Delta v}{2N}(-N + 2i) \right] \right\} \tag{4.26}$$

$$G_2(v) = \frac{P_{sr}}{\Delta v} \sum_{k=1}^{K} d_k \sum_{i=1}^{N} c_k(i)c_l(i) \left\{ u\left[ v - v_0 - \frac{\Delta v}{2N}(-N + 2i - 2) \right] \right.$$

$$\left. -u\left[ v - v_0 - \frac{\Delta v}{2N}(-N + 2i) \right] \right\} \quad (4.27)$$

And then:

$$\int_0^{\infty} G_1(v)dv = \frac{1}{P}\frac{P_{sr}}{\Delta v} \sum_{k=1, k \neq l}^{K} \left[ P\frac{\Delta v}{N}d_k \right] = \frac{P_{sr}}{N} \sum_{k=1, k \neq l}^{K} d_k \quad (4.28)$$

$$\int_0^{\infty} G_2(v)dv = \frac{P_{sr}}{\Delta v} \left\{ (P+1)d_l\frac{\Delta v}{N} + \sum_{k=1, k \neq l}^{K} \left[ \frac{\Delta v}{N}d_k \right] \right\}$$

$$= \frac{P_{sr}}{N}(P+1)d_l + \frac{P_{sr}}{N} \sum_{k=1, k \neq l}^{K} d_k \quad (4.29)$$

To calculate the integrals of $G_1^2(v)$ and $G_2^2(v)$ given in Equations (4.28) and (4.29), let us first consider an example of the spectral density function, $G'(v)$ of the received superposed signal, which is shown in Figure 4.10, where $a(i)$ is the amplitude of the $i$th spectral slot with width $(\Delta v/N)$.

The integral of $G'^2(v)$ can be expressed as [10]:

$$\int_0^{\infty} G'^2(v)dv = \frac{\Delta v}{N} \sum_{i=1}^{N} a^2(i) \quad (4.30)$$

Therefore, using the code properties as expressed in Equations (4.22) and (4.23) we have:

$$\int_0^{\infty} G_1^2(v)dv = \frac{1}{P^2}\frac{P_{sr}^2}{N\,\Delta v} \sum_{i=1}^{N} \left\{ \overline{c}_l(i) \cdot \left[ \sum_{k=1}^{K} d_k c_k(i) \right] \cdot \left[ \sum_{m=1}^{K} d_m c_m(i) \right] \right\} \quad (4.31)$$

$$\int_0^{\infty} G_2^2(v)dv = \frac{P_{sr}^2}{N\,\Delta v} \sum_{i=1}^{N} \left\{ c_l(i) \cdot \left[ \sum_{k=1}^{K} d_k c_k(i) \right] \cdot \left[ \sum_{m=1}^{K} d_m c_m(i) \right] \right\} \quad (4.32)$$

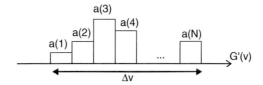

**Figure 4.10** Graphic representation of the spectral density of the received signal $r(v)$ (© 2001 IEEE. Reprinted with permission from Modified quadratic congruence codes for fiber Bragg-grating-based spectral-amplitude-coding optical CDMA systems, Z. Wei, H.M.H. Shalaby and H. Ghafouri-Shiraz, *J. Lightw. Technol.*, 19 (9), 2001.)

In the above equations $d_k$ is the data bit of the $k$th user that is either 1 or 0. Consequently, the photocurrent $I$ can be expressed as:

$$I = I_2 - I_1 = \Re \int_0^\infty G_2(v)dv - \Re \int_0^\infty G_1(v)dv = \Re \frac{P_{sr}}{P} d_l \qquad (4.33)$$

where $\Re$ is the responsivity of the photodetector given by $\Re = (\eta e)/(h v_c)$. Here $\eta$ is the quantum efficiency, $e$ is the electron's charge, $h$ is the Planck's constant, and $v_c$ is the central frequency of the original broadband optical pulse. Since the noises in $PD_1$ and $PD_2$ are independent, the power of noise sources that exist in the photocurrent can be written as [10,18–22]:

$$\langle I^2 \rangle = \langle I_1^2 \rangle + \langle I_2^2 \rangle + \langle I_{th}^2 \rangle = 2eB(I_1 + I_2) + BI_1^2 \tau_{c1} + BI_2^2 \tau_{c2} + 4K_b T_n B/R_L \qquad (4.34)$$

Therefore,

$$\langle I^2 \rangle = 2eB\Re \left[ \int_0^\infty G_1(v)dv + \int_0^\infty G_2(v)dv \right]$$

$$+ B\Re^2 \int_0^\infty G_1^2(v)dv + B\Re^2 \int_0^\infty G_2^2(v)dv + 4K_b T_n B/R_L \qquad (4.35)$$

When all the users are transmitting a bit 1, using the average value as $\sum_{k=1}^K c_k(i) \approx K/P$ and the code correlation properties, the noise power can be written as [10, 18]:

$$\langle I^2 \rangle = \frac{B\Re^2 P_{sr}^2 K}{\Delta v (P+1)P^2} \left( \frac{K-1}{P} + P + K \right) + 2eB\Re P_{sr} \left[ \frac{P-1+2K}{P^2+P} \right] + 4K_b T_n B/R_L \qquad (4.36)$$

Noting that the probability of sending a bit 1 at any time for each user is 50%, the above equation becomes:

$$\langle I^2 \rangle \cong \frac{B\Re^2 P_{sr}^2 K}{2\Delta v (P+1)P^2} \left( \frac{K-1}{P} + P + K \right) + eB\Re P_{sr} \left[ \frac{P-1+2K}{P^2+P} \right] + 4K_b T_n B/R_L \qquad (4.37)$$

From Equations (4.37) and (4.33) we can get the average SNR due to intensity, shot and thermal noises as:

$$SNR = \frac{(I_2 - I_1)^2}{\langle I^2 \rangle} = \frac{\Re^2 P_{sr}^2/P^2}{\substack{B\Re^2 P_{sr}^2 K[(K-1)/P+P+K]/[2\Delta v(P+1)P^2] \\ + eB\Re P_{sr}(P-1+2K)/(P^2+P)+4k_b T_n B/R_L}} \qquad (4.38)$$

Consequently by using Gaussian approximation, the BER is expressed as [10]:

$$P_e = \frac{1}{2} erfc\left( \sqrt{SNR/8} \right) \qquad (4.39)$$

### 4.4.4  Numerical Results

The results discussions and analysis here are based on the ones in [10]. Figure 4.11 shows the relation between the number of simultaneous users and the SNR when the prime number used for MQC code construction is $P = 7$, 11 and 13, respectively. The parameters used in this analysis are listed in Table 4.1. The SNR curve of the former system using Hadamard codes (with the same pulse power and optical bandwidth) is also shown for a comparison purpose. In this figure, the received signal power from each user is $-10$ dBm and the intensity noise is considered to be the main noise source. It should be noted that the SNR values plotted in this figure are all average values and each curve end at the point where the number of simultaneous users is equal to the code size. It has been shown that MQC codes give a much higher SNR when the received signal power is high. With a larger prime number $P$, not only the number of accommodated simultaneous users is significantly increased, but also results in a higher SNR. Therefore, MQC codes can effectively suppress the effect of intensity noise and hence result in a much better performance. This suppression in fact comes from the lower in-phase cross-correlation of MQC codes than that of Hadamard codes under the same code-length. Although MQC codes have lower auto-correlation due to its lower code-weight than Hadamard codes, its in-phase auto-correlation-to-cross-correlation ratio is still larger. However, this lower auto-correlation will cause performance degradation when other noise sources are considered especially at low signal power, as discussed and analyzed later on in this section.

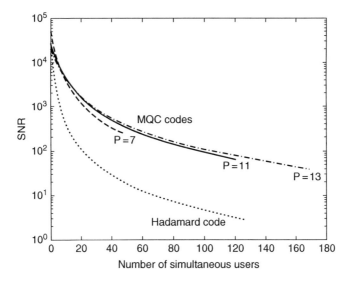

**Figure 4.11**  SNR against the number of simultaneous users when $P_{sr} = -10$ dBm (© 2001 IEEE. Reprinted with permission from Modified quadratic congruence codes for fiber Bragg-grating-based spectral-amplitude-coding optical CDMA systems, Z. Wei, H.M.H. Shalaby and H. Ghafouri-Shiraz, *J. Lightw. Technol.*, **19** (9), 2001.)

**Table 4.1** Typical parameters used in the calculation
(© 2001 IEEE. Reprinted with permission from
Modified quadratic congruence codes for fiber
Bragg-grating-based spectral-amplitude-coding optical
CDMA systems, Z. Wei, H.M.H. Shalaby and
H. Ghafouri-Shiraz, *J. Lightw. Technol.*, **19** (9), 2001.)

| Definitions | Symbols |
|---|---|
| PD Quantum Efficiency | $\eta = 0.6$ |
| Thermal Source Linewidth | $\Delta\nu = 3.75\,\text{THz}$ |
| Operating Wavelength | $\lambda_0 = 1.55\,\mu\text{m}$ |
| Electrical Bandwidth | $B = 80\,\text{MHz}$ |
| Data Bit Rate | $R_b = 155\,\text{Mbps}$ |
| Receiver Noise Temperature | $T_r = 300\,\text{K}$ |
| Receiver Load Resistor | $R_L = 1030\,\Omega$ |

Figure 4.12 shows variations of the BER with the number of simultaneous users when $P_{sr} = -10\,\text{dBm}$. For comparison purposes, BER variations of the former system that uses Hadamard codes are also included. In the former system, since we also send a pulse when data bit is 0, the formula used to calculate the BER becomes $P_e = \frac{1}{2}erfc(\sqrt{SNR/2})$. It has been clearly shown that the system using MQC codes has a much lower BER than the one that uses Hadamard codes. For example, when the required BER is $10^{-9}$, the system using Hadamard codes can only support about 38 users whereas the system using MQC codes can support about 80 simultaneous users when $P = 11$ and more than 86 simultaneous users when $P = 13$. When $P = 7$, the number of possible users is only limited by the code size (i.e. the number of available code sequences which is equal to 49).

Figure 4.13 depicts the BER variations with the received signal power $P_{sr}$ when $P = 7$ and the number of simultaneous users is 49. The solid lines represent the BER where the effects of intensity, shot and thermal noise have been taken into account. The dashed lines indicate the BER performance when the effects of only intensity and shot noise are considered. And the dotted lines indicate the system BER performance when only intensity and thermal noise sources are considered. It is shown that when $P_{sr}$ is large, both the shot and thermal noise are negligibly small compared with the intensity noise, which becomes the main limiting factor of the system performance. However, when $P_{sr}$ is low, the effect of intensity noise becomes minimal, and hence the thermal noise source becomes the main factor that limits the system performance. It is also shown that thermal noise is much more influential than shot noise on system performance.

Figure 4.14 illustrates the BER variations with $P_{sr}$ when the number of simultaneous users is 49 and $P = 7$, 11 and 13. The performance of the former system using Hadamard code is also shown for comparison. Figure 4.15 shows variations of the BER against the number of simultaneous users, $K$, for different values of $P_{sr}$. In Figures 4.14 and 4.15, the effects of the intensity, shot and thermal noise sources have been taken into account.

It is shown that when $P_{sr}$ is less than $-25\,\text{dBm}$, the former system with Hadamard codes will have better performance than the system using MQC codes. This is because the large value of prime number $P$ causes a large power loss ($=1/P$) in the transmitter,

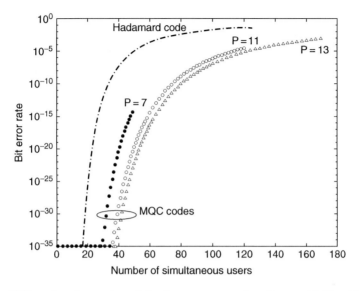

**Figure 4.12** BER against the number of simultaneous users when $P_{sr} = -10$ dBm (© 2001 IEEE. Reprinted with permission from Modified quadratic congruence codes for fiber Bragg-grating-based spectral-amplitude-coding optical CDMA systems, Z. Wei, H.M.H. Shalaby and H. Ghafouri-Shiraz, *J. Lightw. Technol.*, **19** (9), 2001.)

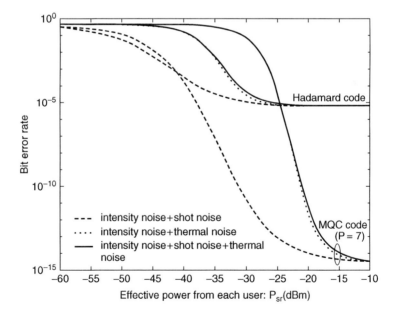

**Figure 4.13** BER against the received signal power $P_{sr}$ when $P = 7$ and full-load (© 2001 IEEE. Reprinted with permission from Modified quadratic congruence codes for fiber Bragg-grating-based spectral-amplitude-coding optical CDMA systems, Z. Wei, H.M.H. Shalaby and H. Ghafouri-Shiraz, *J. Lightw. Technol.*, **19** (9), 2001.)

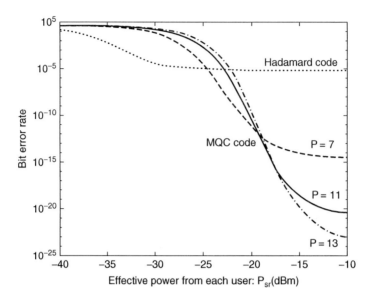

**Figure 4.14**   BER against the received signal power $P_{sr}$ when $K = 49$ and considering the intensity noise, shot noise and thermal noise (© 2001 IEEE. Reprinted with permission from Modified quadratic congruence codes for fiber Bragg-grating-based spectral-amplitude-coding optical CDMA systems, Z. Wei, H.M.H. Shalaby and H. Ghafouri-Shiraz, *J. Lightw. Technol.*, **19** (9), 2001.)

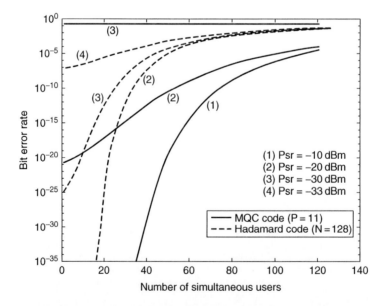

**Figure 4.15**   BER against the number of simultaneous users under various $P_{sr}$ (© 2001 IEEE. Reprinted with permission from Modified quadratic congruence codes for fiber Bragg-grating-based spectral-amplitude-coding optical CDMA systems, Z. Wei, H.M.H. Shalaby and H. Ghafouri-Shiraz, *J. Lightw. Technol.*, **19** (9), 2001.)

and hence the shot and thermal noise sources will severely affect the system performance especially when $P_{sr}$ is not sufficiently large. However, the power loss incurred by Hadamard encoding is always half regardless of the code-length.

If an array of $P + 1$ tuneable lasers is used, large optical power even after MQC encoder can be provided. In this case, the required tuneable range of each laser is only $1/(P + 1)$ of the total optical bandwidth. This method does not have any limitation on the shape of power spectral density as in the case of thermal source and we can also enlarge the optical encoded bandwidth freely to get a higher SNR. Therefore, although the coherent interference between the overlapped optical pulses with the same wavelength may cause degradation, a better performance is promised in such a laser-array-based system.

## 4.5    Negative Binomial Approach Analysis

### 4.5.1    NB Distribution with Hadamard Encoding

In this section, we will analyse the system shown in Figure 4.16 by negative binomial (NB) distribution and compare the results with those obtained from Gaussian approximation. Hadamard code and complementary encoding/decoding scheme are applied in the analysed system.

#### 4.5.1.1    Hadamard Code Property

A Hadamard code is obtained by selecting as code words the rows of a Hadamard matrix as explained in Chapter 2. A Hadamard matrix $M_n$ is a $N \times N$ matrix ($N$ is an even integer) of 1s and 0s with the property that any row differs from any other row in exactly $N/2$ positions. One row of the matrix contains all zeros. The other rows contain $N/2$ zeros and $N/2$ ones. For example when $N = 2$, the Hadamard matrix is:

$$\mathbf{M_2} = \begin{bmatrix} 0 & 0 \\ 0 & 1 \end{bmatrix} \tag{4.40}$$

Furthermore, from $M_n$ a Hadamard matrix $M_{2n}$ can be generated as:

$$\mathbf{M_{2n}} = \begin{bmatrix} \mathbf{M_n} & \mathbf{M_n} \\ \mathbf{M_n} & \overline{\mathbf{M_n}} \end{bmatrix} \tag{4.41}$$

where $\overline{\mathbf{M_n}}$ denotes the complement of $M_n$. Hence, Equations (4.40) and (4.41) can be used to generate Hadamard codes of length $N = 2^m$ with $m$ being a positive integer. It should be noted that Hadamard codes of other block lengths are possible, but they are not linear [23]. The all-zero row is generally deleted from the codeset and therefore the size of a Hadamard code from $N \times N$ Hadamard matrix is $N - 1$. If the unipolar Hadamard code with length $N$ and size $N - 1$ is composed of (0, 1) elements and $c_n(i)$ is the $i$th element of the $n$th Hadamard code sequence, the in-phase correlation between any two sequences is given by:

$$\sum_{i=1}^{N} c_l(i)c_m(i) = \begin{cases} N/2, & m = l \\ N/4, & m \neq l \end{cases} \tag{4.42}$$

Hence,

$$\sum_{i=1}^{N} c_l(i)\overline{c}_m(i) = \begin{cases} 0, & m = l \\ N/4, & m \neq l \end{cases} \tag{4.43}$$

where $m, l \in \{2, 3, 4, \ldots, N\}$.

### 4.5.1.2  System Description

The block diagram of the complementary encoding/decoding SAC-OCDMA system is shown in Figure 4.16. When the data bit is 1, the emitted broadband pulse from the super-fluorescent fibre source (SFS) launches into the upper optical spectral encoder where the PSD of the pulse follows the shape of $A(v)$. When the data bit is 0, the broadband pulse is switched into the complementary encoder which results in a PSD $\overline{A}(v)$.

After a $2 \times 1$ combiner, the encoded optical pulses are sent to a star coupler where optical signals from all users are superimposed. At the receiver side, this superimposed signal is split into two equal parts. Then each part is input into a decoder. The two decoders at the receiver are also complementary to each other and they have the same functions as encoders in the transmitter. According to the property of Hadamard code, signals from interfering users will result in the same amount of MUI in the upper and lower decoders. Hence, after balanced detection the desired signal can be retrieved. The low-pass filter is used to recover the original data stream. Since Hadamard sequence $c_{\mathbf{n}}$ is used as the desired address, $A(v)$ and $\overline{A}(v)$ can be expressed as:

$$A(v) = \sum_{i=1}^{N} c_n(i) \left\{ u \left[ v - v_0 - \frac{\Delta v}{2N}(-N + 2i - 2) \right] - u \left[ v - v_0 - \frac{\Delta v}{2N}(-N + 2i) \right] \right\}$$

$$\tag{4.44}$$

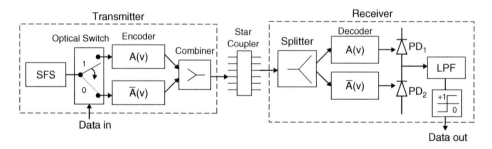

**Figure 4.16**  Block diagram of a complementary-coding SAC-OCDMA system (© 2001 IEEE. Reprinted with permission from Modified quadratic congruence codes for fiber Bragg-grating-based spectral-amplitude-coding optical CDMA systems, Z. Wei, H.M.H. Shalaby and H. Ghafouri-Shiraz, *J. Lightw. Technol.*, **19** (9), 2001.)

and,

$$\overline{A}(v) = \sum_{i=1}^{N} \overline{c}_n(i) \left\{ u\left[ v - v_0 - \frac{\Delta v}{2N}(-N + 2i - 2) \right] - u\left[ v - v_0 - \frac{\Delta v}{2N}(-N + 2i) \right] \right\}$$

(4.45)

where $\Delta v$ is the encoded bandwidth, $v_0$ is the central frequency and $u(v)$ is the unit step function.

### 4.5.1.3  NB Distribution of Hadamard Encoding

The light from a broadband thermal source (including SFS, SLD, LED, etc.) is generated by spontaneous emission. Hence all phase components exist. Therefore light intensity from a thermal source obeys negative exponential distribution [15]. When both the random intensity corrugation and Poisson photon arrival process are taken into account, the photon-count arrived at photodetectors $PD_1$ or $PD_2$ in Figure 4.16 has a NB distribution whose probability mass function is given by [15]:

$$\Pr(\Lambda = k) = \frac{\Gamma(k + M)}{\Gamma(k + 1)\Gamma(M)} \left( \frac{E(\Lambda)}{M + E(\Lambda)} \right)^k \left( \frac{M}{M + E(\Lambda)} \right)^M$$

(4.46)

where $\Lambda$ is the random variable representing the arrived photon-count (from 0 to infinity), $M$ is the degree of freedom of the broadband thermal light, $\Gamma(k)$ is the gamma function of $k$ given by $\Gamma(k) = (k = 1)!$ and $E(\Lambda)$ is the mean of arrived photon number. The variance of this distribution is given by:

$$\sigma_\Lambda^2 = E(\Lambda) + E^2(\Lambda)/M$$

(4.47)

When the data bit period $T \gg \tau_c$, ($\tau_c$ is the coherence time given by Equation (4.11)) the degree of freedom $M$ is equal to $U$ for a polarized thermal source and is equal to $2U$ for an ideally unpolarized thermal source, where $U$ denotes $U \cong T/\tau_c$.

### 4.5.1.4  Analysis of NB Distribution of Hadamard Encoding

Based on the assumptions in Section 4.3, the PSD at the input of photodetectors $PD_1$ and $PD_2$ (see Figure 4.8) of the $l$th receiver over one data bit period can be expressed as:

$$G_1(v) = \frac{P_s}{\Delta v} \sum_{k=1}^{K} \sum_{i=1}^{N} \left\{ \left[ c_k(i)d_k + \overline{c}_k(i)\overline{d}_k \right] \cdot c_l(i) \cdot rect(i) \right\}$$

(4.48)

$$G_2(v) = \frac{P_s}{\Delta v} \sum_{k=1}^{K} \sum_{i=1}^{N} \left\{ \left[ c_k(i)d_k + \overline{c}_k(i)\overline{d}_k \right] \cdot \overline{c}_l(i) \cdot rect(i) \right\}$$

(4.49)

where $K$ is the number of simultaneous users, $d_k$ is the data bit (0 or 1) sent by the $k$th user and $\overline{d}_k$ is its complement; $P_s$ is the received signal power of single users after the $1 \times 2$ splitter at the receiver. The $rect(i)$ function in Equations (4.48) and (4.49) is given by:

$$rect(i) = u\left[v - v_0 - \frac{\Delta v}{2N}(-N + 2i - 2)\right] - u\left[v - v_0 - \frac{\Delta v}{2N}(-N + 2i)\right] \quad (4.50)$$

Using the code properties as expressed in Equations (4.42) and (4.43), then:

$$\int_0^\infty G_1(v)dv = \frac{P_s}{4}(K - 1 + 2d_l) \quad (4.51)$$

$$\int_0^\infty G_2(v)dv = \frac{P_s}{4}(K - 1 + 2\overline{d}_l) \quad (4.52)$$

$$\int_0^\infty G_1^2(v)dv = \frac{P_s^2}{N\Delta v}\sum_{i=1}^N\left\{c_l(i)\cdot\left[\sum_{k=1}^K d_k c_k(i)\right]\cdot\left[\sum_{m=1}^K d_m c_m(i)\right]\right\} \quad (4.53)$$

$$\int_0^\infty G_2^2(v)dv = \frac{P_s^2}{N\Delta v}\sum_{i=1}^N\left\{\overline{c}_l(i)\cdot\left[\sum_{k=1}^K d_k c_k(i)\right]\cdot\left[\sum_{m=1}^K d_m c_m(i)\right]\right\} \quad (4.54)$$

Using the approximation $\sum_{m=1}^K d_m c_m(i) \approx K/2$ along with the code properties, Equations (4.53) and (4.54) become:

$$\int_0^\infty G_1^2(v)dv = \frac{P_s^2}{N\Delta v}\frac{K}{2}\left[d_l\frac{N}{2} + (k-1)\frac{N}{4}\right] = \frac{P_s^2}{8\Delta v}K(K - 1 + 2d_l) \quad (4.55)$$

$$\int_0^\infty G_2^2(v)dv = \frac{P_s^2}{8\Delta v}K(K - 1 + 2\overline{d}_l) \quad (4.56)$$

When the number of simultaneous users $K > 1$, the light coherence times at the inputs of $PD_1$ and $PD_2$ of the $l$th receiver are, respectively, given by:

$$\tau_{c1} = \frac{K(K - 1 + 2d_l)/8\Delta v}{(K - 1 + 2d_l)^2/16} = \frac{2K}{(K - 1 + 2d_l)\Delta v} \quad (4.57)$$

$$\tau_{c2} = \frac{K(K - 1 + 2\overline{d}_l)/8\Delta v}{(K - 1 + 2\overline{d}_l)^2/16} = \frac{2K}{(K - 1 + 2\overline{d}_l)\Delta v} \quad (4.58)$$

Let us denote the number of arrived photons at photodetectors $PD_1$ and $PD_2$ as $k_1$ and $k_2$, respectively. Both of them have NB distribution profiles with the mean values as $E(k_1), E(k_2)$ and degrees of freedom as $M_1$ and $M_2$, respectively. When the received data bit is 1, using the correlation property of Hadamard code, the means of $k_1$ and $k_2$ can be expressed as

$$E[k_1] = P_s T/2hf + (K - 1)P_s T/4hf = (K + 1)P_s T/4hf \quad (4.59)$$

$$E[k_2] = (K - 1)P_s T/4hf \quad (4.60)$$

where $h$ is the Planck's constant, $f$ is the optical source frequency and $T$ is the data bit period. The degrees of freedom are given by $M_1 = 2T/\tau_{c1}$ for $k_1$ and $M_2 = 2T/\tau_{c2}$ for $k_2$ where $\tau_{c1}$ and $\tau_{c2}$ are given by Equations (4.57) and (4.58) respectively. Thus, the two parameters of NB distribution when bits 0 or 1 are received can be obtained and we can calculate the probability mass function in detail.

The number of generated electrons $k_{e1}$ after balanced detection is given by $k_{e1} = \eta(k_1 - k_2) = \eta k_{d1}$, where $\eta$ is the photodetector quantum efficiency and $k_{d1} = k_1 - k_2$. Because the used threshold is zero, the effect of quantum efficiency $\eta$ can be neglected in the analysis of BER. In the following it is assumed that $\eta = 1$ and only the random variable $k_{d1}$, whose probability can be expressed as follow requires consideration:

$$P[k_{d1} = k] = \sum_{i=-\infty}^{\infty} \{P[k_1 = i + k] \cdot P[k_2 = i]\} \qquad (4.61)$$

where $P[k_1]$ and $P[k_2]$ can be obtained from Equation (4.46). To calculate the BER using Equation (4.61), we need the following definition. Given a random variable $\xi$, we define its probability vector $\overline{P}_\xi$ as $\overline{P}_\xi = \{\dots a_{i-1}, a_i, a_{i+1}, \dots\}$, where each element $a_i$ is the probability of $\xi = i$, i.e. $a_i = P[k = i]$. Assuming that the probability vectors of $k_1$, $k_2$ and $k_{d1}$ are denoted as $\overline{P}_{k1}$, $\overline{P}_{k2}$ and $\overline{P}_{d1}$ respectively, the inverse sequence of $\overline{P}_{k2}$ will be $inv(\overline{P}_{k2})$, i.e. $inv(\overline{P}_{k2}) = \{\dots a_{k2,i+1}, a_{k2,i}, a_{k2,i-1}, \dots\}$ where $a_{k2,i} = P[k_2 = i]$. If $conv(x, y)$ represents convolution of sequences $x$ and $y$, Equation (4.61) can be transformed into the following expression:

$$\overline{P}_{d1} = conv[\overline{P}_{k1}, inv(\overline{P}_{k2})] \qquad (4.62)$$

Consequently, the BER when data bit is 1 can be written as:

$$P_{e1} = \left\{ \sum_{i=-\infty}^{-1} P[k_{d1} = i] \right\} + P[k_{d1} = 0]/2 \qquad (4.63)$$

When the data bit is 0, we can calculate bit error rate $P_{e2}$ by swapping $\overline{P}_{k1}$ and $\overline{P}_{k2}$ in Equation (4.62), which results in the same BER. Assuming that the probabilities of sending data bits 1 and 0 are equal, the total BER is expressed as $P_e = (P_{e1} + P_{e2})/2$. Because $P_{e1} = P_{e2}$ therefore $P_e = P_{e1}$. This implies that the total BER can be obtained by using Equation (4.63).

When the number of users $K = 1$, there's no other active user. Hence, if the data bit is 1, the coherence time at $PD_1$ is given by:

$$\tau_{c1} = \frac{(P_s/\Delta v)^2 (N/2)(\Delta v/N)}{[(P_s/\Delta v)(N/2)(\Delta v/N]^2} = \frac{2}{\Delta v} \qquad (4.64)$$

and $E[k_1] = P_s T/2hf$. No optical power exists at $PD_2$. In this case the BER can be obtained by using the following expression:

$$P_{e1} = \left\{ \sum_{i=-\infty}^{-1} P[k_1 = i] \right\} + P[k_1 = 0]/2 \qquad (4.65)$$

Also the same BER can be obtained when the data bit is 0 and similarly the total BER can be expressed as Equation (4.65).

### 4.5.1.5   Numerical Results

In the BER calculation, the effects of both PIIN and shot noise have been considered. It can be also assumed that $\Pr(\Lambda = k) = 0$ when $|k - E[\Lambda]| \geq 4\sigma_\Lambda$, as it is negligibly small. The system operates at 1550 nm wavelength with data bit rate of 622 Mbps. The encoded bandwidth is 3.75 THz wide (which is equal to 30 nm linewidth) and the quantum efficiency is assumed to be 1.

Figure 4.17 compares the BER variations against the number of simultaneous users of both NB (dots) and Gaussian (lines) distributions when the received signal power is $-27$, $-40$ and $-52$ dBm. As expected in both cases, the BER increases with the number of simultaneous users $K$. For a given received signal power of single users after splitter $P_s$, when $K$ is small, the BERs calculated through NB distribution are significantly lower than those calculated by Gaussian approximation. These lower BERs are actually caused by the assumption of $\Pr(\Lambda = k) = 0$ when $|k - E[\Lambda]| \geq 4\sigma_\Lambda$. Accordingly, the BERs approach the corresponding Gaussian curve as the probability calculation range is enlarged, i.e. it is assumed that $\Pr(\Lambda = k) = 0$ when $|k - E[\Lambda]| \geq v\sigma_\Lambda$ where $v > 4$.

However, as the number of users $K$ increases, the BERs calculated through both methods are very close. For example, when $P_{sr} = -27$ dBm and $K > 20$ in Figure 4.17 both methods almost give the same results.

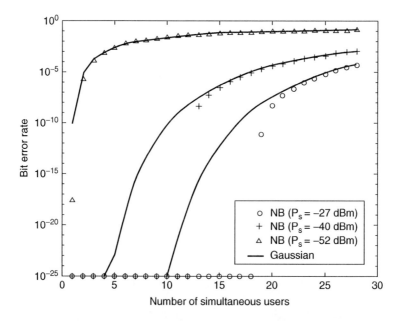

**Figure 4.17**   Variations of the BER against the number of simultaneous users using NB distribution (dots) and Gaussian approximation (lines) (© 2001 IEEE. Reprinted with permission from Performance analysis of optical spectral-amplitude-coding CDMA systems using superfluorescent fiber source, Z. Wei, H. Ghafouri-Shiraz and H.M.H. Shalaby, *Photonics Tech. Letters*, **13** (8), 2001.)

## 4.5.2   NB Distribution with MQC Encoding

In this section, we analyse the performance of an SAC-OCDMA system based on discrete NB distribution when MQC code is employed for users' addresses. The results obtained from this approach are compared with those from Gaussian approximation. It has been shown that when the number of interfering users increases, the NB method results in almost the same BERs as Gaussian approximation. Whereas in the case of $k = 0$, i.e. only one user exists, the NB-distribution method gives a more precise performance estimation. It should be noted that only intensity noise and shot noise are considered in the following analysis.

### 4.5.2.1   System Description

The block diagram of a SAC-OCDMA system applying MQC code is shown in Figure 4.18. The output of the broadband thermal source (BTS) is intensity-modulated at the transmitter. When the data bit is 1, the emission from the BTS launches into the optical spectral encoder where the PSD of the broadband pulse follows the shape of $A(v)$, which is decided by the desired MQC code sequence. When the data bit is 0, the BTS emission is cut off by the modulator and hence nothing is sent. The encoded optical pulses are sent to a star coupler where optical signals from all users are superimposed and then sent back to them. At each receiver, this superimposed signal is split into two equal parts.

Both parts are input into decoders separately. The two decoders in the receiver are complementary to each other and they are configured based on the receiver's address sequence. Therefore they can help the receiver to retrieve the desired signal assigned to it. Then, after balanced detection and low-pass filter (LPF), the original data can be recovered. In this system, because the MQC codes have ideal in-phase cross-correlation, not only the MUI can be cancelled, but also the effect of PIIN is suppressed.

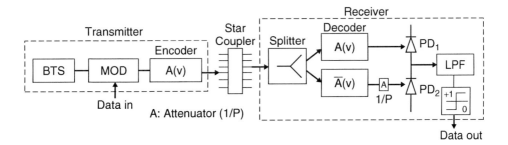

**Figure 4.18**   Block diagram of a spectral-amplitude-coding optical CDMA system (© 2001 IEEE. Reprinted with permission from Performance analysis of optical spectral-amplitude-coding CDMA systems using super-fluorescent fiber source, Z. Wei, H. Ghafouri-Shiraz and H.M.H. Shalaby, *Photonics Tech. Letters*, **13** (8), 2001.)

### 4.5.2.2   MQC Code Property

The MQC code employed in the analysis is a sequence of (0, 1) elements constructed by a prime number $P$, see Chapter 2. If $c_n(i)$ is the $i$th element of the $n$th code sequence and $\bar{c}_n(i)$ is its complement, the in-phase correlation between any two MQC sequences will be given by:

$$\sum_{i=1}^{N} c_l(i)c_m(i) = \begin{cases} P+1, & m=l \\ 1, & m \neq 1 \end{cases} \tag{4.66}$$

where $m, l \in \{1, 2, 3, \ldots, N\}$ and $N = P^2 + P$ is the code-length of MQC. Hence:

$$\sum_{i=1}^{N} c_l(i)\bar{c}_m(i) = \begin{cases} 0, & m=l \\ P, & m \neq 1 \end{cases} \tag{4.67}$$

### 4.5.2.3   Performance Analysis

When MQC sequence $c_n$ is used as the desired address, $A(v)$ and its complement $\bar{A}(v)$ can be expressed as:

$$A(v) = \sum_{i=1}^{N} c_n(i)rect(i) \tag{4.68}$$

and,

$$\bar{A}(v) = \sum_{i=1}^{N} \bar{c}_n(i)rect(i) \tag{4.69}$$

where $rect(i)$ is given by Equation (4.50). According to Figure 4.18, the PSD at the input of PD$_1$ and PD$_2$ (denoted as $G_1(v)$ and $G_2(v)$, respectively) of the first receiver over one data bit period can be expressed as:

$$G_1(v) = \frac{P_{sr}}{\Delta v}d_1 \sum_{i=1}^{N} c_1(i)rect(i) + \frac{P_{sr}}{\Delta v} \sum_{k=2}^{K+1} \sum_{i=1}^{N} c_k(i)c_1(i)rect(i) \tag{4.70}$$

$$G_2(v) = \frac{1}{P}\frac{P_{sr}}{\Delta v} \sum_{k=2}^{K+1} \sum_{i=1}^{N} c_k(i)\bar{c}_1(i)rect(i) \tag{4.71}$$

where $K$ is the number of active interfering users who are sending data bit 1, $d_1$ is the data bit (0 or 1) sent by the first user. By taking $P_s$ as the received signal power of a single user after the $1 \times 2$ splitter at the receiver and using the code correlation properties, Equations (4.66) and (4.67), the following equations are obtained:

$$\int_0^{\infty} G_1(v)dv = \frac{P_s}{N}[K + (P+1)d_1] \tag{4.72}$$

$$\int_0^{\infty} G_2(v)dv = \frac{KP_s}{N} \tag{4.73}$$

$$\int_0^\infty G_1^2(v)dv = \frac{P_s^2}{N\Delta v} \sum_{i=1}^{N} \left\{ c_1(i) \cdot \left[ \sum_{k=1}^{K+1} c_k(i) \right] \cdot \left[ \sum_{m=1}^{K+1} c_m(i) \right] \right\} \tag{4.74}$$

$$\int_0^\infty G_2^2(v)dv = \frac{1}{P^2} \frac{P_s^2}{N\Delta v} \sum_{i=1}^{N} \left\{ \bar{c}_1(i) \cdot \left[ \sum_{k=1}^{K+1} c_k(i) \right] \cdot \left[ \sum_{m=1}^{K+1} c_m(i) \right] \right\} \tag{4.75}$$

Using the approximation $\sum_{m=1}^{K} c_m(i) \approx K/P$ and the code properties, Equations (4.74) and (4.75) become:

$$\int_0^\infty G_1^2(v)dv = \frac{P_s^2}{N\Delta v} \frac{K+1}{P}[K + (P+1)d_1] \tag{4.76}$$

$$\int_0^\infty G_2^2(v)dv = \frac{P_s^2}{N\Delta v} \frac{K+1}{P^2}K \tag{4.77}$$

Therefore, when the number of interfering users $K > 0$, the coherence times of the thermal light at the input of $PD_1$ and $PD_2$ in the first receiver are, respectively, given by:

$$\tau_{c1} = \frac{N(K+1)}{P\Delta v[K + (P+1)d_1]} \tag{4.78}$$

$$\tau_{c2} = \frac{N(K+1)}{P^2 \Delta v K} \tag{4.79}$$

Let us denote the number of arrived photons at photodetectors $PD_1$ and $PD_2$ as $k_1$ and $k_2$. Both of them have NB distribution profiles with mean values $E(k_1), E(k_2)$ and degrees of freedom $M_1, M_2$ respectively. When the received data bit is 1, based on the correlation property of MQC, the means of $k_1$ and $k_2$ can be expressed as:

$$E[k_1] = d_1 P_s T/Phf + KP_s T/Nhf \tag{4.80}$$

$$= [K + (P+1)d_1]P_s T/Nhf$$

$$E[k_2] = KP_s T/Nhf \tag{4.81}$$

where $h$ is the Planck's constant, $f$ is the optical source central frequency and $T$ is the data bit period. The degrees of freedom are given by $M_1 = 2T/\tau_{c1}$ for $k_1$ and $M_2 = 2T/\tau_{c2}$ for $k_2$ where $\tau_{c1}$ and $\tau_{c2}$ are calculated by Equations (4.78) and (4.79), respectively. Thus the detailed probability distribution of the arrived photons at both PDs can be obtained and denoted by $P[k_1]$ and $P[k_2]$ respectively.

The number of generated electrons $k_{e1}$ after balanced detection is given by $k_{e1} = \eta(k_1 - k_2) = \eta k_{d1}$, where $\eta$ is the photodetector quantum efficiency and $k_{d1} = k_1 - k_2$. For simplicity, the effect of quantum efficiency $\eta$ is neglected in the analysis of the BER. In the following it is assumed that $\eta = 1$ and only the random variable $k_{d1}$ requires consideration. The probability of $k_{d1}$ can be expressed as:

$$P[k_{d1} = k] = \sum_{i=-\infty}^{\infty} \{P[k_1 = i + k] \cdot P[k_2 = i]\} \tag{4.82}$$

Using the same method and definition introduced in previous section, Equation (4.82) can be transformed into the following expression:

$$\overline{P}_{d1} = conv[\overline{P}_{k1}, \, inv(\overline{P}_{k2})] \tag{4.83}$$

Given a threshold as:

$$Th = \frac{1}{2}\left(\int_0^\infty G_2(v)dv - \int_0^\infty G_1(v)dv\right) = \frac{P_s}{2P} \tag{4.84}$$

the BER when data bit is 1 can be written as:

$$P_{e1} = \left\{\sum_{i=-\infty}^{Th-1} P[k_{d1} = i]\right\} + P[k_{d1} = Th]/2 \tag{4.85}$$

When data bit is 0, we can calculate BER $P_{e2}$ similarly. It should be noted that the means of photons in this case should be $E[k_1] = E[k_2] = KP_sT/Nhf$ and the BER should be calculated by:

$$P_{e2} = \left\{\sum_{i=Th}^{\infty} P[k_{d1} = i]\right\} + P[k_{d1} = Th]/2 \tag{4.86}$$

Assuming that the probabilities of sending data bit 1 and 0 are equal then the overall BER becomes:

$$P_e = (P_{e1} + P_{e2})/2 \tag{4.87}$$

When there is only one user ($K = 0$), if data bit is 1, the coherence time at $PD_1$ is given by:

$$\tau_{c1} = \frac{(P_s/\Delta v)^2(N/P)(\Delta v/N)}{[(P_s/\Delta v)(N/P)(\Delta v/N)]^2} = \frac{P}{\Delta v} \tag{4.88}$$

Moreover $E[k_1] = P_sT/Phf$ and no optical power exists at $PD_2$. In this case the BER can be obtained by using the following expression:

$$P_{e1} = \left\{\sum_{i=0}^{Th-1} P[k_1 = i]\right\} + P[k_1 = Th]/2 \tag{4.89}$$

Because nothing is received when data bit is 0, and hence no error is generated, the overall BER can be given by $P_e = P_{e1}/2$.

#### 4.5.2.4 Numerical Results

In the calculation of BER, the effects of both PIIN and shot noise based on NB distribution have been considered. It can be assumed that $\Pr(\Lambda = k) = 0$ when $|k - E[\Lambda]| \geq 12\sigma_\Lambda$, as it is negligibly small. The system operates at 1550 nm wavelength with data bit rate of

622 Mbps. The encoded bandwidth is 3.75 THz wide (which is equal to 30 nm linewidth) and the quantum efficiency is assumed to be 1.

Figure 4.19 compares the BER variations against the number of interfering users of both NB distribution and Gaussian approximation when $P = 7$ and the received signal power after splitter, $P_s$ is $-22$, $-35$ and $-42$ dBm. It has been found that in both cases the BER increases with the number of simultaneous interfering users $K$. We have also compared the BERs calculated by both methods with different $P_s$ when there is only one user present. The results are listed in Table 4.2. It has been shown that NB distribution precisely gives a lower BER than Gaussian approximation in this case.

Figure 4.20 compares the BERs as a function of the number of simultaneous users when received signal power after the splitter is $-35$ dBm for $P = 5$ and 13. It can be observed from Figure 4.20 that the BER is enhanced by the increase in $P$ value. This is due to the increase in the code-length and greater number of accommodated users in the network. It is also apparent that the multiple user interferences degrade the overall performance when increasing the number of simultaneous users results in lower BER.

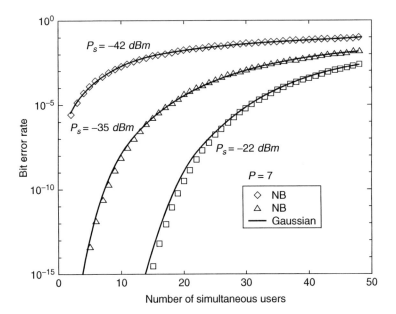

**Figure 4.19** Variation of the BER against the number of interfering users of Gaussian approximation (lines) and NB distribution (others) when $P = 7$

**Table 4.2** Single-user BER comparisons

|          | $P_s = -40$ dBm       | $P_s = -50$ dBm       | $P_s = -60$ dBm       |
|----------|------------------------|------------------------|------------------------|
| NB       | $1.6 \times 10^{-13}$  | $3.9 \times 10^{-3}$   | $1.0 \times 10^{-1}$   |
| Gaussian | $9.7 \times 10^{-11}$  | $1.7 \times 10^{-2}$   | $2.4 \times 10^{-1}$   |

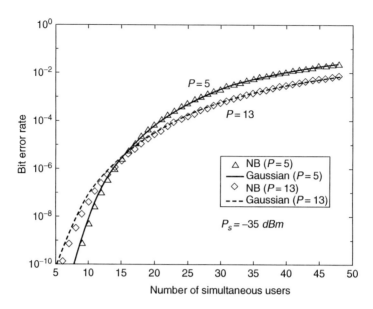

**Figure 4.20** Variation of the BER against the number of interfering users of Gaussian approximation (lines) and NB distribution (others) when $P_{sr} = -35$ dBm

It can be seen from Figures 4.19 and 4.20 that both methods result in almost the same BERs when $K > 0$. And when $K = 0$, a more precise BER can be obtained from the NB distribution.

## 4.6 Spectral-Phase-Coding Schemes

The overall system performance has been analysed theoretically for SAC-OCDMA introducing various encoding schemes and technologies. Another interesting encoding scheme is spectral-phase-coding (SPC) OCDMA. We briefly introduce this scheme in this section, including recent updates and progress in this method.

Theoretical analysis of OCDMA performance has been performed for both spectral and temporal phase coding schemes in [24] and [25] respectively; however, it is not known whether any theoretical performance comparison of these methods has been reported. The schemes have some common features, including the following:

- Performance is analysed by the ability to cancel the multiple user interference (MUI). MUI includes both temporal (i.e. overlapped signals in time) and spectral interference referred to as beat noise [24]. Nevertheless, the theories fully account for both types of interference. The temporally overlapped interference is generally the more serious interference and needs to be kept an eye on. Here we generally use MUI to refer to both types of interference.
- BER performance degrades with increasing number of active users for a given code-length. Hence there is a soft-limited trade-off between user count and error-rate performance.

- Code properties, e.g. code-length and correlation values play a significant role in overall system performance. For a given number of active users, longer codes allow significantly better suppression of interference and significantly lower error rate. Equally for a fixed error rate, longer code-length can accommodate a greater number of active users in a network. However, increase in the code-length brings more complexity in encoder/decoder structure, lower spectral efficiency and possibly more power consumption. As seen, there is a trade-off between performance, complexity and efficiency.
- SPC-OCDMA also requires error correction techniques and code-user translation (i.e. fully asynchronous in time and fully overlapped spectrally) [26] to achieve high spectral efficiency.

Recently, time-spread phase-encoded OCDMA has been demonstrated in a truly asynchronous and spectrally overlapped manner in multi-gigabits per second [27]. It is mainly feasible due to ultralong code-length using 511-chip superstructure fibre Bragg grating (SSFBG) encoders/decoders in which the interference is suppressed reasonably. On the other hand, this SSFBG technology is not programmable and the very long code-length obstructs high bit rates. In OCDMA, there are synchronism issues at both receiver and transmitter sides. The CDMA technique supports burst traffics in the networks, and data transmission in the asynchronous regime can simplify and decentralize the network control and management. On the other hand, full asynchronism is difficult to implement in practice while maintaining sufficient MUI suppression at the same time. Therefore, some level of synchronism is built into many two-user OCDMA schemes, and particularly in most studies employing the SPC method. A fully asynchronous two-user SPC OCDMA has been demonstrated experimentally at low bit rates in [28].

The uncoded or properly decoded signal duration is the chip duration, $t_{chip}$. The larger duration of the encoded or improperly decoded pseudo-noise-like signals is referred to as the slot duration $t_{slot}$. In the SPC scheme introduced in [26] and explained here, the individual features in the pseudo-noise signals have characteristic durations equal to $t_{chip}$, and the number of independent features equal to the code-length, $t_{slot}/t_{chip} = N$. In fully asynchronous systems, signals are transmitted by any user at anytime within bit duration, $t_{bit}$, with no need for user coordination. In a fully synchronous system, the signals are transmitted at specific user coordination within the transmission duration with a timing precision below $t_{chip}$. Besides, a synchronous receiver would require optical clock recovery for gating with timing precision below $t_{chip}$. To overcome this issue, the concept of slot-level coordination [26] has been proposed that relaxes the timing requirements of the synchronous approach. The concept works as the transmission time of a user is controlled on the timescale of the slot duration $t_{slot}$, but without the need for chip-level timing control [29, 30]. OCDMA can be classified based on different levels of synchronism requirements. At the transmitter, no timing coordination (full asynchronism) between slot-level ($t_{slot}$), and chip-level ($t_{chip}$) is distinguished. At the receiver, asynchronous and synchronous detections are distinguished by requiring chip-level ($t_{chip}$) precision by using either ultrafast electronics or optical gating with clock recovery. The true clock recovery in the presence of multiple overlapping SPC-OCDMA users is still a challenge for practical application of synchronous gating in SPC-OCDMA. Therefore, a self-gating approach via nonlinear optics, which avoids the OCDMA clock recovery issues has been employed in [26] by deploying asynchronous detection.

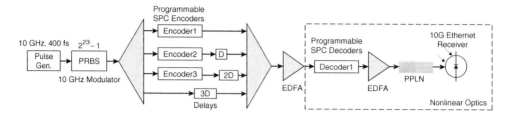

**Figure 4.21**    Four-user programmable SPC OCDMA set up

The experimental set-up of the four-user 10 Gbps OCDMA is shown in Figure 4.21 [26]. An actively mode-locked fibre laser is used as a pulsed laser source followed by a dispersion-decreasing fibre soliton compressor producing nearly transform-limited 0.4 ps pulses at 10 GHz centred at 1542 nm. A 10 Gbps $2^{23} - 1$ pseudo-random binary sequence (PRBS) data stream encodes the pulses. For three users, the modulated ultrashort pulses are input into fibre-coupled Fourier-transform pulse shapers [31] to spectrally phase-encode the spectrum of the source laser. The decoder at the receiver is based on pulse shapers and an ultrasensitive nonlinear optical intensity discriminator (i.e. thresholder) based on second-harmonic generation (SHG) in a periodically poled lithium niobate (PPLN) waveguide. The receiver consists of a fibre-coupled Fourier-transform pulse shaper used to select the user channel to decode, an optical amplifier, a highly sensitive fibre pigtailed PPLN waveguide chip to perform the nonlinear discrimination function [32] and a 10 Gbps Ethernet photoreceiver (3 dB bandwidth of 7.5 GHz), operating at the second-harmonic wavelength of 0.77 μm.

Both SPC encoding and decoding were performed by the ultrashort pulse shaping techniques using a fibre-coupled Fourier-transform pulse shaper. This incorporates a $2 \times 128$ pixel liquid-crystal modulator (LCM) array to spectrally phase-encode the spectrum of the source laser. The individual pixels of the LCM can be electronically controlled independently to give an arbitrary phase shift in the range of to $2\pi$ with 12-bit resolution [29, 30]. As a result, discrete frequency components comprising the short input pulses are horizontally diffracted by a grating and controlled by the LCM. Other spectral phase encoding–decoding devices are also available now, including AWG [1] and ring resonators [33].

This spectrally phase-encoded OCDMA offers full programmability, whereas it has to employ comparatively short codes (e.g. length 31 is typical). For this reason some form of timing coordination is generally employed in order to obtain good error rate performance in the experiments. In addition to programmability, SPC simplifies the code translation. Since multiple users share the same channel in OCDMA networks, a code assigned to a specific user might have already been used between certain optical nodes due to the limited code space. Code translation or random access protocol (RAP) is a solution to avoid double-coding an encoded user by another unused code. As a result, RAP is an efficient algorithm to increase user counts and/or reduce the number of required codes by reusing codes in OCDMA networks. All-optical CDMA RAP has been pursued in time-domain coding [34, 35] or 2-D time/wavelength coding [36]. However, complicated nonlinear optical processing schemes with short codelengths (4–8 chips) have been considered and the code translators have not supported dynamic reconfiguration.

## 4.7 Summary

In this chapter, the spectrally encoded OCDMA schemes and systems have been introduced regarding different technologies and transceiver architecture. The spectral-amplitude-coding OCDMA system has been analysed in detail considering two statistical approaches including Gaussian and negative binomial distribution for photon-counts at the photodetectors, while the photon arrival follows the Poisson behaviour. The performance of systems using MQC and Hadamard codes has also been studied when taking the effects of the intensity, shot and thermal noise sources into account. It has been shown that the spreading codes with more flexible length and correlation constraints such as the MQC families can suppress phase-induced intensity noise more effectively and improve the overall system performance significantly. When the received signal power of single user $(P_{sr})$ at the receiver is high enough, the phase-induced intensity noise is the main factor which limits the system performance. However, when $P_{sr}$ is not sufficiently high, thermal and shot noise sources become the main limiting factors and the effect of thermal noise is much larger than that of shot noise.

The bit error rate of an SAC-OCDMA system using discrete NB distribution has been analysed directly in this section. Hadamard code and complementary encoding/decoding have been used in the analysis. It has been shown that the bit error rate obtained by NB distribution is very close to that calculated using Gaussian approximation, especially when the number of simultaneous users is large. When the system capacity is considered, there are generally lots of simultaneous users. Therefore Gaussian approximation can provide an accurate estimation as well as NB distribution for the capacity of SAC-OCDMA systems. Modified quadratic congruence codes are also used as user addresses in this analysis. It has been shown that when interfering users are present, i.e. $K > 0$, the encoding/decoding behaviour results in almost the same bit error rates as the Gaussian approximation. Whereas when there is no interference involved, i.e. $K = 0$, the NB-distribution-based behaviour gives more precise performance estimation.

Also in this chapter, the recent progress in spectrally phase-coded OCDMA systems including 10 Gbps multiuser system demonstrations and code translations required in a multipoint network implementation have been reviewed. It is clear that OCDMA remains of significant research interest in the research community around the world, especially for applications in optical access networks, and motivates development of optical signal processing technologies. However, there are still challenges for future research to focus on, among them (i) complexity and cost reduction of the entire system, including the broadband laser source, encoding–decoding devices, detection system and simplified timing control for multiuser OCDMA systems; and (ii) overall system performance improvement of multiuser OCDMA systems, including increasing the user count and network capacity while maintaining error-free operation for all users present sharing network resources.

## References

1. Tsuda, H. *et al*. (1999) Photonic spectral encoder/decoder using an arrayed-waveguide grating for coherent optical code division multiplexing. In: *Proc. WDM Components, Trends in Optics and Photonics*, San Diego, USA.
2. Wang, S.Y. *et al*. (2006) The experimental demonstration of MW-OCDMA system based on FBG en/decoder. In: *SPIE*, **6025**, 60251A.

3. Yan, M. *et al*. (2008) En/decoder for spectral phase-coded OCDMA system based on amplitude sampled FBG. *IEEE Photonics Tech. Letters*, **20** (10), 88–90.

4. Hinkov, I. *et al*. (1995) Feasibility of optical CDMA using spectral encoding by acoustically tunable optical filters. *Electronics Letters*, **31** (5), 384–386.

5. Laakkonena, P., Kuittinen, M. and Turunena, J. (2001) Coated phase masks for proximity printing of Bragg gratings. *Optics Communications*, **192** (3–6), 153–159.

6. Li, H. *et al*. (2006) Optimization of a continuous phase-only sampling for high channel–count fiber Bragg gratings. *Optics Express*, **14** (8), 3152–3160.

7. Othonos, A. and Kalli, K. (1999) *Fiber Bragg gratings: fundamentals and applications in telecommunications and sensing*. Artech House, Boston, USA.

8. Cherin, A.H. (1983) *An introduction to optical fibres*. McGraw-Hill, USA.

9. Keiser, G. (1983) *Optical fiber communications*. McGraw-Hill, USA.

10. Wei, Z., Shalaby, H.M.H. and Ghafouri-Shiraz, H. (2001) Modified quadratic congruence codes for fiber Bragg-grating-based spectral-amplitude-coding optical CDMA systems. *J. Lightw. Technol.*, **19** (9), 1274–1281.

11. Fathallah, H., Rusch, L.A. and LaRochelle, S. (1999) Passive optical fast frequency-hop CDMA communication system. *J. Lightw. Technol.*, **17** (3), 397–405.

12. Huang, J.F. and Hsu, D.Z. (2000) Fiber-grating-based optical CDMA spectral coding with nearly orthogonal *m*-sequence codes. *IEEE Photonics Tech. Letters*, **12** (9), 1252–1254.

13. Smith, E.D.J., Blaikie, R.J. and Taylor, D.P. (1998) Performance enhancement of spectral-amplitude-coding optical CDMA using pulse position modulation. *IEEE Trans. on Comm.*, **46** (9), 1176–1185.

14. Zhou, X. *et al*. (2000) Code for spectral amplitude coding optical CDMA systems. *Electronics Letters*, **36** (8), 728–729.

15. Goodman, J.W. (2000) *Statistical optics*. New York: Wiley.

16. Nguyen, L., Young, J.F. and Aazhang, B. (1996) Photoelectric current distribution and bit error rate in optical communication systems using a superfluorescent fiber source. *J. Lightw. Technol.*, **14** (6), 1455–1466.

17. Dennis, T. and Young, J.F. (1999) Measurements of BER performance for bipolar encoding of an SFS. *J. Lightw. Technol.*, **17** (9), 1542–1546.

18. Wei, Z., Ghafouri-Shiraz, H. and Shalaby, H.M.H. (2001) Performance analysis of optical spectral-amplitude-coding CDMA systems using super-fluorescent fiber source. *IEEE Photonics Tech. Letters*, **13** (8), 887–889.

19. Wei, Z. and Ghafouri-Shiraz, H. (2002) IP transmission over spectral-amplitude-coding CDMA links. *J. Microw. & Opt. Tech. Let.*, **33** (2), 140–142.

20. Wei, Z. and Ghafouri-Shiraz, H. (2002) IP routing by an optical spectral-amplitude-coding CDMA network. *IEE Proc. Communications*, **149** (5), 265–269.

21. Wei, Z. and Ghafouri-Shiraz, H. (2002) Proposal of a novel code for spectral amplitude coding optical CDMA systems *IEEE Photonics Tech. Letters*, **14** (3), 414–416.

22. Wei, Z. and Ghafouri-Shiraz, H. (2002) Codes for spectral-amplitude-coding optical CDMA systems. *J. Lightw. Technol.*, **20** (8), 1284–1291.

23. Prucnal, P.R. (2005) *Optical code division multiple access: fundamentals and Applications*. CRC Taylor & Francis Group, Florida, USA.

24. Heritage, J.P., Salehi, J.A. and Weiner, A.M. (1990) Coherent ultrashort light pulse code-division multiple access communication systems. *J. Lightw. Technol.*, **8** (3), 478–491.

25. Wang, X. and Kitayama, K. (2004) Analysis of beat noise in coherent and incoherent time-spreading OCDMA. *J. Lightw. Technol.*, **22** (10), 2226–2235.

26. Weiner, A.M., Jiang, Z. and Leaird, D.E. (2007) Spectrally phase-coded O-CDMA. *J. Optical Networking*, **6** (6), 728–755.

27. Wang, X. *et al*. (2005) 10-user, truly-asynchronous OCDMA experiment with 511-chip SSFBG en/decoder and SC-based optical thresholder. In: *OFC*, Anaheim, CA, USA.

28. Shen, S. *et al*. (2000) Bit error rate performance of ultrashort-pulse optical CDMA detection under multi-access interference. *Electronics Letters*, **36** (21), 1795–1797.

29. Jiang, Z. *et al*. (2005) Four-user, 10 Gb/s spectrally phase coded O-CDMA system operating at 30 fJ/bit. *IEEE Photonics Tech. Letters*, **17** (3), 705–707.

30. Jiang, Z. *et al*. (2005) Four-user, 2.5-Gb/s, spectrally coded OCDMA system demonstration using low-power nonlinear processing. *J. Lightw. Technol.*, **23** (1), 143–158.

31. Weiner, A.M. (1995) Femtosecond optical pulse shaping and processing. *Progress in Quantum Electronics*, **3** (9), 161.
32. Parameswaran, K.R. *et al.* (2002) Highly efficient second-harmonic generation in buried waveguides formed by annealed and reverse proton exchange in periodically poled lithium niobate. *Optics Letters*, **27** (3), 179–181.
33. Agarwal, A. *et al.* (2006) Spectrally efficient six-user coherent OCDMA system using reconfigurable integrated ring resonator circuits. *IEEE Photonics Tech. Letters*, **18** (18), 1952–1954.
34. Kitayama, K., Wada, N. and Sotobayashi, H. (2000) Architectural considerations for photonic IP router based on upon optical code correlation. *J. Lightw. Technol.*, **18** (12), 1834–1844.
35. Kamath, P., Touch, J.D. and Bannister, J.A. (2004) The need for medium access control in optical CDMA networks. In: *IEEE InfoCom*, Hong Kong.
36. Gurkan, D. *et al.* (2003) All-optical wavelength and time 2-D code converter for dynamically reconfigurable O-CDMA networks using a PPLN waveguide. In: *OFC*, CA, USA.

# 5

# Incoherent Temporal OCDMA Networks

## 5.1 Introduction

On-off keying (OOK) and pulse position modulation (PPM) OCDMA are two popular modulation schemes for incoherent OCDMA networks. PPM, as an energy efficient modulation, excels over OOK if the average power rather than chip time is the constraining factor [1]; however, in practical OCDMA systems the chip time is important, whereas power issues become critical in mobile and personal devices. In this chapter, we have also analysed Manchester codes which give further improvement to OCDMA system performance when they are assigned to users. In the analysis we have assumed that the multiple access interference (MAI) is the dominant noise in the OCDMA system.

Here incoherent synchronous OCDMA employing PPM signalling alongside its transmitters and receivers architectures are analysed. The analysis leads to the system bit error rate (BER) with respect to the group-padded modified prime code (GPMPC). This spreading code was introduced in Section 2.4.5.

The PPM-OCDMA system without an interference canceller becomes unreliable as the number of simultaneous subscribers increases. The reason is that MAI increases rapidly as the number of active users increases. Increasing the multiplicity $M$ and the prime number $P$ is helpful for improving the overall system performance, but it is not realistic to increase $M$ and $P$ continuously. Moreover, increasing $M$ and $P$ will increase the system complexity. If the amount of MAI can be reduced or removed, the system performance can be improved remarkably and more active users can be accommodated in the network. Hence, in this chapter, systems using both Manchester encoding and MAI canceller are studied. Also, in order to realize the preference of GPMPC, the performance of various codes including new-modified prime code (n-MPC) and modified prime code (MPC) that were introduced in Chapter 2, are demonstrated and compared for better understanding.

Although PPM is a power-efficient modulation scheme, it is not convenient for a PPM system to achieve high throughput due to the requirement of bandwidth expansion [2].

*Optical CDMA Networks: Principles, Analysis and Applications*, First Edition. Hooshang Ghafouri-Shiraz and M. Massoud Karbassian.
© 2012 John Wiley & Sons, Ltd. Published 2012 by John Wiley & Sons, Ltd.

In the past few years, interest has been given to overlapping PPM (OPPM), which is as an alternative signalling format to the conventional PPM in incoherent optical channels. OPPM can be considered as a type of PPM signalling format, where overlapping is allowed between pulse positions. Later on in the chapter (Section 5.10) we will discuss how an OPPM system can achieve higher throughput than a PPM system without the need of bandwidth expansion due to the fixed-assigned time-slots. Moreover, OPPM retains the advantages of PPM in terms of power efficiency and implementation simplicity.

Also in this chapter, a synchronous OPPM-OCDMA signalling format and system is analysed where GPMPC has been employed as a spreading code. Then, the transmitter and receiver models for different architectures are investigated in detail in terms of bit error rate (BER) performance. Based on the properties of the GPMPC, the MAI cancellation technique is also introduced and analysed for the OPPM-OCDMA network, unlike the conventional method that uses hard-limiter(s) before and after the OCDMA correlators to eliminate the floor interference at the receivers [3]. Since the Manchester codes have the capability to further improve the system performance, they are applied to OPPM-OCDMA transceivers as well to reduce the inter-symbol interference. As the overlapping is allowed, there is self-interference at the synchronization instants at OPPM-OCDMA receivers, thus the entire process is also taken into account including self-interference. The numerical results are obtained on the assumption that the dominant noise source in the network is MAI and that both photodiode dark current noise and thermal noise contributions are negligible. Finally, the throughputs for both OPPM- and PPM-OCDMA networks are discussed and evaluated.

## 5.2   PPM-OCDMA Signalling

The $M$-ary PPM-OCDMA signalling format is shown in Figure 5.1. One frame with a duration of $T$ seconds consists of $M$ time-slots, each of which is $\tau$ seconds wide, where $T = M \cdot \tau$. There are $P^2 + 2P$ chips ($P$ is a prime number) in a GPMPC sequence

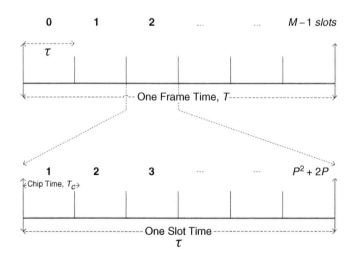

**Figure 5.1**   $M$-ary PPM-OCDMA signalling format with GPMPC

with chip time $T_c$. Each symbol is represented by a train of optical pulses placed in one of $M$ adjacent time-slots. Therefore, for an $M$-ary PPM-OCDMA communication system, there are $M$ possible pulse positions within the symbol frame $T$. In a single time frame, each user is allowed to transmit only one of the $M$ symbols. A pre-assigned unique spreading sequence can be used to distinguish different users, although several of them can transmit the same symbol in a frame. When a user transmits a symbol, the unique spreading sequence of the desired user will occupy the corresponding time-slot. For a proper spread, the spreading sequence with length $L_c$ must be exactly fitted into time-slot $\tau$ (also called the spreading interval), where $\tau = T_c \cdot L_c$.

## 5.3 PPM-OCDMA Transceiver Architecture

### 5.3.1 PPM-OCDMA Transmitter Architectures

#### 5.3.1.1 Simple Transmitter

A typical transmitter model for an incoherent PPM-OCDMA system is shown in Figure 5.2, which consists of information source, optical PPM and OCDMA encoders [4]. In the following, the main function of each transmitter block is described.

*Information Source*

The total number of information sources depends on the total number of available codes (code cardinality), where $N$ out of the total number of users are the active ones and each user transmits continuous data symbols. As Figure 5.3 indicates for GPMPC with $P = 3$ and $M = 3$, the maximum number of active users is $P^2 = 9$. Assuming that active

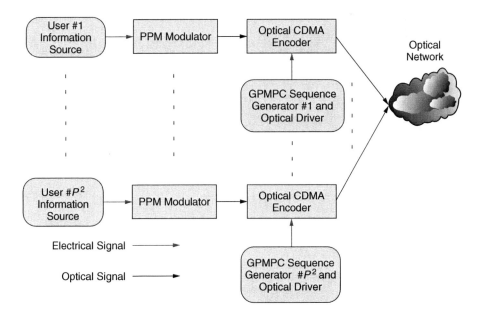

**Figure 5.2**   Incoherent PPM-OCDMA transmitters structure

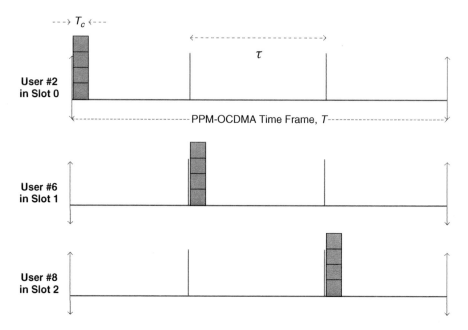

**Figure 5.3** Example of PPM signals for users #2, #6 and #8 when $M = 3$

intended user #2 sends data in slot 0, user #6 sends data in slot 1 and user #8 also sends data in slot 2.

### PPM Modulator

The output symbol of each information source is modulated into one of $M$ time-slots, and generates an optical PPM signal with the tall narrow shape using a laser pulse of width $T_c$ and certain time delay referenced to a certain point. The time delay depends on the amplitude of the data symbol transmitted from the user. The position of laser pulses for intended users #2, #6 and #8 are displayed in Figure 5.3, which are the output waveforms of PPM encoder.

### Optical CDMA Encoder

The optical PPM signal is then passed to an OCDMA encoder, where it is spread into a train of shorter laser pulses with chip width $T_c$. The train of shorter laser pulses is the spreading sequence of the desired user. The spreading sequence is one of GPMPC sequences in the analysis. An OCDMA encoder can be implemented by using optical tapped-delay lines (OTDL), which includes delayers, combiners and a splitter. The encoder consists of a splitter $1:w$ where $w$ is the code-weight (number of ones in the code). Then the delay-lines with corresponding delay time with respect to the position of ones within the spreading code. And finally, the pulses are combined in a combiner $w:1$ to shape the OCDMA-encoded signal.

Unless otherwise specified, in the remaining part of this chapter we assume the arbitrary signature sequences assigned to the intended users #2, #6 and #8 are, respectively,

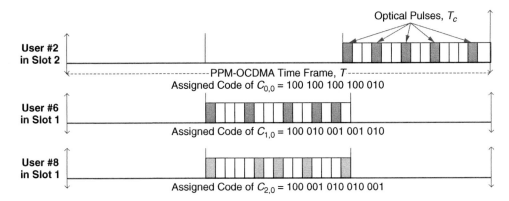

**Figure 5.4**  Example of signalling model for 3-ary PPM-OCDMA, e.g. three users #2, #6 and #8 have signature codes: 100100100100010, 100010001001010 and 100001010010001, respectively

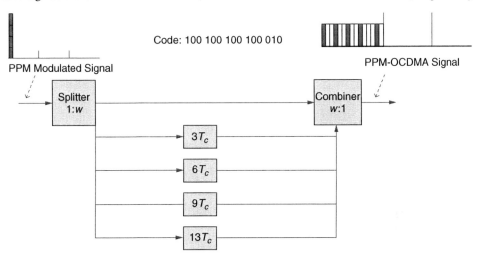

**Figure 5.5**  Example of OTDLs for encoding 100 100 100 100 010 as a signature code

$C_{0,0} = 100\ 100\ 100\ 100\ 010$, $C_{1,0} = 100\ 010\ 001\ 001\ 010$ and $C_{2,0} = 100\ 001\ 010\ 010\ 001$. Figure 5.4 illustrates the OCDMA-encoded signals of users #2, #6 and #8 which are the outputs of their corresponding encoder as in Figure 5.5, while Figure 5.6 shows the combined signals at the entering point to the 'optical network' in Figure 5.2 through a star coupler for example. Figure 5.6 clearly shows that these signals are overlapped and cause interference.

### 5.3.1.2   Transmitter with MAI Cancellation

The transmitter model for PPM-CDMA system with interference cancellation is similar to that shown in Figure 5.2. The only difference is that the maximum number of accommodated users will be $P^2 - P$, not $P^2$. Since the last sequence code of each group is

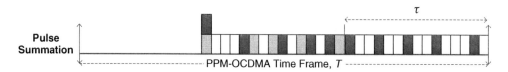

**Figure 5.6**   Example of the combination of PPM-OCDMA signals in an optical channel

reserved as the reference correlator at the receiver, they cannot be allocated to any user. Let us assume $N$ active users are in the system, and each user transmits continuous data symbols. Then the number of idle users is $P^2 - P - N$.

### 5.3.1.3   Transmitter with MAI Cancellation and Manchester Encoding

Manchester encoding is a type of data communications in which: (i) data and clock signals are combined to form a single self-synchronizing data stream, (ii) each encoded bit contains a transition at the midpoint of a bit period, (iii) the direction of transition is determined by whether the bit is a zero or one and (iv) the first half is the true bit value whereas the second half is the complement of the true bit value. Therefore, the rules of Manchester encoding are as follows: (a) if the original data is a logic zero, the Manchester code is a transition of zero to one at one time period; (b) if the original data is a logic one, the Manchester code is a transition of one to zero at one time period. Since, Manchester coding can also be interpreted as a return-to-zero coding, the analysis includes both return-to-zero (RZ) and non-return-to-zero (NRZ) signalling formats [5–7].

In the analysis, we define both first- and second-half time periods denoting true bit values. Manchester encoding is systematically allocated to different users in the system; it is indicated that the first half users, i.e. $(P + 1)/2$ groups (out of $P$ groups) are assigned to transmit data by using the first half-chip interval $[0, T_c/2]$ while the half users from the remaining $(P - 1)/2$ groups share the second half-chip interval $[T_c/2, 0]$. This coding scheme ensures that the two groups of users do not interfere with each other and thus will help to reduce multiuser interference [7].

Referring to the signal model example for $M = 3$ illustrated in Figures 5.4 and 5.6; Figures 5.7 and 5.8 illustrate the signal formats for the system with Manchester coding.

The spreading sequences $C_{0,0}$ and $C_{1,0}$ for users #2 and #6 are, respectively, located in the first half-chip interval, while spreading sequence $C_{2,0}$ for user #8 is allocated in the second half-chip interval. Now, it can be seen from Figure 5.8 that the signals are not overlapped and do not cause interference. It should be noted that the sequences on the figures are arbitrarily selected.

## 5.3.2   PPM-OCDMA Receiver Architectures

### 5.3.2.1   Simple Receiver

The receiver of the incoherent PPM-OCDMA without interference cancellation and Manchester codes is shown in Figure 5.9. The main function of each block is explained in details as follows.

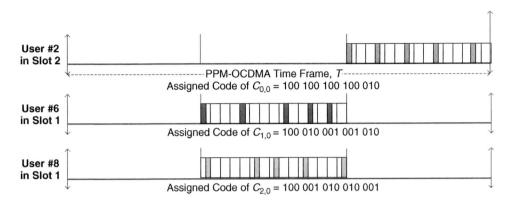

**Figure 5.7** Signalling model for 3-ary PPM-OCDMA system with Manchester codes, the three active users #2, #6 & #8 have signature codes: 100100100100010, 100010001001010 and 100001010010001 respectively, for example

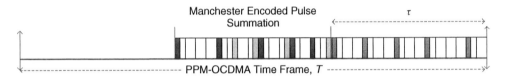

**Figure 5.8** Example of the combination of Manchester-coded PPM-OCDMA signals in an optical channel

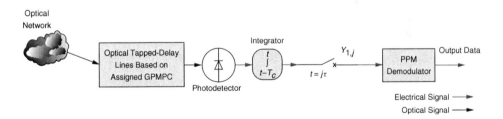

**Figure 5.9** Incoherent PPM-OCDMA receiver model

### *Optical Tapped-Delay Line (OTDL)*

The received PPM-OCDMA signal from optical fibre, which includes all the users' information and noise, is correlated by its own spreading sequence by OTDL. An OTDL could be regarded as an optical matched filter. The **mark positions** denoting the ones in the spreading sequence determine the structure of the OTDL as also shown in Figure 5.5. The amount of delay is not only dependent on the spreading sequence but also on the mark positions within the chip intervals. The correlated spreading sequence is the same as the pre-assigned at the transmitter. If the incoming signal is encoded with the correct address, the output of the optical matched filter will yield an auto-correlation peak, otherwise smaller cross-correlation amplitude is generated.

### Photodetector

The photodetector is used to convert the demultiplexed optical signal into an electrical signal, which is proportional to the photon-counts and lightwave intensity.

### Integrator

The integration is performed over the entire chip duration $T_c$. Sampling the integrated signal is done at the moment of $j \cdot \tau$ only, where $j \in \{1, 2, \ldots M\}$. It should be mentioned that in a direct-detection PPM-OCDMA system, the synchronization is carried out at the end of each time-slot. Then the signal is sampled at the last chip position, covering the entire spreading code-length where the maximum and accurate auto-correlation can be computed within the spreading slot $\tau$ of the time frame $T$.

### PPM Detector

Thereafter, the photon-count of each time-slot is sent to the decision mechanism. The obtained $M$-ary samples are passed through the PPM decoder, which is a comparator over the $M$ samples. The slot containing the maximum photon-counts is declared as the true bit value.

#### 5.3.2.2   Receiver with MAI Cancellation

The incoherent PPM-OCDMA receiver model with MAI cancellation is shown in Figure 5.10.

The MAI cancellation technique has been proposed in [8] based on the correlation properties of MPC. The same scheme is used here while GPMPC is employed as spreading code instead. The idea is to pre-reserve a code to provide interference estimation. The estimated interference is then subtracted from the received signal after photodetection.

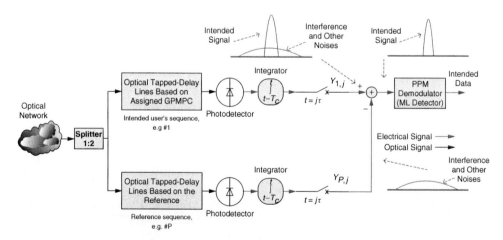

**Figure 5.10**   Incoherent PPM-OCDMA receiver structure with MAI cancellation

As can be seen from Figure 5.10, the received signal, which consists of intended data, MAI and noise, is divided into two equal parts by a $1 \times 2$ optical splitter, then the split signals are fed into two optical matched filters. The upper portion, referred to as the main branch, is used to extract useful signal while the lower portion, called the reference branch, is used to estimate the amount of MAI. In the main branch the injected signal is correlated with its own spreading sequence, while in the reference branch, the fed signal is correlated with the reference spreading sequences which are the last sequence code from each group preserved at the beginning. The mark positions of the desired user's spreading sequence determine the structure of the OTDL in the main branch, and the mark positions of the reference spreading sequence determine the structure of the OTDL in the reference branch. In practice, since the output of each photodetector follows the Poisson process and the signals in both braches suffer from the same fibre length loss and dispersions, and the photodetectors in both branches presumably have the same characteristics, the MAI and noise, e.g. shot noise, thermal noise and amplified spontaneous emission (ASE), are cancelled out after subtraction. If the incoming signal is encoded with the correct address, the output of the optical matched filter will yield an auto-correlation peak. Otherwise, the cross-correlation amplitude is generated. Moreover, based on the correlation properties of GPMPC, the detection will outperform where the correlation values differ remarkably.

All of the correlation outputs are then converted to electrical signals using a photodetector. Integration is performed over the entire chip duration, while synchronization is applied at the moment of $j \cdot \tau$ only, where $j \in \{1, 2, \ldots M\}$. They are sampled at the last chip position.

Based on the group correlation property of GPMPC, the photon-count $Y_{p,j}$ is only composed of MAI which is the same interference as in the main branch. The MAI cancellation is achieved by subtracting the photon-count $Y_{p,j}$ in the reference branch from the photon-counts $Y_{1,j}$ in the main branch at the moment $j \cdot \tau$. Then $M$ subtractions are passed through a PPM decoder, where comparison is performed. The slot (corresponding to a particular symbol) with the maximum subtraction value is declared to be the transmitted true bit value.

### 5.3.2.3  Receiver with MAI Cancellation and Manchester Encoding

Figure 5.11 displays the incoherent PPM-OCDMA receiver model with interference cancellation and Manchester encoding.

The essential principle of the receiver model with interference cancellation and Manchester codes is similar to the system with interference cancellation but the only notable difference is the range of integrations.

The electrical signals (after photodetection) of the users in group 1 to group $(P + 1)/2$ will be integrated over the first half-chip intervals for both main and reference branches, while the integration is performed over the second half-chip intervals for the remaining users from group $(P - 1)/2$ to group $P$. In Figure 5.11 the integration is either from $t - T_c$ to $t - T_c/2$, which is the first half-chip duration, or from $t - T_c/2$ to $t$, which makes the second half-chip duration. The integrated signal is sampled at the moment of $j \cdot \tau$ only, where $j \in \{1, 2, \ldots M\}$.

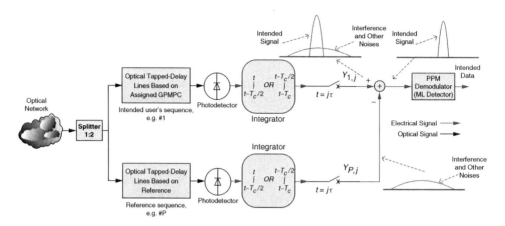

**Figure 5.11** Incoherent Manchester-encoded PPM-OCDMA receiver structure with MAI cancellation

## 5.4 PPM-OCDMA Performance Analysis

Using GPMPC as the spreading sequences, the network described in Figure 5.2 for prime number $P$ and multiplicity $M$ is analysed. Since the entire available sequence code is $P^2$, the total number of subscribers equals $P^2$. It is assumed that $N$ out of $P^2$ users are active and the remaining users are idle. We define a random variable $\gamma_n, n \in \{1, 2, \ldots, P^2\}$ as follows:

$$\gamma_n = \begin{cases} 1, & \text{if user \# } n \text{ is Active} \\ 0, & \text{if user \# } n \text{ is Idle} \end{cases} \tag{5.1}$$

Thus $\sum_{n=1}^{P^2} \gamma_n = N$.

### 5.4.1 Analysis of Simple Receiver

Assuming that user #2 is the intended user. If the random variable $T$ represents the number of active users in the $1^{\text{st}}$ group and variable $t$ is the realization of $T$.

$$T = \sum_{n=1}^{P^2} \gamma_n \tag{5.2}$$

The probability distribution of $T$, given that user #2 is active for any $t \in \{t_{\min}, t_{\min+1}, \ldots, t_{\max}\}$, where $t_{\min} = \max(N + P - P^2, 1)$ and $t_{\max} = \min(N, P)$ is expressed as [9]:

$$P_T(t) = \frac{\binom{P^2 - P}{N - t} \cdot \binom{P - 1}{t - 1}}{\binom{P^2 - 1}{N - 1}} \tag{5.3}$$

where $C_b^a = \dbinom{a}{b} = \dfrac{a!}{(a-b)! \cdot b!}$, which is the combination of $b$ out of $a$. Let the collection of the photon-counts $(Y_{n,0}, Y_{n,1}, \ldots, Y_{n,M-1})$ be denoted by the Poisson random vector $Y_n$ for user #$n$. $\mathbf{Q}$ denotes the average photon-counts per pulse where $\mathbf{Q} = \mu \cdot \ln M / (P+2)$ and $\mu$ is a parameter proportional to the received signal power [10]. We define an interference random vector $k = (k_0, k_1, \ldots, K_{M-1})^T$ of size $M$, where the random variable $k_j$ represents the number of optical interference pulses introduced to time-slot $j$. The vector $u = (u_0, u_1, \ldots, u_{M-1})^T$ is the realization of vector $k$. Given, $T = t, k$ is a multinomial random vector with probability:

$$P_{k|T}(u_0, u_1, \ldots, u_{M-1}|t) = \frac{1}{M^{N-t}} \cdot \frac{(N-t)!}{u_0! . u_1! . \ldots . u_{M-1}!} \tag{5.4}$$

where $\sum_{j=0}^{M-1} u_j = N - t$.

The bit-error probability can be lower-bounded depending on the PPM modulation scheme [11] as follows:

$$P_b = \frac{M}{2(M-1)} \sum_{t=t_{\min}}^{t_{\max}} P_E \cdot P_T(t) \tag{5.5}$$

Taking $\mathbf{Q} \to \infty$, by modifying and rewriting the probability according to the GPMPC properties, the lower-bounded BER is derived as [11]:

$$P_E \geq \sum_{u_1=P+3}^{N-t} \binom{N-t}{u_1} \frac{1}{M^{u_1}} \cdot \left(1 - \frac{1}{M}\right)^{N-t-u_1} \cdot \sum_{u_0}^{\min(u_1-P-3, N-t-u_1)} \binom{N-t-u_1}{u_0}$$

$$\cdot \frac{1}{(M-1)^{u_0}} \cdot \left(1 - \frac{1}{M-1}\right)^{N-t-u_0-u_1} + 0.5 \sum_{u_1=P+2}^{\frac{N-t+P+2}{2}} \binom{N-t}{u_1} \frac{1}{M^{u_1}} \tag{5.6}$$

$$\cdot \left(1 - \frac{1}{M}\right)^{N-t-u_1} \cdot \binom{N-t-u_1}{u_1-P-2} \cdot \frac{1}{(M-1)^{u_1-P-2}} \cdot \left(1 - \frac{1}{M-1}\right)^{N-t-2u_1+P+2}$$

### 5.4.2 Analysis of Receiver with MAI Cancellation and Manchester Encoding

As discussed earlier, the last sequence code in each group is preserved as the reference sequence, thus the total number of reference codes is $P$. Then the entire available spreading sequences are $P^2 - P$ so the idle users are $P^2 - P - N$. In this system, Equation (5.3) for any $t \in \{t_{\min}, t_{\min+1}, \ldots, t_{\max}\}$ is rewritten as follows:

$$P_T^1(t) = \frac{\dbinom{P^2-2P+1}{N-t} \cdot \dbinom{P-2}{t-1}}{\dbinom{P^2-P-1}{N-1}} \tag{5.7}$$

where $t_{\min} = \max(N + 2P - P^2 - 1, 1)$ and $t_{\max} = \min(N, P-1)$.

For the interference cancellation with Manchester codes shown in Figure 5.11, a new random variable $R$ for the number of active users from group 2 to group $(P + 1)/2$ is defined. If we denote $r$ as the realization of $R$, then the probability of any $r \in \{r_{\min}, r_{\min+1}, \ldots, r_{\max}\}$, given $T = t$, can be written as:

$$P_{R|T}(r|t) = \frac{\binom{\frac{P^2 - 2P + 1}{2}}{r} \cdot \binom{\frac{P^2 - 2P + 1}{2}}{N - t - r}}{\binom{P^2 - 2P + 1}{N - 1}} \tag{5.8}$$

where [4]: $r_{\min} = \max\left(0, N - t - \frac{P^2 - 2P + 1}{2}\right)$, $r_{\max} = \min\left(\frac{P^2 - 2P + 1}{2}, N - t\right)$

Given $T = t$, and $R = r$, the probability of interference vector $k$ is then given as:

$$P_{k|(T,R)}(u_0, u_1, \ldots u_{M-1}|t, r) = \frac{1}{M^r} \cdot \frac{r!}{u_0! . u_1! . \ldots . u_{M-1}!} \tag{5.9}$$

An upper-bounded bit-error probability based on the PPM modulation scheme can be derived as:

$$P_b = \frac{M}{2(M - 1)} \sum_{t=t_{\min}}^{t_{\max}} \sum_{r=r_{\min}}^{r_{\max}} P_E \cdot P_{R|(t,r)} \cdot P_T^1(t) \tag{5.10}$$

where the upper-bounded BER can be modified according to GPMPC coding as:

$$P_E \leq (M - 1) \sum_{u_1=0}^{r} \binom{r}{u_1} \cdot \left(\frac{1}{M}\right)^{u_1} \cdot \left(1 - \frac{1}{M}\right)^{r-u_1} \cdot \sum_{u_0=0}^{r-u_1} \binom{r - u_1}{u_0}$$

$$\cdot \left(\frac{1}{M - 1}\right)^{u_0} \cdot \left(1 - \frac{1}{M - 1}\right)^{r-u_0-u_1} \cdot \exp\left[-\mathbf{Q} \cdot \frac{(P + 2)^2}{4.(P + 2 + u_0 + u_1)}\right] \tag{5.11}$$

It is important to note that, if $\mathbf{Q} \to \infty$, then $P_E = 0$.

### 5.4.3 Analysis of a Receiver with MAI Cancellation

The system with only MAI cancellation is very similar to the structure discussed in the previous section. The integration in this system is carried out over the entire chip time rather than half chip time. The MAI contribution comes from the users in group 2 to group $P$. Following the BER probability given in Equation (5.5), the upper-bounded BER can be expressed as:

$$P_b = \frac{M}{2(M - 1)} \sum_{t=t_{\min}}^{t_{\max}} P_E \cdot P_T^1(t) \tag{5.12}$$

where the error probability is modified to GPMPC coding as:

$$P_E \leq (M - 1) \sum_{u_1=0}^{N-t} \binom{N - t}{u_1} \cdot \left(\frac{1}{M}\right)^{u_1} \cdot \left(1 - \frac{1}{M}\right)^{N-t-u_1} \cdot \sum_{u_0=0}^{N-t-u_1} \binom{N - t - u_1}{u_0}$$

$$\cdot \left(\frac{1}{M - 1}\right)^{u_0} \left(1 - \frac{1}{M - 1}\right)^{N-t-u_0-u_1} \cdot \exp\left[-\mathbf{Q} \cdot \frac{(P + 2)^2}{4.(P + 2 + u_0 + u_1)}\right] \tag{5.13}$$

## 5.5 Discussion of Results

The GPMPC is applied to the following three different optical receiver structures and the overall performance of each receiver is discussed in detail. These optical receivers are (i) simple receiver, (ii) receiver with only MAI cancellation and (iii) receiver with both MAI cancellation and Manchester encoding. In order to compare the GPMPC performance, both MPC and n-MPC (Chapter 2) performances are obtained when they are employed in the above three receiver structures. Therefore, when GPMPC is employed, Equations (5.3)–(5.6) are applicable for performance analysis of receiver structure (i), Equations (5.7), (5.12) and (5.13) for receiver structure (ii) and Equations (5.7)–(5.11) for receiver structure (iii). The corresponding equations for n-MPC and MPC can be found in [9, 12] respectively.

### 5.5.1 BER Against Received Signal Power

Figure 5.12 illustrates the BER performances of all three codes in three introduced receiver structures against the average photons per pulse ($\mu$) which is a parameter proportional to received signal power. The full-load communication has been assumed in the analysis which means that the total number of active users are present $N = P^2 - P$. The prime number $P$ and multiplicity $M$ are set to 11 and 8, respectively, thus the total number of active users is 110. The lower-bounded BER for receiver (i) is considered when $\mu = \infty$ (i.e. relatively high power), therefore a constant BER is achieved for the lower-bounded BER as shown in Figure 5.12. It is obvious that a lower BER is achieved for higher received signal power. The BER = $10^{-9}$ is also depicted as a reference for better explanation.

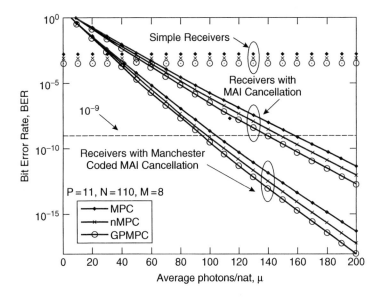

**Figure 5.12**  Performance of the PPM-OCDMA receivers using different codes against the average number of photons per pulse $\mu$, when $M = 8$, $P = 11$ and $N = 110$

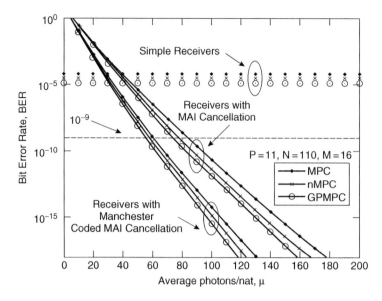

**Figure 5.13** Performance of the PPM-OCDMA receivers using different codes against the average number of photons per pulse $\mu$, when $M = 16$, $P = 11$ and $N = 110$

As Figure 5.12 shows, when $\mu = 100$, the lower-bounded error rates obtained for simple receiver structures are $1.5 \times 10^{-3}$ (MPC), $7.2 \times 10^{-4}$ (n-MPC) and $3.3 \times 10^{-4}$ (GPMPC), which is the lowest error rate. Also, the upper-bounded error rates in receivers with MAI cancellation are $1.7 \times 10^{-6}$ (MPC), $7.1 \times 10^{-7}$ (n-MPC) and $3 \times 10^{-7}$ (GPMPC). Furthermore, the upper-bounds of error rates of Manchester-coded receivers with MAI cancellation is expected to be $2.2 \times 10^{-9}$ (MPC), $1 \times 10^{-9}$ (n-MPC) and $2.8 \times 10^{-10}$ (GPMPC). The above results indicate that the GPMPC performance is considerably better.

As $\mu$ increases, the optical power increases. Thus, the ones and zeros of a PPM pulse will be more distinguishable. Results also indicate that the overall performance with the MAI canceller improves significantly as compared with the systems without having the MAI cancellers.

It can be seen that, there is no advantage with the MAI canceller for smaller value of received signal power ($\mu < 40$). Furthermore, the BER performance can be improved further by applying Manchester codes in the PPM-OCDMA cancellation system.

Also, Figure 5.13 shows the performance analysis in a higher multiplicity of PPM modulation (i.e. $M = 16$). It is apparent that lower BER is achieved with higher multiplicity and less received power $\mu$. Higher multiplicity offers more pulse positions to expand the capacity whereas it restricts the system complexity.

## 5.5.2 BER Against Number of Active Users

Figure 5.14 plots the BER performances of various receivers using different coding schemes when $\mu = 100, P = 11$ and $M = 8$ against the number of active users in the

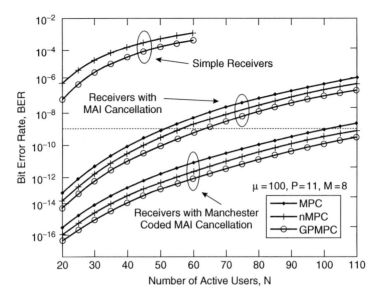

**Figure 5.14** Performance of the PPM-OCDMA receivers using different codes against the number of users, $N$, when $\mu = 100$, $P = 11$ and $M = 8$

network. In this analysis, the upper-bounds of BER for receivers with MAI cancellation and Manchester-coded MAI cancellation are considered when $\mu = 100$, while the lower-bound of BER for simple receiver when $\mu = \infty$ (i.e. relatively high received signal power) is taken into account for only n-MPC and GPMPC. The results indicate that the GPMPC can accommodate a greater number of users. To examine the results, when $N = 60$, the error rate at the simple receiver using n-MPC is $1.1 \times 10^{-3}$ while using GPMPC it is $4.4 \times 10^{-4}$, as shown in Figure 5.14. However, simple receiver structures are unable to support users greater than 60, as the BER soars very high (BER $\approx 1$).

Furthermore in Figure 5.14, the BER at the receivers with MAI cancellation have been improved remarkably. The error rate at the receiver with MAI using n-MPC is $1.77 \times 10^{-9}$, whereas the one using GPMPC has a BER of $6.6 \times 10^{-10}$. The scheme with Manchester coding has been much enhanced to have BER of less than $10^{-11}$ regardless of coding scheme, but among them GPMPC is still the best performer.

It can be noticed that the bit-error probability increases as the number of subscribers increases in these three receivers. The reason is that the interference increases as the number of active users increases. The interference canceller can effectively remove the MAI and improve the BER performance. It can also be seen that the error rate can become better when Manchester codes are applied to the system as well as the interference cancellation. It is also seen in Figure 5.14 that the simple receivers are unable to support more than 60 users under this given condition due to the interference growth.

The same results are expected from the receivers when multiplicity increases (i.e. $M = 16$) as presented in Figure 5.15. The BER $= 10^{-9}$ is also depicted to assist the eye. The higher multiplicity makes the system implementation difficult. It has been found that when the system design limits the multiplicity, Manchester encoding offers its benefits to enhance the outcomes.

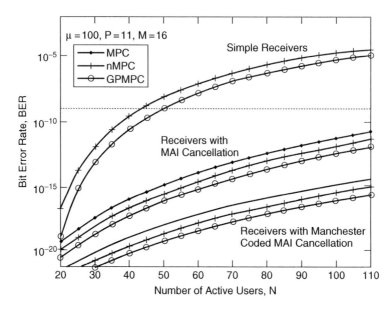

**Figure 5.15** Performance of the three PPM-CDMA receivers using different codes against the number of users, $N$, when $\mu = 100$, $P = 11$ and $M = 16$

### 5.5.3 BER Against Prime Number

The BER performance of receivers with MAI cancellation and Manchester-coded receivers with MAI cancellation are compared against prime number $P$ in Figure 5.16. Due to the outperformance of the other two receivers as compared with the simple receivers, only receivers (ii) and (iii) are presented. In this analysis, the receivers are evaluated in the case of full-load, i.e. $N = P^2 - P$, $M = 8$ and $\mu = 100$. It is apparent that the error rate is very low when using GPMPC as compared to the other coding schemes, especially when $P$ is small.

We now examine the BER of various receivers with different coding schemes. At the receivers with MAI cancellation when $P = 13$, the error rates are $8.6 \times 10^{-6}$ (MPC), $4.3 \times 10^{-6}$ (n-MPC) and $2.1 \times 10^{-6}$ (GPMPC). The improvement is very notable for example at the Manchester-coded receivers with MAI cancellation, the bit error rates are $1.1 \times 10^{-8}$ (MPC), $4.9 \times 10^{-9}$ (n-MPC) and finally $2.1 \times 10^{-9}$ (GPMPC).

It is also expected that the performance will become better with higher multiplicity as presented in the previous figures. However, in practice, the implementations may be restricted by accurate time-slots in high value multiple array PPM signalling. It is important to mention that the time-slots cannot be increased arbitrarily due to the limited electronic circuitry. Therefore, when we are limited to increasing multiplicity, Manchester coding can be a good option to enhance the performance, although at the cost of network bandwidth.

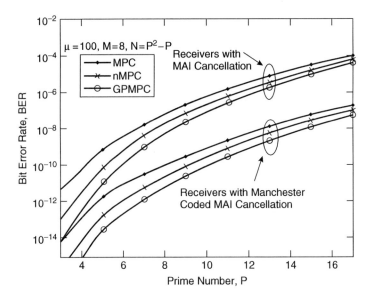

**Figure 5.16** Performance of the PPM-OCDMA receivers using different codes against prime number $P$ when $\mu = 100$, $N = P^2 - P$ and $M = 8$

## 5.6 Overlapping PPM-OCDMA Signalling

In this section we investigate the OCDMA transceivers with overlapping PPM (OPPM) signalling. An $M$-ary OPPM modulation deploys $M$ time-slots. A $T$ time frame of OPPM modulation comprises $M$ time-slots. The modulated signal is permitted to spread over a spreading interval with $\tau$ slot duration, which is again subdivided into $P$ smaller subintervals each of width $\tau/P$ where $P$ is a prime number. An overlap of $(M - \gamma) \cdot \tau$, where $\gamma \in \{1, 2 \dots M\}$, is the overlapping index [13] that is allowed between any two adjacent spreading intervals. To encode an optical OPPM signal, the spreading sequence of length $L$ (taking the GPMPC as $L = P^2 + 2P$) must be exactly fitted into the time-slot $\tau$; thus $\tau/P$ can be appropriate as chip duration of $T_c$. If a wrapped signal is allowed, the time frame $T$ must satisfy the following condition:

$$T = M\tau = \frac{M}{\gamma}L \cdot T_c \qquad (5.14)$$

Figure 5.17 illustrates the 8-ary OPPM-OCDMA signalling format of an arbitrary sequence of $C_{0,0}$ with $\gamma = 5$ (i.e. $\gamma = L/P = P + 2$). Shifting the position within the time frame $T$ for different numbers of subintervals can form different slots. Let us assume that the initial position represents slot 0. Right-shifting the time-slot for one subinterval forms slot 1 and so on. When the time-slot reaches the end of the time frame, it is broken into two blocks. The right-hand block is wrapped back to the beginning of the time frame, while the left-hand block is put at the end of the time frame. Those slots with characteristic that have part of their signals wrapped back to the beginning of the time frame are called *wrapped slots* while the slots without that property are referred

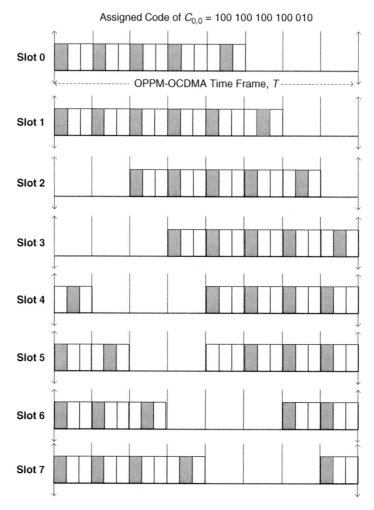

**Figure 5.17** OPPM-OCDMA signalling for $\gamma = 5$ and $M = 8$ for a signature of $C_{0,0} = 100\ 100$ 100 100 010

to as *unwrapped slots*. It is apparent from Figure 5.17 that the unwrapped slots are 0 to 3 and the wrapped slots are 4 to 7. When the OPPM signal is wrapped, the spreading sequence also follows the OPPM signal.

## 5.7 OPPM-OCDMA Transceiver Architecture

### 5.7.1 OPPM-OCDMA Transmitter Architectures

#### 5.7.1.1 Simple Transmitter

A transmitter model for incoherent OPPM-OCDMA network is shown in Figure 5.18. It consists of information source, OPPM modulator and OCDMA encoder [11]. The main

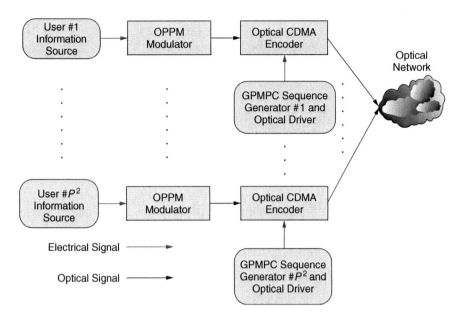

**Figure 5.18**   Incoherent OPPM-OCDMA transmitters structure

function of each block is similar to the transmitter model for PPM-OCDMA network explained earlier.

### Information Source

The information source is responsible for providing the optical pulses representing the data stream from users. The total number of information sources depends on the total number of available sequences $P^2$, where $N$ users out of them are active users. In Figure 5.18, for GPMPC with $P = 3$ and $M = 8$, the maximum number of active users is $P^2 = 9$.

### OPPM Modulator

Each data stream is then fed into OPPM modulators, where a tall narrow laser pulse of width $T_c$ is generated and time delayed in accordance with the data symbol to generate $M$-ary OPPM signalling. The time delay depends on the amplitudes of the data symbol transmitted from the information source. Figure 5.19 shows the outputs of optical OPPM modulator for users #2, #5 and #7 in slots 0, 2 and 4, respectively.

### Optical CDMA Encoder

The outputs of the modulator are then passed to the OCDMA encoder, where it is spread into a shorter optical pulse with the same width $T_c$, with reference to the signature sequence and it is only allowed to occur within the spreading time interval $\tau$. The structure of the optical tapped-delay line (OTDL) and the OCDMA encoder for both unwrapped and wrapped signals is shown in Figures 5.20(a) and 5.20(b), respectively.

For example as shown in Figure 5.21, user #2 sends data in slot 0 encoded with arbitrary signature code $C_{0,0} = 100\ 100\ 100\ 100\ 010$; user #5 transmits data in slot 2 with arbitrary

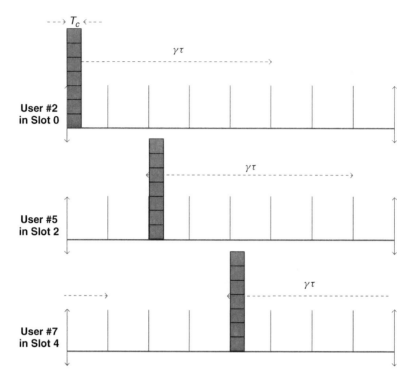

**Figure 5.19**   OPPM signalling for users #2, #5 and #7 at $P = 3$, $\gamma = 5$ and $M = 8$

signature code $C_{1,2} = 001\ 100\ 010\ 010\ 100$; and user #7 sends data in slot 4 which is encoded by arbitrary signature code $C_{2,1} = 001\ 010\ 100\ 100\ 010$.

Finally, all the signals are combined together to form an optical signal, which is transmitted across the optical channels to the receivers, which are illustrated in Figure 5.22. It can be seen from Figure 5.22 that the signals are overlapped and caused interference.

### 5.7.1.2   Transmitter with MAI Cancellation

The transmitter model of OPPM-CDMA system with interference cancellation is similar to that discussed in Figure 5.18. The last spreading sequence of each GPMPC group is again preserved as a reference at the receiver, so the total available signature codes becomes $P^2 - P$. Assuming $N$ active users are in the network, and each user transmits $M$-ary continuous data symbols. Therefore, the idle users are $P^2 - P - N$.

### 5.7.1.3   Transmitter with MAI Cancellation and Manchester Encoding

It should be noted that Manchester encoding is introduced to further improve the system performance, although it extends the bandwidth. As mentioned, the first half users from group 1 to group $(P + 1)/2$ are assigned the first half-chip interval $[0, T_c/2]$, while the

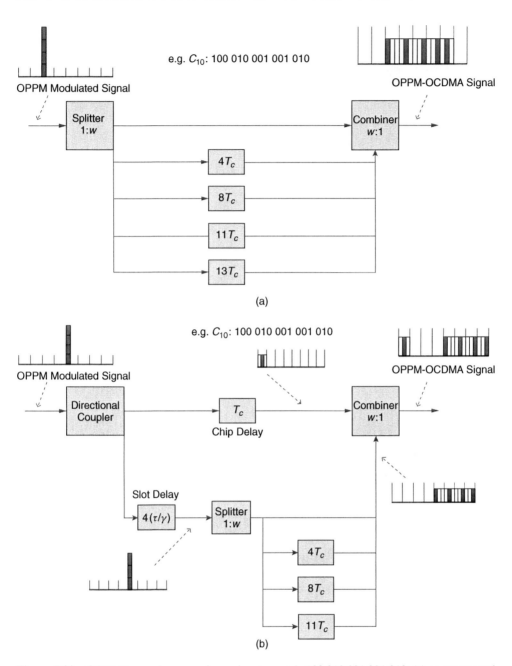

**Figure 5.20** OCDMA encoder, assuming a signature code 100 010 001 001 010: (a) an unwrapped signal and (b) a wrapped signal

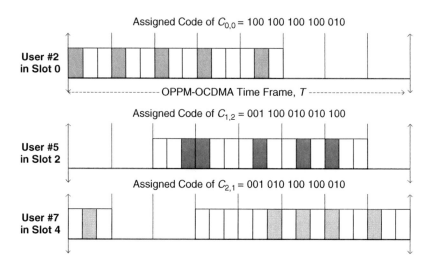

**Figure 5.21**   OPPM-OCDMA signalling for $\gamma = 5$ and $M = 8$ with assigned arbitrary codes

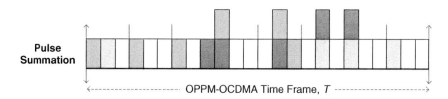

**Figure 5.22**   OPPM-OCDMA signalling combination in the optical channel

rest of the users share the second half-chip interval $[T_c/2, T_c]$. This ensures that the two groups of users do not interfere with each other and thus will help to reduce MAI among users from different groups. As a comparison with the signals model for $\gamma = 5$ and $M = 8$ in Figures 5.21 and 5.22, Figures 5.23 and 5.24 demonstrate the signals model for the transmitter with Manchester-coded MAI cancellation. The spreading sequence $C_{0,0}$ for user #2 is assigned in the first half-chip interval, while spreading sequence $C_{1,2}$ and $C_{2,1}$ for users #5 and #7 are using the second half-chip intervals. Note that the signalling can also be interpreted as return-to-zero (RZ) and non-return-to-zero (NRZ) formats. It can still be seen in Figure 5.24 that the signals are overlapping in this example to show that carefully designing and allocating Manchester codes to users is very significant.

## 5.7.2   OPPM-OCDMA Receiver Architectures

### 5.7.2.1   Simple Receiver

A receiver model for incoherent OPPM-OCDMA is presented in Figure 5.25 as introduced in [13].

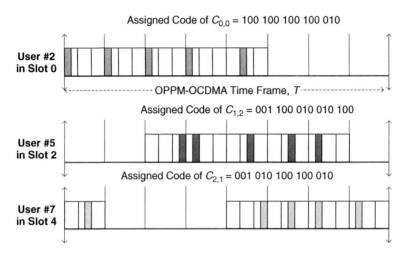

**Figure 5.23** Manchester-coded OPPM-OCDMA signalling format with assigned arbitrary sequences

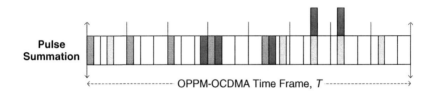

**Figure 5.24** Manchester-coded OPPM-OCDMA signalling combination in the optical channel

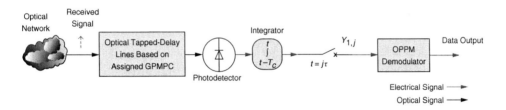

**Figure 5.25** Incoherent OPPM-OCDMA receivers simple architecture

## Optical Tapped-Delay Line (OTDL)

The received signal from $N$ active users is correlated with the unique spreading sequence by the OTDL, acting as an optical correlator [4]. The mark positions of the spreading sequence determine the structure of the OTDL. The number of delays is not only dependent on the spreading sequence but also on the positions of the marks within the chip intervals as shown in Figure 5.20 for the encoding process. If it is encoded with its own spreading sequence of the optical correlator, the output of the optical correlator will yield an auto-correlation peak. Otherwise a cross-correlation value is obtained and the signal is further spread.

### *Photodetector*

The output is then converted to an electrical signal by the photodetector. The electrical signal is proportional to the collected photon-counts and lightwave intensity.

### *OPPM Demodulator*

This part demodulates the electric signal containing the photon-counts collected over the chip duration from the integrator. Finally the data is extracted or demodulated based on the maximum likelihood detection rule with respect to the signal power within the time-slot.

### 5.7.2.2   Receiver with MAI Cancellation

The MAI cancellation technique is similar to the one introduced in Section 5.3.2.2. One sequence from each GPMPC group is preserved as a reference to estimate the MAI noise. The estimated interference is then subtracted from the received signal after photodetection [8]. Figure 5.26 shows the OPPM-OCDMA receiver structure with MAI cancellation.

The received signal, consisting of a desired signal, MAI and noise, is fed into a $1 \times 2$ optical splitter where it is divided into two equal signals. Like the PPM-OCDMA receiver model, the upper portion, the main branch, is used to extract intended data, while the lower portion, the reference branch, is used to estimate the MAI. The signal in the main branch is correlated with the same signature sequence which characterizes the desired user, while in the reference branch the signal is correlated with the reference sequence, which is the last sequence in each group preserved initially. If the injected signal is encoded with the correct address code, the output of the OTDL will yield an auto-correlation peak.

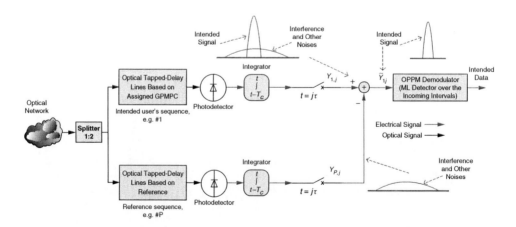

**Figure 5.26**   Incoherent OPPM-OCDMA receiver architecture with MAI cancellation

Otherwise, cross-correlation value is generated. In the photodetector, optical signals are converted to electrical signals. The electric signals will be integrated over the entire chip duration, synchronized at each bit interval. Sampling is performed at the end of each mark position of each slot. According to the correlation properties of the GPMPC, the photon-count $Y_{P,j}$ of the reference branch mainly consists of MAI as the received signal is correlated with the reference spreading code, generating the cross-correlation value. In the main branch, the intended signal and MAI constitute the photon-counts $Y_{1,j}$ as shown in Figure 5.26.

Interference cancellation is carried out by subtracting $Y_{P,j}$ from $Y_{1,j}$. In practice, since the output of the photodetectors follows the Poisson process, and the photodetectors used in the branches have the same characteristics, the MAI and noise, e.g. shot and thermal noise, are cancelled out after subtraction as seen in Figure 5.26. All the subtraction is then passed to the decision unit based on maximum likelihood detection which is a comparator circuit selecting the interval which contains the greatest power among all the $M$ intervals.

### 5.7.2.3 Receiver with MAI Cancellation and Manchester Encoding

A block diagram of receiver model for the incoherent OPPM-OCDMA with Manchester-coded MAI cancellation is presented in Figure 5.27.

Apart from the fact that the integration range is different, the main function of each device for the system with interference cancellation and Manchester codes is the same as the receiver with interference only.

The optical signals are converted to electrical signals by the photodetectors. Following this, the electric signals will be integrated over the first half-chip or the second half-chip which are determined by the users in each group. The active users from group 1 to group $(P + 1)/2$ will be integrated over the first half-chip intervals $[t - T_c, t - T_c/2]$ for both the main and reference branches, while the integration is performed over the second half-chip intervals $[t - T_c/2, t]$ for the remaining subscribers from group $(P - 1)/2$ to group $P$. Later, the integrated signal is sampled at the end of each subinterval.

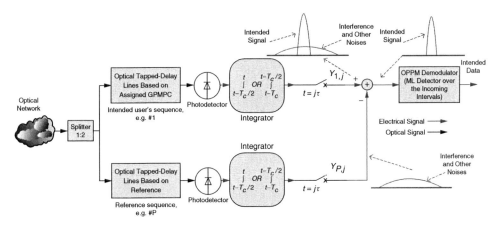

**Figure 5.27** Incoherent OPPM-OCDMA receivers architecture with MAI cancellation and Manchester encoding

## 5.8 OPPM-OCDMA Performance Analysis

Here, by employing GPMPC in an incoherent OPPM-OCDMA system, the receivers' performance in terms of bit error rate will be derived. We define the random Poisson vector $Y_{1,j}$, denoting the photon-counts collected at the receiver, where $j \in \{0, 1, 2, \ldots M - 1\}$. $Y_{1,j}$ consists of two parts: the intended signals and the interference introduced by other users. Index $i$ is declared to be the true one if $Y_{1,i} > Y_{1,j}$ for every $i \neq j$. $S_{1,i} = 1$ represents the intended user, e.g. #1, transmitting signal in slot $i$. The probability of error is then:

$$P[E\,|i] = \Pr\{Y_{1,j} \geq Y_{1,i}, \ some \ j \neq i | S_{1,i} = 1\} \tag{5.15}$$

Hence,

$$P[E] = \Pr\{Y_{1,j} \geq Y_{1,0}, \ some \ j \neq 0 | S_{1,0} = 1\} \tag{5.16}$$

Then the BER based on the $M$-ary modulation format becomes [9]:

$$P_b = \frac{M}{2(M - 1)} P_E \tag{5.17}$$

### 5.8.1 Analysis of Simple Receiver

The codeset cardinality is $P^2$, assuming that the $N$ out of $P^2$ are active users, $P^2 - N$ will be idle users. Let us denote the random variable $T$ as the number of active users in the first group and parameter $t$ as its realization. The probability density function (PDF) of $T$ for any $t \in \{t_{\min}, t_{\min+1}, \ldots, t_{\max}\}$, where $t_{\min} = \max(N + P - P^2, 1)$ and $t_{\max} = \min(N, P)$ can be written as:

$$P_T(t) = \frac{\binom{P^2 - P}{N - t} \cdot \binom{P - 1}{t - 1}}{\binom{P^2 - 1}{N - 1}} \tag{5.18}$$

Since different users may contribute different numbers of interfering pulses to user #1, assuming that user #1 is encoded within first group sequence codes, the $N$ active users are divided into two categories to further investigate: (i) the first group active users whose signature codes are in the same group as the desired user and (ii) the non-first group users whose signature codes are in the other groups.

#### (i) Interference Due to the First Group Users

The number of interfering users in the first group is denoted by a random variable $H$, with $h$ being its realization. Since the overlapped symbols are allowed, the interfering slots are $\kappa = |M - \gamma|$ out of $M$ slots (see Figure 5.21 when $M = 8$ and overlapping index $\gamma = 5$) which can introduce an interfering pulse to user #1 in the first group, and the conditional PDF of interference in first group can be expressed as:

$$P_{H|T}(h|t) = \binom{t - 1}{h} \cdot \left(\frac{\kappa}{M}\right)^h \cdot \left(1 - \frac{\kappa}{M}\right)^{t-1-h} \tag{5.19}$$

where $h \in \{0, 1, \ldots t - 1\}$.

When the interfering signature codes are left-rotated (or right-rotated) for $j \cdot P$ chips from the code of the desired user, the number of interfering pulses introduced to the desired users are then either $j$ or $\gamma - j$, where $j \in \{1, 2, 3, \ldots, \gamma\}$ [13]. Denote the number of interfering pulses caused by the first group users by $L$, and use variable $l$ as the realization of $L$. The conditional PDF of $L$ is then:

$$P_{L|HT}(l|h,t) = \frac{1}{\kappa^h} \tag{5.20}$$

And $l \in \{l_{min}, l_{min+1}, \ldots, l_{max}\}$ where $l_{min} = \sum_{j=1}^{h} \min(\gamma - j, j)$ and $l_{max} = \sum_{j=1}^{h} \max(\gamma - j, j)$, in other words:

$$l \equiv l_{min} + \left\lfloor \frac{l_{max} - l_{min}}{\kappa^h} \times \rho \right\rfloor \tag{5.21}$$

where $\rho \in \{1, \ldots, \kappa^h\}$ and $\lfloor x \rfloor$ returns the integer of $x$.

*(ii) Interference Due to the Non-First Group Users*

The non-first group users introduce interfering pulses to the desired user with the probability of $\gamma/M$ ($\gamma$ is the overlapping index). A random variable $U$ denotes the number of non-first-group interfering users, and the variable $u$ is its realization. The conditional PDF of $U$ is then written as:

$$P_{U|T}(u|t) = \binom{N-t}{u} \cdot \left(\frac{\gamma}{M}\right)^u \cdot \left(1 - \frac{\gamma}{M}\right)^{N-t-u} \tag{5.22}$$

where $u \in \{0, 1 \ldots, N - t\}$.

Thus $P_E$ is described based on the interferences and modulation scheme as:

$$P_E = P_e\{Y_{1,j} \geq Y_{1,0}, \ some \ j \neq 0 | S_{1,0} = 1, T = t, U = u, H = h\} \cdot P(H = h) \cdot$$
$$P(U = u) \cdot P(T = t)$$
$$\geq (M - 1) \cdot P_e\{Y_{1,1} \geq Y_{1,0} | S_{1,0} = 1, T = t, U = u, H = h\} \cdot P(H = h) \cdot \tag{5.23}$$
$$P(U = u) \cdot P(T = t)$$
$$= (M - 1) \cdot P_1 \cdot P_{L|HT} \cdot P_{H|T}(h|t) \cdot P_{U|T}(u|t) \cdot P_T(t)$$

where $P_1$ is defined – from [2] – as:

$$P_1 = \sum_{y_1}^{\infty} e^{-\mathbf{Q}(u+1)} \cdot \frac{[\mathbf{Q} \cdot (u + 1)]^{y_1}}{y_1!} \cdot \sum_{y_0}^{y_1} e^{-\mathbf{Q}(u+P+2)} \cdot \frac{[\mathbf{Q} \cdot [u + P + 2)]^{y_0}}{y_0!} \tag{5.24}$$

where $\mathbf{Q}$ denotes the average photon-counts per pulse, $\mathbf{Q} = \mu \cdot \log \dfrac{M}{P+2}$ [10].

The BER can finally be expressed based on the modulation format and the above interference analysis as:

$$P_b = \frac{M}{2} \sum_{t_{min}}^{t_{max}} \sum_{u=0}^{N-t} \sum_{h=0}^{t-1} \sum_{\rho=0}^{2^h} P_1 \cdot P_{L|HT} \cdot P_{H|T} \cdot P_{U|T} \cdot P_T \tag{5.25}$$

## 5.8.2  Analysis of Receiver with MAI Cancellation

As mentioned, the last sequence code from each group is initially preserved as the reference sequence; thus the total number of reference codes is $P$. Then the entire available spreading sequences are $P^2 - P$. The idle users are $P^2 - P - N$. In this case, Equation (5.18) for any $t \in \{t_{\min}, t_{\min+1}, \ldots t_{\max}\}$ where $t_{\min} = \max(N + 2P - P^2 - 1, 1)$ and $t_{\max} = \min(N, P - 1)$ is rewritten as:

$$P_T^1(t) = \frac{\binom{P^2 - 2P + 1}{N - t} \cdot \binom{P - 2}{t - 1}}{\binom{P^2 - P - 1}{N - 1}} \tag{5.26}$$

A Poisson random vector $Y_1$ is used to represent the photon-counts collected from the main branch, while $Y_P$ denotes the photon-counts received from the reference branch. Then vector $\tilde{Y}_1$ is defined as $\tilde{Y}_1 = Y_1 - Y_P$. Hence, we have [2]:

$$P_E = P_r\{\tilde{Y}_{1,j} \geq \tilde{Y}_{1,0}, \ some \ j \neq 0 | S_{1,0} = 1\}$$

$$\geq (M - 1) \cdot P_r\{\tilde{Y}_{1,1} \geq \tilde{Y}_{1,0} | S_{1,0} = 1, T = t, U = u, H = h\} \cdot P(H = h)$$

$$\cdot P(U = u) \cdot P(T = t)$$

$$= (M - 1) \cdot P_1 \cdot P_{L|HT} \cdot P_{H|T}(h|t) \cdot P_{U|T}(u|t) \cdot P_T^1(t) \tag{5.27}$$

where $P_1 \leq \exp\left[-\mathbf{Q}\dfrac{(P + 2)^2}{4.(P + 2 + 2u + l)}\right]$.

Thus the BER can be written based on the interference analysis and modulation format as:

$$P_b = \frac{M}{2} \sum_{t_{\min}}^{t_{\max}} \sum_{u=0}^{N-t} \sum_{h=0}^{t-1} \sum_{\rho=0}^{2^h} \exp\left[-\mathbf{Q}\frac{(P + 2)^2}{4.(P + 2 + 2u + l)}\right] \cdot P_{L|HT} \cdot P_{H|T} \cdot P_{U|T} \cdot P_T^1 \tag{5.28}$$

## 5.8.3  Analysis of Receiver with MAI Cancellation and Manchester Encoding

In the receiver with Manchester-coded MAI cancellation shown in Figure 5.27, a new random variable $W$ for the number of active users from group 2 up to group $(P + 1)/2$ is defined. Variable $w$ is the realization of $W$ for any $w \in \{w_{\min}, \ldots, w_{\max}\}$ and can be written as [7]:

$$P_{W|T}(w|t) = \frac{\binom{(P^2 - 2P - 1)/2}{w} \cdot \binom{(P^2 - 2P + 1)/2}{N - t - w}}{\binom{P^2 - 2P + 1}{N - t}} \tag{5.29}$$

where $w_{\min} = \max\{0, N - t - (P^2 - 2P + 1)/2\}$, $w_{\max} = \min\{N - t, (P^2 - 2P + 1)/2\}$.

Then, the conditional PDF of variable $U$, non-first-group interferers, is written as:

$$P^1_{U|T}(u|t) = \binom{w}{u} \cdot \left(\frac{\gamma}{M}\right)^u \cdot \left(1 - \frac{\gamma}{M}\right)^{w-u} \tag{5.30}$$

where $u \in \{0, 1, 2 \ldots w\}$.

Hereafter, the BER based on the modulation scheme and interference analysis is expressed as:

$$P_b = \frac{M}{2} \sum_{t_{\min}}^{t_{\max}} \sum_{w_{\min}}^{w_{\max}} \sum_{u=0}^{N-t} \sum_{h=0}^{t-1} \sum_{\rho=0}^{2^h} \exp\left[-Q\frac{(P+2)^2}{4.(P+2+2u+l)}\right]$$

$$\cdot P_{L|HT} \cdot P_{H|T} \cdot P^1_{U|T} \cdot P^1_T \tag{5.31}$$

### 5.8.4 Analysis of Self-Interferences (SI)

Since the synchronization is performed at the last chip of a code sequence, the self-interference arises due to incomplete orthogonal code sequences. This subsection is devoted to investigating the effect of SI at synchronous incoherent OPPM-OCDMA transceivers. The SI and MAI as dominant degrading factors have been considered at the receivers. Referring to Equation (5.17), the BER is given by:

$$P_b = \frac{M}{2(M-1)} P_E \tag{5.32}$$

#### 5.8.4.1 Analysis of SI at Simple Receiver

Recall that the random Poisson vector $Y_{1,j}$ denotes the photon-counts collected by receiver #1, where $j \in \{0, 1, \ldots M - 1\}$. Index $i$ is declared to be the true one if $Y_{1,j} > Y_{1,j}$ for every $i \neq j$. $S_{1,i} = 1$ represents the user #1 transmitting data at slot $i$. When the spreading sequence shifts, the probability of SI is $q = 1/P^2$. Hence, the probability of error $P_E$ can be, as discussed previously:

$$P_E = P_1 \cdot P\{T = t\} \cdot P\{U = u\} \cdot P\{H = h\} \tag{5.33}$$

where $P\{T = t\}, P\{U = u\}$ and $P\{H = h\}$ have been introduced earlier in Sections 5.8.1–5.8.3. The PDF of $P_1$ is then introduced as:

$$P_1 = \sum_{j=0}^{M-1} P_r\{Y_{1,j} \geq Y_{1,0}, \; some \; j \neq 0 | S_{1,0} = 1, T = t, U = u, H = h, L = l\}$$

$$\leq (M - \gamma) \cdot P_r\{Y_{1,1} \geq Y_{1,0} | S_{1,0} = 1, T = t, H = h, U = u, L = l, v_0 = 0\} \tag{5.34}$$

$$+ \sum_{j=1}^{\gamma-1} P_r\{Y_{1,j} \geq Y_{1,0} | S_{1,0} = 1, T = t, H = h, U = u, L = l, v_j = k_j\}$$

For any $j \in \{1, 2, \ldots, M - 1\}$, $v_j \in \{0, 1\}$ denotes the number of pulses that cause SI in slot $j$ due to the transmitted data at slot 0 by the intended user. In the first term of the above $P_1$ expression, there is no self-interference due to $v_0 = 0$, $P_r = [v_0 = 0] = 0$, while the second term causes the actual SI due to the remaining $\gamma - 1$ interfering slots [14]. These slots interfere with slot 0 with a probability of $P_r\{v_j = 1\} > 0$. Hence, $P_1$ can be expressed as:

$$P_1 = (M - \gamma) \cdot P_r\{Y_{1,1} > Y_{1,0}|S_{1,0} = 1, T = t, U = u, H = h, L = l, v_1 = 0\}$$
$$+ (\gamma - 1) \cdot q \cdot P_r\{Y_{1,j} > Y_{1,0}|S_{1,0} = 1, T = t, U = u, H = h, L = l, v_1 = 1\}$$
$$+ (\gamma - 1) \cdot (1 - q) \cdot P_r\{Y_{1,j} > Y_{1,0}|S_{1,0} = 1, T = t, U = u, H = u, L = l, v_1 = 1\}$$
$$= (M - \gamma) \cdot P_{r1} + (\gamma - 1) \cdot q \cdot P_{r2} + (\gamma - 1) \cdot (1 - q) \cdot P_{r1}$$
$$= (M - 1) \cdot P_{r1} + (\gamma - 1) \cdot q \cdot (P_{r2} - P_{r1}) \tag{5.35}$$

where

$$P_{r1} = P_r\{Y_{1,1} > Y_{1,0}|S_{1,0} = 1, T = t, U = u, H = h, L = l, v_1 = 0\} \tag{5.36}$$

$$= \sum_{y_1=0}^{\infty} e^{-\mathbf{Q}(u+l)} \cdot \frac{[\mathbf{Q} \cdot (u + l)]^{y_1}}{y_1!} \cdot \sum_{y_0=0}^{y_1} e^{-\mathbf{Q}(u+P+2)} \cdot \frac{[\mathbf{Q} \cdot (u + P + 2)]^{y_0}}{y_0!}$$

and

$$P_{r2} = P_r\{Y_{1,1} > Y_{1,0}|S_{1,0} = 1, T = t, U = u, H = h, L = l, v_1 = 1\} \tag{5.37}$$

$$= \sum_{y_1=0}^{\infty} e^{-\mathbf{Q}(u+l+1)} \cdot \frac{[\mathbf{Q} \cdot (u + l + 1)]^{y_1}}{y_1!} \cdot \sum_{y_0=0}^{y_1} e^{-\mathbf{Q}(u+P+2)} \cdot \frac{[\mathbf{Q} \cdot (u + P + 2)]^{y_0}}{y_0!}$$

Therefore, the error probability $P_b$ can be obtained as:

$$P_b = \frac{M}{2(M - 1)} P_E$$

$$= \frac{M}{2(M - 1)} \cdot P_1 \cdot P\{T = t\} \cdot P\{U = u\} \cdot P\{H = h\}$$

$$= \frac{M}{(2(M - 1)} \cdot \sum_{t=t_{\min}}^{t_{\max}} \sum_{u=0}^{N-t} \sum_{h=0}^{t-1} \sum_{\rho=0}^{\kappa^h} [(M - 1) \cdot P_{r1} + (\gamma - 1) \cdot q \cdot (P_{r2} - P_{r1})]$$

$$\cdot P_{L|HT} \cdot P_{H|T} \cdot P_{U|T} \cdot P_T \tag{5.38}$$

### 5.8.4.2   Analysis of SI at Receiver with MAI Cancellation

Since the last signature codes from GPMPC groups are preserved as the reference correlation, the total number of available spreading sequence becomes $P^2 - P$. Consequently, the probability of self-interference at this receiver becomes $q = 1/(P^2 - P)$. A Poisson random vector $Y_1$ is used to represent the photon-counts collected from the main branch, while $Y_P$ denotes the photon-counts received from the reference branch. The vector $\tilde{Y}_1$ is defined as $\tilde{Y}_1 = Y_1 - Y_P$. Then, error probability $P_E$ is defined as:

$$P_E = \sum_{j=0}^{M-1} P_r\{\tilde{Y}_{1,1} > \tilde{Y}_{1,i}, \text{ some } j \neq i | S_{1,i} = 1, T = t, U = u, H = h, L = l\}$$

$$\cdot P(T = t) \cdot P(U = u) \cdot P(H = h)$$

$$\leq \sum_{j=0}^{M-1} P_r\{\tilde{Y}_{1,1} > \tilde{Y}_{1,0}, \text{ some } j \neq 0 | S_{1,0} = 1, T = t, U = u, H = h, L = l\}$$

$$\cdot P(T = t) \cdot P(U = u) \cdot P(H = h) \tag{5.39}$$

By further analysis, we have:

$$P_E = \sum_{j=0}^{M-1} P_r\{\tilde{Y}_{1,1} > \tilde{Y}_{1,0}, \text{ some } j \neq 0 | S_{1,0} = 1, T = t, U = u, H = h, L = l\}$$

$$\leq (M - \gamma) \cdot P_r\{\tilde{Y}_{1,1} > \tilde{Y}_{1,0}, \text{ some } j \neq 0 | S_{1,0} = 1, T = t, U = u, H = h, L = l, v_1 = 0\}$$

$$+ (\gamma - 1) \cdot q \cdot P_r\{\tilde{Y}_{1,1} > \tilde{Y}_{1,0}, \text{ some } j \neq 0 | S_{1,0} = 1, T = t, U = u, H = h, L = l,$$

$$v_1 = 1\} + (\gamma - 1) \cdot (1 - q) \cdot P_r\{\tilde{Y}_{1,1} > \tilde{Y}_{1,0}, \text{ some } j \neq 0 | S_{1,0} = 1, T = t, U = u,$$

$$H = h, L = l, v_1 = 0\}$$

$$= (M - \gamma) \cdot P_{r1}^1 + (\gamma - 1) \cdot q \cdot P_{r2}^1 + (\gamma - 1) \cdot (1 - q) \cdot P_{r1}^1$$

$$= (M - (\gamma + 1) \cdot q) \cdot P_{r1}^1 + (\gamma - 1) \cdot q \cdot P_{r2}^1 \tag{5.40}$$

The term in the right-hand side of the last inequality is due to the $M - 1 - (\gamma - 1)$ slots that do not cause SI with slot 0, where:

$$P_{r1}^1 = P_r\{\tilde{Y}_{1,1} > \tilde{Y}_{1,0}, \text{ some } j \neq 0 | S_{1,0} = 1, T = t, U = u, H = h, L = l, v_1 = 0\} \tag{5.41}$$

$$P_{r2}^1 = P_r\{\tilde{Y}_{1,1} > \tilde{Y}_{1,0}, \text{ some } j \neq 0 | S_{1,0} = 1, T = t, U = u, H = h, L = l, v_1 = 1\} \tag{5.42}$$

Since the slots are uniformly distributed, these probabilities are the same and then we define $\theta(u, t)$ as:

$$\theta(u, t) = P_{r2}^1 = P_{r1}^1 = P_r\{\tilde{Y}_{1,1} > \tilde{Y}_{1,0}, \text{ some } j \neq 0 | S_{1,0} = 1, T = t, U = u,$$

$$H = h, L = l, v_1 = 1\} \tag{5.43}$$

We can further simplify the calculations by using the Chernoff Bound [4], and then we have:

$$\theta(u, t) = P_r\{Y_{1,1} - Y_{P,1} > Y_{1,0} - Y_{P,0} | S_{1,0} = 1, T = t, U = u, H = h, L = l, v_1 = 1\}$$

$$\leq E[z^{[Y_{1,1} - Y_{P,1} - Y_{1,0} - Y_{P,0}]} | S_{1,0} = 1, T = t, U = u, H = h, L = l, v_1 = 1] \tag{5.44}$$

where $E[]$ is the conditional expectation operator and $Z > 1$ is the number of interfering slots. The natural logarithm computation is carried out on $\theta(u,t)$ and the expectation term is performed as:

$$Ln\theta(u,t) \leq \mathbf{Q} \cdot (u + l + 1) \cdot (z - 1) - \mathbf{Q} \cdot (u + 1) \cdot (1 - z^{-1})$$
$$- \mathbf{Q} \cdot (P + 2 + u) \cdot (1 - z^{-1}) - \mathbf{Q} \cdot u(1 - z) \tag{5.45}$$

Now, by setting $z - 1 = \delta, \delta > 0$ and integer, we have $1 - z^{-1} \leq 1$, whereas $\delta - \delta^2 \leq 0$. Thus by considering new boundaries, we obtain the lower-bounded [2]:

$$1 - z^{-1} \geq \delta - \delta^2 \tag{5.46}$$

By substituting Equation (5.46) in Equation (5.45), we have:

$$Ln\theta(u,t) = \mathbf{Q}(u + l + 1)(\delta) - \mathbf{Q}(u + l)(\delta - \delta^2) - \mathbf{Q}(u + P + 2)(\delta - \delta^2) - \mathbf{Q}(u)(-\delta)$$
$$= \mathbf{Q}(2u + l + P + 2)(\delta^2) - \mathbf{Q}(P + 2)\delta \tag{5.47}$$

By minimizing Equation (5.47) regarding $\delta$, it is obtained as:

$$\delta = \frac{P + 2}{2(P + 2 + 2u + l)} \tag{5.48}$$

Therefore, by substituting Equation (5.48) into Equation (5.47), we have:

$$Ln\theta(u,t) = \mathbf{Q}(2u + l + P + 2)\left(\frac{P + 2}{2(P + 2 + 2u + l)}\right)^2$$
$$- \mathbf{Q}(P + 2)\left(\frac{P + 2}{2(P + 2 + 2u + l)}\right) \tag{5.49}$$

and bringing Equation (5.49) back to the exponential format, we obtain:

$$\theta(u,t) \leq \exp\left[-\mathbf{Q}\frac{(P + 2)^2}{4(P + 2 + 2u + l)}\right] \tag{5.50}$$

Hereafter,

$$P_{r2}^1 = \theta(u,t) \leq \exp\left[-\mathbf{Q}\frac{(P + 2)^2}{4(P + 2 + 2u + l)}\right] \tag{5.51}$$

Similarly for $P_{r1}^1$:

$$P_{r1}^1 \leq \exp\left[-\mathbf{Q}\frac{(P + 2)^2}{4(P + 2 + 2u + l)}\right] \tag{5.52}$$

Hence, the upper-bounded BER of the receiver with MAI and SI becomes:

$$P_b = \frac{M}{2(M - 1)} \cdot P_E$$
$$= \frac{M}{2(M - 1)} \cdot P_E \cdot P\{T = t\} \cdot P\{U = u\} \cdot P\{H = h\} \tag{5.53}$$

$$= \frac{M}{2} \sum_{t=t_{\min}}^{t_{\max}} \sum_{u=0}^{N-t} \sum_{h=0}^{t-1} \sum_{\rho=0}^{\kappa^h} \left[ P_{r1}^1 + (\gamma - 1) \cdot q \cdot \left( P_{r2}^1 - P_{r1}^1 \right) \right] \cdot P_{L|HT} \cdot P_{H|T} \cdot P_{U|T} \cdot P_T^1$$

where the probabilities of other elements in Equation (5.53) have been introduced in Sections 5.8.2 and 5.8.3.

### 5.8.4.3   Analysis of SI at Receiver with MAI Cancellation and Manchester Encoding

This receiver is very similar to the receiver with MAI discussed previously. The only difference is that active users are divided into two groups, referring to Section 5.8.3, and then the upper-bounded BER can be obtained considering MAI and SI as:

$$P_b = \frac{M}{2} \sum_{t=t_{\min}}^{t_{\max}} \sum_{w=w_{\min}}^{w_{\max}} \sum_{u=0}^{N-t} \sum_{h=0}^{t-1} \sum_{\rho=0}^{\kappa^h} \left[ P_{r1}^1 + (\gamma - 1) \cdot q \cdot \left( P_{r2}^1 - P_{r1}^1 \right) \right]$$

$$\cdot P_{L|HT} \cdot P_{H|T} \cdot P_{U|T}^1 \cdot P_{W|T} \cdot P_T^1 \qquad (5.54)$$

## 5.9   Discussion of Results

This section presents the three transceivers' performance based on the above analysis. To better understand the results, the performances are compared with the latest developed prime code families in detail. The analysis for n-MPC in incoherent OPPM-OCDMA system can be found in [12]. We have examined the BER performances of receivers which are affected by (i) MAI only and (ii) MAI as well as SI.

### 5.9.1   BER Performance of Receivers with MAI

The following three structures and their overall performances are discussed in details: (i) simple receiver; (ii) receiver with only MAI cancellation and (iii) receiver with both MAI cancellation and Manchester encoding. This section presents the results of the receivers when MAI has been considered as the dominant degrading interference.

Figure 5.28 shows the bit error rate (BER) evaluation for GPMPC, n-MPC and MPC codes employed in incoherent simple OPPM-OCDMA receivers against the average photons per pulses, $\mu$. Prime number $P$ and number of active users $N$ are set to be 7 and 42 (i.e. full-load), respectively. Two different multiplicities of $M = 8$ and 16 are investigated at the receivers. Figure 5.28 clearly shows that the performance becomes better as $\mu$ increases. GPMPC outperforms other codes due to greater difference correlation values as discussed earlier. It is also seen that multiplicity plays a significant role in improving the system performance but that it compromises the structure complexity, e.g. for $\mu = 70$ and $M = 16$ bit error rate of n-MPC is 0.0081, MPC is 0.0094 while GPMPC's is 0.0065.

The BER performance of simple receivers using GPMPC, MPC and n-MPC, under a given condition, is illustrated on Figure 5.29 against the number of active users.

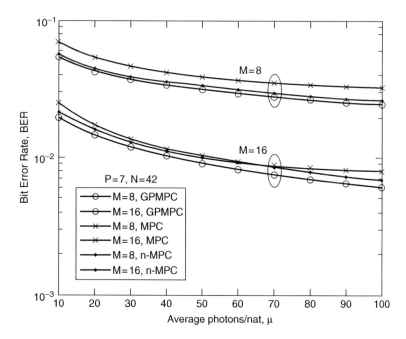

**Figure 5.28** BER performance of OPPM-OCDMA simple receivers using different codes against the average photons per pulse $\mu$, when $P = 7$, $N = 42$ and $M = 8$ and 16

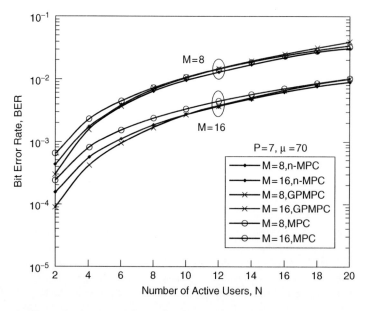

**Figure 5.29** BER performance of OPPM-OCDMA simple receivers using different codes against the number of active users $N$, when $P = 7$, $\mu = 70$ and $M = 8$ and 16

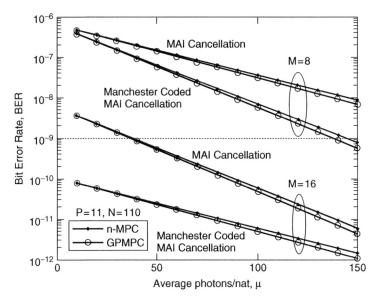

**Figure 5.30** BER performance of OPPM-OCDMA receivers considering MAI using different codes against the average photons per pulse $\mu$, when $P = 11$, $N = 110$ and $M = 8$ and 16

The figure shows that as the number of users increases, the BER increases too, due to the interference enhancement, which makes the system performance unreliable. Again multiplicity has an effective role in the receiver's performance, but the overall BER performance of the simple receiver, shown in Figures 5.28 and 5.29, is so poor, compared with the minimum requirement of $10^{-9}$ used in optical communication systems, that it makes the use of interference cancellation essential.

To examine other receivers, we now only evaluate the performance of GPMPC and n-MPC code families since n-MPC has already outperformed MPC [12]. The BER $= 10^{-9}$ is also drawn on the figures as a reference.

Figure 5.30 shows variations of BER performance with the average photon-count. In this analysis the prime number $P$ and number of active users $N$ are 11 and 110 (i.e. full-load), respectively. Also, since the average photon-count ($\mu$) is directly proportional to signal power, the BER decreases as the received signal's power increases. The results shown in Figure 5.30 clearly show that, as expected, reduction of interference together with Manchester encoding has improved the system performance as compared with simple receivers and receivers with only MAI cancellation. The results indicate that, by optimizing both $\mu$ and $M$ a more reliable communication link can be guaranteed.

Figure 5.31 illustrates the performance of the incoherent OPPM-OCDMA receivers with MAI cancellation and Manchester-coded against the number of active users when $P = 7, M = 8$, 16 and $\mu = 100$. It is observed from Figure 5.31 that there is valley (minimum) in the BER values of the receivers with MAI cancellation with Manchester coding schemes. By studying the minimum value of BER with respect to the number of active users, it is observable that the value is minimum at around 50–60 per cent of the total number of users, i.e. $N = P^2 - P$. Table 5.1 shows a few examples based on

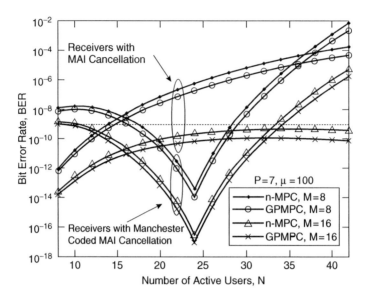

**Figure 5.31** BER performance of OPPM-OCDMA receivers considering MAI using different codes against the number of active users $N$, when $P = 7$, $\mu = 100$ and $M = 8$ and 16

**Table 5.1** Minimum BER for the number of active users considering only MAI at receivers with Manchester-coded MAI cancellation, when $M = 16$ and $\mu = 100$

| $P$ | $BER_{min}$ of n-MPC | $BER_{min}$ of GPMPC | $N_{eff}$ | $N_{Full}$ | Supported No. of Active Users $(N_{eff}/N_{Full})\%$ |
|---|---|---|---|---|---|
| 5 | $1.84 \times 10^{-12}$ | $8.51 \times 10^{-13}$ | 12 | 20 | 60 |
| 7 | $9.26 \times 10^{-17}$ | $3.46 \times 10^{-17}$ | 24 | 42 | 57.1 |
| 11 | $6.85 \times 10^{-22}$ | $4.8 \times 10^{-22}$ | 60 | 120 | 50 |
| 13 | $5.17 \times 10^{-27}$ | $3.25 \times 10^{-27}$ | 84 | 156 | 53.8 |
| 17 | $6.23 \times 10^{-35}$ | $7.23 \times 10^{-36}$ | 140 | 272 | 51.47 |

various prime numbers $P$ minimum BER of different spreading codes, effective number of users denoting the minimum BER, $N_{eff}$ and total number of users $N_{Full}$.

Accordingly, if the number of maximum supported active users is set to 55 per cent (as an average $N_{eff}$) of the total users in the access network, then the transceivers will apparently have much enhanced performance. It is noted that the maximum number of supported active users in the current networks is usually set to 10–20 per cent of total users due to capital expenditure cost of the network design and implementation [15].

By comparing the BER values at the effective points, $N_{eff}$, it can be observed that the BER differences are remarkable and it can compensate for the network capacity.

Furthermore, the multiplicity $M$ has an independent effect on improving the BER, but the implementation will be complicated due to the precise timing and switching design. The performance behaviour, shown in Figures 5.30 and 5.31 when $M = 8$ and 16, indicated that by increasing $M$ itself the network capacity is recovered as well as the overall performance under a given condition especially with GPMPC.

## 5.9.2 BER Performance of Receivers with MAI and SI

In this section, the receivers are examined against the number of active users $N$, and the average number of photons per pulse $\mu$, when both the multiple-access and self-interferences are taken into account. Obviously, simple receivers will degrade even more due to higher level of interferences, so it is important to investigate the performance of the receivers of following two transmission systems: (i) receivers with only MAI cancellation and (ii) receivers with Manchester-coded MAI cancellation. The evaluations for different receivers are based on the analysis mentioned in Section 5.8.4 for GPMPC.

Figure 5.32 shows the receivers BER, when $P = 11, N = 110$ (i.e. full-load) and $M = 8$ and 16, against the average number of photons per pulse $\mu$. As can be seen, the higher the received signal power, the more reliable is the communication. At the beginning both code performances are very good and GPMPC bit error rate decreases as $\mu$ increases.

It is seen that the BER difference between two spreading codes can be traded in with system capacity and/or power consumption. This implies that a system employing GPMPC can have the same BER as a system employing n-MPC, while supporting a greater number of users and/or consuming less power.

For example, when $\mu = 50$, the BER performance of a GPMPC receiver is the same as that of an n-MPC receiver having $\mu = 55$. Moreover, based on the results shown in Figure 5.32 when $M = 16$ and $\mu = 100$ the BER of a receiver employing GPMPC

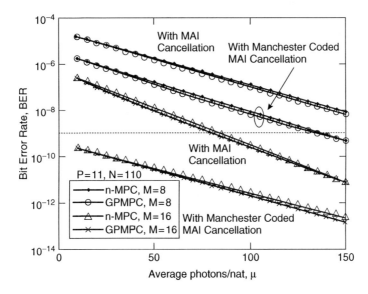

**Figure 5.32** BER performance of OPPM-OCDMA receivers considering MAI and SI using different codes against the average photons per pulse $\mu$, when $P = 11$, $N = 110$ and $M = 8$ and 16

is $1.8 \times 10^{-10}$ whereas that of n-MPC is $2.9 \times 10^{-10}$, which suggests that the BER of n-MPC is 1.6 times higher than that of GPMPC.

As discussed before, the multiplicity factor $M$ provides more possible positions within a time-slot to accommodate a higher number of signals. Figure 5.32 clearly shows that superior performance is achieved by a higher level of multiplicity factor $M$. However, the higher multiplicity factor makes the system implementation complex, so applying Manchester encoding is an excellent choice to virtually increase the signal multiplicity and accordingly the system capacity.

Finally, Figure 5.33 illustrates the BER performance of incoherent OPPM-OCDMA receivers against the number of active users $N$, when $P = 7, \mu = 100, M = 8$ and 16. Here we see a similar behaviour to Figure 5.31 with a minimum BER valley. The effective number of users $N_{eff}$ shows where the minimum error rate occurs at the receivers with Manchester-coded MAI reduction. Table 5.2 shows the example results of $N_{eff}$ with respect to various prime number $P$.

It is seen that by increasing the number of active users, the interference increases. Obviously, the receiver will not be able to support a great number of users under reasonable bit error rate. Yet GPMPC is seen to have superior performance compared with n-MPC, especially in Manchester-encoded MAI cancellation receiver structure. For example, with the receiver with Manchester-coded MAI cancellation, when $M = 8$ at $N_{eff}$, from Figure 5.33, the BER for the receiver employing n-MPC is $9.8 \times 10^{-13}$, while it is $3.7 \times 10^{-13}$ for GPMPC, i.e. 62 per cent improvement.

By comparing the error rates of the receivers considering only MAI in Section 5.8.1 with the receivers considering MAI and self-interference in Section 5.8.4, it is observed that the performance is obviously further degraded by the self-interference.

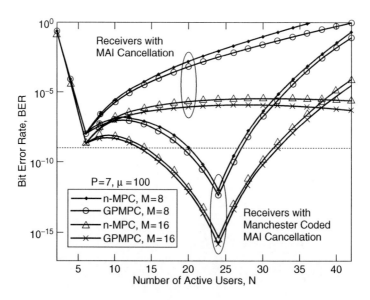

**Figure 5.33** BER performance of OPPM-OCDMA receivers considering MAI and SI using different codes against the number of active users $N$, when $P = 7$, $M = 8$, 16 and $\mu = 100$

**Table 5.2** Minimum BER for the number of active users considering MAI and SI at receivers with Manchester-coded MAI cancellation, when $M = 16$ and $\mu = 100$

| $P$ | $BER_{min}$ of n-MPC | $BER_{min}$ of GPMPC | $N_{eff}$ | $N_{Full}$ | Supported No. of Active Users $(N_{eff}/N_{Full})\%$ |
|---|---|---|---|---|---|
| 5 | $8.6 \times 10^{-12}$ | $1.2 \times 10^{-12}$ | 12 | 20 | 60 |
| 7 | $3.3 \times 10^{-16}$ | $1.3 \times 10^{-16}$ | 24 | 42 | 57.1 |
| 11 | $6.9 \times 10^{-20}$ | $4.5 \times 10^{-20}$ | 60 | 120 | 50 |
| 13 | $5.3 \times 10^{-25}$ | $3.1 \times 10^{-25}$ | 84 | 156 | 53.84 |

## 5.10 Analysis of Throughput

In practice, one of the important parameters for the performance evaluation of an optical CDMA system is the data rate. For a given user, throughput is the rate of data transmission given by the amount of information transmitted per second by the user.

### 5.10.1 OPPM-OCDMA Throughput

Referring to Section 5.2, $T$ is the duration of each $M$-ary time frame, with chip duration of $T_c$. The spreading sequence of length $L$ must be exactly fitted into time-slot $\tau$, where $\tau = LT_c$. The throughput, $R_{T-OPPM}$, of an OPPM-OCDMA is defined as data packet per time-slot [2, 14]:

$$R_{T-OPPM} = \frac{\ln M}{T} = \frac{\ln M}{M\tau/\gamma} = \frac{\gamma \ln M}{MLT_c} \tag{5.55}$$

where $\gamma$ is the overlapping index. Since the pulse-width $T_c$ is always fixed, for simplicity the throughput-pulse-width product $R_{O-OPPM}$ is defined as:

$$R_{O-OPPM} = R_{T-OPPM} \cdot T_c = \frac{\gamma \ln M}{MLT_c} \cdot T_c = \frac{\gamma \ln M}{ML} \tag{5.56}$$

The above equation shows that, for a fixed pulse width of $T_c$, the throughput-pulse-width product $R_{O-OPPM}$ is proportional to throughput $R_{T-OPPM}$. In addition, the users-throughput product, $NR$, is defined as the product of the number of users and $R_{O-OPPM}$, that is:

$$NR_{OPPM} = N \cdot R_{O-OPPM} = N \cdot \frac{\gamma \ln M}{ML} \tag{5.57}$$

$NR$ is a measure of the total data rate from all users transmitted within the channel. In practice, we are interested in characterizing the maximum throughput that can be achieved when keeping the bit-error probability below a prescribed threshold. Therefore, the parameters $\gamma$, $M$ and $L$ are allowed to vary in order to optimize the throughput under the constraint that $P_b \leq \varepsilon$ where $\varepsilon$ is a precision reference for BER, for example $10^{-9}$. In doing so, we have:

$$R_{O-OPPM,\max} = \max_{\substack{\gamma,M,L \\ P_b \leq \varepsilon}} R_{O-OPPM}, NR_{OPPM,\max} = \max_{\substack{\gamma,M,L \\ P_b \leq \varepsilon}} NR_{OPPM} \qquad (5.58)$$

### 5.10.2 PPM-OCDMA Throughput

Under a given condition in an incoherent PPM-OCDMA transceiver, the throughput $R_{T-PPM}$ is given by [14]:

$$R_{T-PPM} = \frac{\ln M}{T} = \frac{\ln M}{M\tau} = \frac{\ln M}{MLT_c} \qquad (5.59)$$

The throughput-pulse-width product $R_{O-PPM}$ in the PPM-OCDMA can be expressed as:

$$R_{O-PPM} = R_{T-PPM} \cdot T_c = \frac{\ln M}{MLT_c} \cdot T_c = \frac{\ln M}{ML} \qquad (5.60)$$

Similarly, in PPM-OCDMA transceiver, the users-throughput product can be written as:

$$NR_{PPM} = N \cdot R_{O-PPM} = N \cdot \frac{\ln M}{ML} \qquad (5.61)$$

Now, by considering GPMPC in the PPM- and OPPM-OCDMA systems, the throughput-pulse-width product could be rewritten as:

$$R_{O-OPPM} = \frac{\gamma \ln M}{ML} = \frac{(P+2) \cdot \ln M}{M \cdot (P^2 + 2P)} \qquad (5.62)$$

$$R_{O-PPM} = \frac{\ln M}{ML} = \frac{\ln M}{M \cdot (P^2 + 2P)} \qquad (5.63)$$

Hence, the users-throughput products can be expressed as:

$$NR_{OPPM} = N \cdot \frac{\gamma \ln M}{ML} = \frac{N \cdot (P+2) \cdot \ln M}{M \cdot (P^2 + 2P)} \qquad (5.64)$$

$$NR_{PPM} = N \cdot \frac{\ln M}{ML} = \frac{N \cdot \ln M}{M \cdot (P^2 + 2P)} \qquad (5.65)$$

Equations (5.63) and (5.64) indicate that, under a given condition, OPPM system supports $P + 2$ (i.e. $\gamma$) times higher throughput than the PPM scheme. In other words, OPPM is offering the higher throughput without the bandwidth expansion as it is required in the PPM system. However, if the bandwidth increase becomes challenging or impossible due to the components or bandwidth limitations, OPPM modulation is able to provide a higher throughput by introducing an overlapping index in which the spreading is performed over a number of slots rather than only one slot as in PPM modulation. This overlapping index feature of OPPM brings a huge advantage into the system capacity enhancement with no need for bandwidth expansion as it is usually expensive and complex.

## 5.11   Summary

In this chapter, GPMPC as the latest advanced spreading code from prime code families suitable for synchronous incoherent OCDMA has been employed in pulse-position and overlapping pulse-position modulations in the OCDMA networks. Three types of transceivers, including (i) simple transceiver (ii) transceiver with MAI cancellation and (iii) transceiver with MAI cancellation and Manchester coding, have been analysed and examined. The Manchester coding can be interpreted as a return-to-zero signalling format. Accordingly, the analysis included both return-to-zero and non-return-to-zero signalling at the receivers. The lower-bounded BER for receiver (i) and the upper-bounded BER for receivers (ii) and (iii) have been derived. In the analysis, the MAI was assumed to be the dominant noise while background and photodiode dark current noises were ignored. However, the self-interference has been taken into account.

The results have indicated that receivers employing an appropriate coding scheme with multiple access interference cancellation are able to accommodate a greater number of users, while still maintaining low BER and low power consumption. It has been indicated that Manchester encoding plays a significant role in data formats leading to enhanced interference cancellation. However, it should be noted that bandwidth expansion is always necessary for Manchester encoded signals. On the other hand, to overcome this limitation, overlapping PPM signalling has been analysed. The overlapping index feature of OPPM provides improvement in the system capacity without bandwidth expansion.

## References

1. Shalaby, H.M.H. (2002) Complexities, error probabilities and capacities of optical OOK-CDMA communication systems. *IEEE Trans on Comm.*, **50** (12), 2009–2017.
2. Shalaby, H.M.H. (1999) Direct-detection optical overlapping PPM-CDMA communication systems with double optical hard-limiters. *J. Lightw. Technol.*, **17** (7), 1158–1165.
3. Ohtsuki, T. (1999) Performance analysis of direct-detection optical CDMA systems with optical hard-limiter using equal-weight orthogonal signaling. *IEICE Trans. Comm.*, **E82-B** (3), 512–520.
4. Lee, T.S., Shalaby, H.M.H. and Ghafouri-Shiraz, H. (2001) Interference reduction in synchronous fiber optical PPM-CDMA systems *J. Microw. Opt. Tech. Let.*, **30** (3), 202–205.
5. Ghafouri-Shiraz, H., Karbassian, M.M. and Liu, F. (2007) Multiple access interference cancellation in Manchester-coded synchronous optical PPM-CDMA network. *J. Optical Quantum Electronics*, **39** (9), 723–734.
6. Jau, L.L. and Lee, Y.H. (2004) Optical code-division multiplexing systems using Manchester-coded Walsh codes. *IEE Optoelectronics*, **151** (2), 81–86.
7. Lin, C.L. and Wu, J. (2000) Channel interference reduction using random Manchester codes for both synchronous and asynchronous fiber-optic CDMA systems. *J. Lightw. Technol.*, **18** (1), 26–33.
8. Liu, M.Y. and Tsao, H.W. (2000) Cochannel interference cancellation via employing a reference correlator for synchronous optical CDMA system. *J. Microw. Opt. Tech. Let.*, **25** (6), 390–392.
9. Shalaby, H.M.H. (1995) Performance analysis of optical synchronous CDMA communication systems with PPM signaling. *IEEE Trans. Comm.*, **43** (2/3/4), 624–634.

10. Shalaby, H.M.H. (1998) Chip-level detection in optical code division multiple access. *J. Lightw. Technol.*, **16** (6), 1077–1087.
11. Shalaby, H.M.H. (1998) Cochannel interference reduction in optical PPM-CDMA systems. *IEEE Trans. Comm.*, **46** (6), 799–805.
12. Liu, F. and Ghafouri-Shiraz, H. (2005) Analysis of PPM-CDMA and OPPM-CDMA communication systems with new optical code. In: *SPIE Proc.*, vol. 6021.
13. Shalaby, H.M.H. (1999) A performance analysis of optical overlapping PPM-CDMA communication systems. *J. Lightw. Technol.*, **19** (2), 426–433.
14. Karbassian, M.M. and Ghafouri-Shiraz, H. (2007) Fresh prime codes evaluation for synchronous PPM and OPPM signaling for optical CDMA networks. *J. Lightw. Technol.*, **25** (6), 1422–1430.
15. Mestdagh, D.J.G. (1995) *Fundamentals of multiaccess optical fiber networks*. Artech House Inc, Boston, USA.

# 6

# Coherent Temporal OCDMA Networks

## 6.1 Introduction

Recently, coherent time-spreading OCDMA employing either direct time-spreading using super structures such as fibre Bragg grating [1] and arrayed waveguide grating [2] or spectrally phase coding time-spreading by spatial lightwave modulator [3] has attracted a lot of attention because of its overall superior performance over incoherent schemes. Several architectures have been considered for the use of CDMA within optical fibre communication systems. The most common system uses direct-detection with bipolar codes such as Gold sequences [4], while unipolar prime code families, particularly the latest ones introduced in Chapter 2, have more flexible code-length and are partially orthogonal. To retain the advantage of $\{0, 1\}$ codes as a power saving option, in this chapter we consider unipolar signalling and coherent binary phase shift keying (BPSK) modulation. The capacity of the system using prime codes is limited by the maximum achievable bit rate of the electronic circuitry that generates the pseudo-noise (PN) sequences. Here, we concentrate on a maximum attainable chip-rate of 10 Gchip/s and a desired bit rate of hundreds of Mbps. This imposes a limit on the length of the spreading sequences which must be in the order of hundreds of chips per bit.

In this chapter, the signal-to-noise ratio (SNR) of coherent homodyne and heterodyne OCDMA architectures employing the group-padded modified prime code (GPMPC), introduced in Section 2.4.5, and two phase modulation methods are analysed. The phase modulation methods include using either a Mach–Zehnder interferometer (MZI) as an external phase modulator or a distributed feedback (DFB) laser diode's driving current as an injection-locking method along with the dual-balanced detection structure. We have also evaluated the performance penalty imposed on the OCDMA system as a result of the limited phase excursion of $\pm 0.42\pi$, imposed by the injection-locking modulation [4]. Furthermore, we also analyse the coherent BPSK-OCDMA scheme with heterodyne

*Optical CDMA Networks: Principles, Analysis and Applications*, First Edition. Hooshang Ghafouri-Shiraz and M. Massoud Karbassian.
© 2012 John Wiley & Sons, Ltd. Published 2012 by John Wiley & Sons, Ltd.

detection. The external phase modulation method is considered for heterodyne detection. In the analysis, the system SNR with respect to the number of simultaneous active users is investigated considering both receiver noise and multiple access interference (MAI).

## 6.2 Coherent Homodyne BPSK-OCDMA Architecture

This section is dedicated to describing the architecture of the coherent BPSK-OCDMA network. In the coherent OCDMA system which is based on the external phase modulation as shown in Figure 6.1, the outgoing data is first BPSK encoded to generate the in-phase and quadrature phase (IQ) signals in the electrical domain. Then, the BPSK-encoded signal phase-modulates the lightwave through MZI as an external phase modulator. The use of an external MZI modulator for this application has been justified experimentally in [5]. Finally, the signals are CDMA encoded by means of GPMPC sequences and multiplexed via star couplers and transmitted over a star passive optical network (SPON) as a network infrastructure. This implies the optical carrier re-use as it is code-multiplexed and distinguishable by different spreading codes. At the intended receiver as shown in Figure 6.1, the inverse process takes place where another MZI is used to demodulate the received signal using the same GPMPC sequence used for encoding at the transmitter end. Referring to the GPMPC correlation properties, in Chapter 2, the signal that has been multiplied by the same code will de-spread, while signals multiplied by different codes will be further spread in the frequency domain and hence removed. In practice, the received signal is required to be filtered to reject unwanted crosstalk outside the receiver bandwidth. The intended demodulated signal and tiny co-channel interference (i.e. reduced through de-spreading process) is photodetected in dual-balanced structure to remove noise and direct current (dc) values to reliably detect IQ signals.

Figure 6.2 illustrates the coherent OCDMA system based on the injection-locking of the DFB laser diodes' driving current for the phase modulation. Two quality current-driven DFB laser diodes are used, one at the transmitter as an optical source and the other at the receiver as a local oscillator. The signal is phase-modulated by controlling the injection current of the DFB laser at the transmitter.

The optical signal is then CDMA encoded by assigned a GPMPC sequence and transmitted over the network. At the receiver end the synchronized local oscillator is combined coherently with the received OCDMA signal. In CDMA detection process, the portion

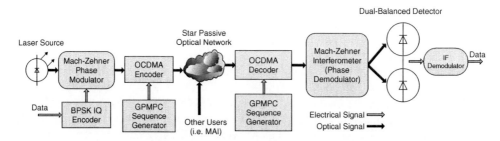

**Figure 6.1** Coherent homodyne BPSK-OCDMA transceiver with MZI phase modulator (© 2008 IEEE. Reprinted with permission from Study of phase modulations with dual-balanced detection in coherent homodyne optical CDMA network, M.M. Karbassian and H. Ghafouri-Shiraz, *J. Lightw. Technol.*, **26** (16), 2008.)

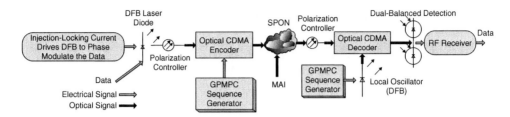

**Figure 6.2** Coherent homodyne BPSK-OCDMA transceiver with injection-locking DFB laser (© 2008 IEEE. Reprinted with permission from Study of phase modulations with dual-balanced detection in coherent homodyne optical CDMA network, M.M. Karbassian and H. Ghafouri-Shiraz, *J. Lightw. Technol.*, **26** (16), 2008.)

of the received signal encoded with the same GPMPC sequence at the transmitter (i.e. the intended data for the intended receiver) is de-spread, whereas signals encoded with other sequences (i.e. MAI) are further spread and reduced. The coherently mixed optical signals are incident on a dual-balanced detector whose electrical output conserves the phase information. The generated bipolar electrical signal is integrated over a bit interval and the result is compared with a threshold to form the final bit estimation based on the maximum likelihood (ML) decision rule.

When phase modulation is achieved by an MZI modulator, the chip-rate can increase to the maximum value of 10 Gchip/s; in contrast, the injection current adjusted to achieve phase modulation has a maximum chip-rate of about 1 Gchip/s due to the limited phase excursion [4, 6–9]. Therefore, there is an apparent limitation on the overall system performance.

### 6.2.1 Analysis of Phase Modulation with MZI

According to Figure 6.1, to extract the information contained in the phase of the optical carrier, coherent detection is employed. For estimation and removal of dc values in the baseband signal, a dual-balanced detector is a reasonable option [8, 10]. The receiver operates under the shot-noise-limited regime by increasing the local oscillator power such that both dark current and receiver thermal noise can become negligibly small.

To obtain an expression for the output of the dual-balanced detector (see Figure 6.2), let $K$ be the number of active users and $S_i(t - \tau_i)$ be the signature sequence of the $i^{\text{th}}$ user, where $\tau_i$ is the relative time delay between the $i^{\text{th}}$ user and the desired user (e.g. user #1). We employ GPMPC spreading sequences and assume that all users have the same polarization and average power given by $\hat{S}^2$. Let $C_i(t)$ denotes the piecewise-constant function which is the product of the data-bit and code sequence bit values of the $i^{\text{th}}$ user at time $t$. The initial phase offset of the $i^{\text{th}}$ user is a random variable $\theta_i$ uniformly distributed over the interval $(0, 2\pi)$.

$C_i(t)$ phase-modulates the lightwave and assumes either 0 or 1 (as product of two bit values). This is the reason that $C_i(t)$ appears in the phase argument showing the lightwave's phase shift. Hence, the received signal is:

$$s(t) = \sqrt{2}\hat{S} \sum_{i=1}^{K} \cos(\omega_c t + C_i(t)(\pi/2) + \theta_i) \tag{6.1}$$

The local oscillator will also be phase-modulated by the spreading sequence of the desired user $S_1(t)$, which is the value of $S_i(t - \tau_i)$ when $i = 1$ and $\tau_i = 0$. This appears in the phase argument as well, thus the local oscillator signal is also given by:

$$l(t) = \sqrt{2}\,\hat{L} \sum_{i=1}^{K} \cos(\omega_c t + S_1(t)(\pi/2) + \theta_{LO}) \tag{6.2}$$

where $\theta_{LO}$ is the initial phase offset of the local oscillator. Now, the product of $l(t)$ and $s(t)$ at the receiver is:

$$l(t).s(t) = 2.\hat{L}.\hat{S}.\sum_{i=1}^{K} \cos(\omega_c t + S_1(t)(\pi/2) + \theta_{LO})\cos(\omega_c t + C_i(t)(\pi/2) + \theta_i) \tag{6.3}$$

Hence:

$$l(t).s(t) = 2.\hat{L}.\hat{S}.\sum_{i=1}^{K} \left[ \frac{1}{2}cos\left[\omega_c t + S_1(t)(\pi/2) + \theta_{LO} + \omega_c t + C_i(t)(\pi/2) + \theta_i\right] \right.$$
$$\left. + \frac{1}{2}cos\left[\omega_c t + S_1(t)(\pi/2) + \theta_{LO} - \omega_c(t) - C_i(t)(\pi/2) - \theta_i)\right] \tag{6.4}$$

The output after the dual-detection becomes:

$$\Re.l(t).s(t) = \Re.\hat{L}.\hat{S}.\sum_{i=1}^{K} \cos\left[(S_1(t) - C_i(t)).(\pi/2) + \theta_{LO} - \theta_i\right] \tag{6.5}$$

where the higher frequency component $2\omega_c$ is filtered by the detection process. In Equation (6.5) $\Re = \eta e/h\nu$ is the photodetectors' responsivity where $\eta$ is the quantum efficiency of each detector, $h$ is Planck's constant, $e$ is the fundamental charge of an electron, and $\nu$ is the employed optical frequency. Note that we assume that the local oscillator tracks the phase of the desired user, thus we use $\theta_{LO} = \theta_1 = 0$. As mentioned earlier, when a relatively high power local oscillator is employed, the receiver operates under the shot-noise-limited regime and its noise has one-sided power spectral density as:

$$N_0 = 2\Re\,\hat{L}^2 \tag{6.6}$$

Integration of the detector output, i.e. Equation (6.5), over one bit interval $T$, results in:

$$S_{out} = \Re \sum_{i=1}^{K} \int_0^T l(t)s(t)dt + \sqrt{N_0} \int_0^T n(t)dt$$
$$= \Re\hat{L}\hat{S}b_0^1 T + \sqrt{\Re T}.\hat{L} + \Re\hat{L}\hat{S}\sum_{i=2}^{K}\left[b_{-1}^i R_{i,1}(\tau_i) + b_0^i \hat{R}_{i,1}(\tau_i)\right]\cos\theta_i \tag{6.7}$$

where $b_0^1$ represents the information bit being detected, $b_{-1}^i$ and $b_0^i$ are overlapping of the previous and the following bits of the $i^{\text{th}}$ user's bit. The continuous-time partial cross-correlation functions $R_{i,j}(\tau)$ and $\hat{R}_{i,j}(\tau)$ are defined as:

$$R_{i,j}(\tau) = \int_o^\tau s_i(t - \tau)s_j(t)dt$$

$$\hat{R}_{i,j}(\tau) = \int_\tau^T s_i(t - \tau)s_j(t)dt \tag{6.8}$$

where $\tau = \tau_i - \tau_j$ and $\tau_1 = 0$. The noise $n(t)$ in Equation (6.7) is assumed to have a Gaussian distribution with zero mean and unit variance [4]; all data bits are independent and equiprobable and the delays are independent and uniformly distributed over a bit interval. The first term in Equation (6.7) is due to the desired user, while the second term is additive white Gaussian noise (AWGN), and the third term is the multiuser interference (i.e. MAI) which has normal distribution when a reasonable number of active users are involved [4].

Finally, the signal-to-noise ratio (SNR) of the coherent homodyne BPSK-OCDMA with MZI phase modulator is derived in terms of the number of active users $K$ as:

$$SNR(K) = \frac{\Re^2 \hat{L}^2 \hat{S}^2 T^2}{\Re^2 \hat{L}^2 \hat{S}^2 T^2 \dfrac{K-1}{3N} + \Re T \hat{L}^2} = \frac{1}{\dfrac{K-1}{3N} + \dfrac{1}{\Re \hat{S}^2 T}} \tag{6.9}$$

In Equation (6.9), note that the numerator contains the signal power of the intended user (Equation (6.7)) and the denominator contains the MAI and noise. The noise term is clear as it came from Equations (6.6) and (6.7). For the MAI term given in Equation (6.7), the power represents the variance of each term inside the sigma as the cross-correlation values, which is $T^2/3N$ [4, 7, 8], where $N$ is the length of the signature sequence. On the other hand, the addition (sigma) is then repeated $K - 1$ times producing the $\frac{(K-1)T^2}{3N}$, first term in Equation (6.9)'s denominator as MAI.

Also note that for single-user $SNR(1) = \Re \hat{S}^2 T = E_b/N_0$ where $E_b$ is the signal energy of one bit. The bit error rate (BER) of BPSK signalling with Gaussian noise, related to the above SNR, is then [11]:

$$BER_{BPSK}(K) = Q\left(\sqrt{2 \times SNR(K)}\right) = \frac{1}{2} erfc\sqrt{SNR(K)} \tag{6.10}$$

where $Q(x)$ and $erfc(x)$ are Q-function and complementary-error-function respectively, where

$$Q(x) = \frac{1}{\sqrt{2\pi}} \int_x^\infty \exp\left(\frac{-u^2}{2}\right) du = \frac{1}{2}\left[1 - erf\left(\frac{x}{\sqrt{2}}\right)\right] \tag{6.11}$$

Figure 6.3 shows the comparison of BER against the single-user SNR for GPMPC sequences. In the analysis we have used a Gold sequence of 511 chips and a GPMPC of $P = 23$ [1–3, 7, 12]. As mentioned earlier, if we pose the very high chip-rate while keeping the bit rate sufficiently high, then obviously a limit is also imposed to the length of the spreading sequence.

Since Gold sequences are $N = 2^n - 1$ chips long, with $n$ being an odd integer, we are limited to $n = 9$, i.e. a length of 511; whereas with GPMPC, $P$ values are 23 and 29

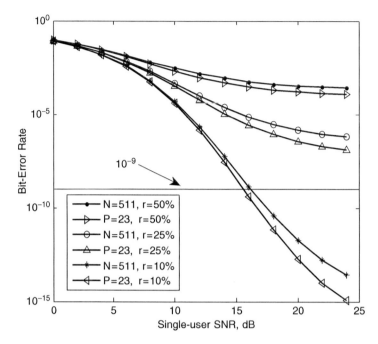

**Figure 6.3** BER performance of homodyne BPSK-OCDMA with MZI against single-user SNR

which correspond to code-lengths ($N = P^2 + 2P$) of 575 and 899, respectively. Also the desired BER threshold of $10^{-9}$ is indicated in Figure 6.3 as a reference.

The parameter $r$ shown in the inset of Figure 6.3, which ranges from 10 to 50 per cent, denotes the percentage of the total possible number of active users involved in the communication system according to the coding scheme. For example, 10 per cent of users (53 users) are accommodated with guaranteed BER of $4 \times 10^{-10}$ at single-user SNR of 15 dB when $P = 23$; while 10 per cent of users (50 users) with a Gold sequence of 511 chips are accommodated with BER of $1.3 \times 10^{-9}$. Hence, for a given value of $r$, when the signal power increases, the system BER decreases; however, for higher values of $r$, the system BER degrades due to the increase in MAI. Also, the performance of the system using the GPMPC, surpasses the overall performance of Gold sequences in different conditions.

Figure 6.4 explains the system behaviour against the number of active users for different single-user SNRs ($E_bN$) and compares GPMPC with $P = 23$ and Gold-sequences with 511 chips. The figure clearly shows that (i) as the single-user SNR increases the BER decreases, (ii) the system performance is limited by MAI and (iii) systems employing GPMPC can accommodate greater number of active users as compared with those employing Gold-sequence. For example, with a GPMPC system having $E_bN = 16$ dB we can have a maximum of 55 simultaneous active users that maintain $BER \leq 10^{-9}$, whereas this figure reduces to 45 users in the case of Gold sequences.

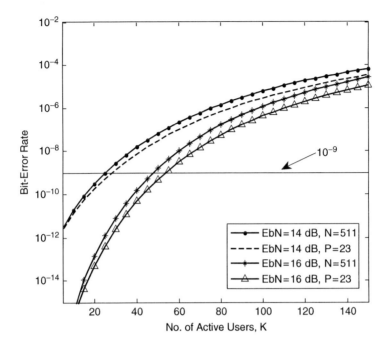

**Figure 6.4**  BER performance of homodyne BPSK-OCDMA with MZI against the number of active users $K$

Now, when code-lengths are kept below 1000, for effective data-rate and encoder/decoder power efficiency, $P = 29$ (i.e. 899) still remains in the scope, whereas the next step for Gold sequences is 1023.

### 6.2.2  Analysis of Phase Modulation with DFB Injection-Locking

In the previous section, we studied matched-filtering during coherent detection with a dual-balanced detector followed by an integrator which leads to a baseband electrical signal consisting of the de-spread signal with AWGN and MAI. In homodyne detection the local oscillator is at the same frequency as that of the carrier and the output electrical signal is at the baseband. Here, we discuss the phase modulation by injection locking of the driving current of a DFB laser based on the architecture shown in Figure 6.2.

To examine the effect of the phase limitation on the modulation process, the injection current of the DFB laser diode is modulated to accomplish PSK signalling at the transmitter with phase excursion being limited to $\pm 0.42\pi$ [4] and modulation speed of 1 Gchip/s. At the receiver the signal is demodulated by the injection-locking current of the second DFB laser, which is used as a local oscillator, as shown in Figure 6.2. Since it is no longer feasible to track the desired user's initial phase offset by the injection-locking method, another tracking method is assumed, therefore $\theta_1 = \theta_{LO} = 0$ and since

$\pm 0.42\pi = \pm \pi/2 \mp 0.08\pi$ the received signal can be expressed as:

$$s(t) = \sqrt{2}\hat{S} \sum_{i=1}^{K} \cos(\omega_c t + C_i(t)(\pi/2 - 0.08\pi) + \theta_i) \qquad (6.12)$$

The local oscillator signal is then:

$$l(t) = \sqrt{2}\hat{L} \sum_{i=1}^{K} \cos(\omega_c t + S_1(t)(\pi/2 - 0.08\pi) + \theta_{LO}) \qquad (6.13)$$

The output of the dual-balanced detector under these circumstances is:

$$\Re l(t)s(t) = 2\Re\hat{L}\hat{S} \sum_{i=1}^{K} \cos\left[(S_1(t) - C_i(t))(\pi/2 - 0.08\pi) + \theta_{LO} - \theta_i\right]$$

$$= \Re\hat{L}\hat{S} \sin^2(0.08\pi) \sum_{i=1}^{K} \cos(\theta_{LO} - \theta_i)$$

$$+ \Re\hat{L}\hat{S} \cos^2(0.08\pi) \sum_{i=1}^{K} \cos((S_1(t) - C_i(t))(\pi/2) + \theta_i)$$

$$+ \Re\hat{L}\hat{S} \sin(0.16\pi) \sum_{i=1}^{K} \sin((S_1(t) - C_i(t))(\pi/2) + \theta_i) \qquad (6.14)$$

Due to the phase limitation, two new terms in the output of the dual-balanced detector appear. The first term in Equation (6.14) is the dc component which can be removed through the dual-balanced detector. The last term in Equation (6.14) involves the weighted-sum of the difference of two codes (essentially a new pseudo-code). The weighted-sum is bounded by the sum over the pseudo-codes, which in turn is negligible as compared with that of the second term because the code-length becomes arbitrarily larger than the number of active users (i.e. $N >> K$). Hence, the output can only be approximated by the second term and additive noise in Equation (6.15) [4, 8, 9]. By integrating the second term and the added noise term over a bit interval $T$, the output signal becomes:

$$S_{out} = \Re\hat{L}\hat{S} \cos^2(0.08\pi) \sum_{i=1}^{K} \int_0^T \cos((S_1(t) - C_i(t))\pi/2 + \theta_i)dt + \sqrt{N_0} \int_0^T n(t)dt$$

$$= \Re\hat{L}\hat{S} b_0^1 T \cos^2 0.08\pi + \sqrt{\Re T}\hat{L}$$

$$+ \Re\hat{L}\hat{S} \cos^2 0.08\pi . \sum_{i=2}^{K} \left[b_{-1}^i R_{i,1}(\tau_i) + b_0^i \hat{R}_{i,1}(\tau_i)\right] \cos\theta_i \qquad (6.15)$$

Using the same Gaussian approximation for the MAI, the SNR becomes:

$$SNR(K) = \frac{(\Re \cos^2 0.08\pi)^2 \hat{L}^2 \hat{S}^2 T^2}{(\Re \cos^2 0.08\pi)^2 \hat{L}^2 \hat{S}^2 T^2 \dfrac{K-1}{3N} + \Re T \hat{L}^2} = \frac{1}{\dfrac{K-1}{3N} + \dfrac{1}{\Re \hat{S}^2 T \cos^4 0.08\pi}}$$

$$(6.16)$$

Comparison of Equations (6.9) and (6.16) indicates that the only difference between them is the presence of a cosine term. As seen, the phase excursion incorporated by the injection-locking method causes a power loss of $\cos^4 0.08\pi$, equivalent to (approximately) 1.2 dB loss over the single-user SNR (i.e. $10 \log \vartheta$, where $\vartheta = 0.42\pi$ is the phase excursion [6]). The phase limitation has been employed to both encoder and decoder and accordingly the power loss of approximately 0.6 dB at each transmitter and receiver is expected for each user.

Figure 6.5 shows the BER against the single-user SNR (i.e. $E_b N = E_b/N_0$) when $r = 10\%$ and $P = 23$ (i.e. 53 users) for both modulation methods. The system with external phase modulation outperforms the other since there is no restriction on phase excursion as well as the chip-rates. Also, for any given BER the difference between the two graphs is 1.2 dB as expected.

The behaviour of these two methods is also illustrated in Figure 6.6 against the number of active users $K$ when $E_b N = 16$ dB and $P = 23$. The external modulation indicates

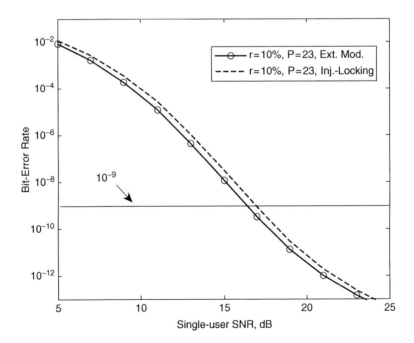

**Figure 6.5** BER comparisons of homodyne BPSK-OCDMA with different phase modulations against single-user SNR, $E_b N$

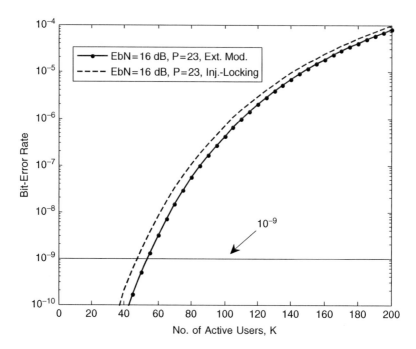

**Figure 6.6** BER comparisons of homodyne BPSK-OCDMA with different phase modulations against the number of active users $K$

remarkably enhanced performance, particularly for the lower number of simultaneous active users. As the figure shows, when $BER = 10^{-9}$ the system employing GPMPC and MZI can support 55 users, whereas the system with injection-locking phase modulation method can accommodate only 50 users, i.e. a 10 per cent capacity degradation.

## 6.3  Coherent Heterodyne BPSK-OCDMA Architecture

In this section, we study the coherent heterodyne BPSK-OCDMA architecture employing GPMPC and compare it with the commonly used Gold sequences. As a reference configuration, we consider a star passive optical network (SPON) with $Z$ transmitters and receivers employing the BPSK modulation as their basic structures as shown in Figure 6.7. Each incoming bit is encoded by means of a GPMPC sequence, acting as the address of the destination.

Let $x_i$ be the GPMPC sequence identifying the $i^{th}$ receiver, and each 1 and 0 value forming the GPMPC sequence is the so-called 'chip'. The following rule is applied to the BPSK scheme: either $x_i$ or $\overline{x}_i$ is transmitted, depending on whether a 1 or a 0 data bit is to be sent, where $\overline{x}_i$ is derived from $x_i$ by inverting each chip in the sequence.

The signals from all the transmitters are then summed up and broadcast to each receiver. The receivers perform a correlation between the received signal and their own prime code sequence (address); all the signals except the properly encoded one, will be decoded as

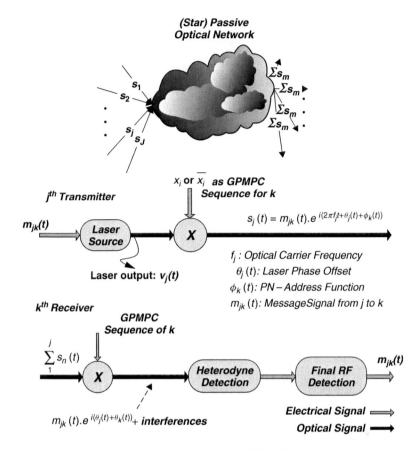

**Figure 6.7** Transceiver architecture for heterodyne BPSK-OCDMA from $j \rightarrow k$ (© 2007 IEEE. Reprinted with permission from Performance analysis of heterodyne detected coherent optical CDMA using a novel prime code family, M.M. Karbassian and H. Ghafouri-Shiraz, *J. Lightw. Technol.*, **25** (10), 2007.)

interfering noise, whereas the latter will give rise to a correlation peak. Hence, several simultaneous transmissions, addressed to different receivers are possible. Because the cross-correlation between GPMPC sequences in different groups is not *zero* (but it is as low as *one*), the interfering signals will reduce the noise margin of the receivers.

The spreading and de-spreading operations can be performed directly on the optical domain by means of a lithium-niobate crystal phase modulator [14], driven by the incoming data and the pseudo-noise sequence. After the de-spreading, the signal is heterodyne-detected and processed based on the decision rule (i.e. ML) according to the chosen modulation scheme. A block diagram of the system is shown in Figure 6.8.

Because of the spreading, the maximum achievable bit rate is limited by the speed of the electronic circuitry which generates the prime code sequences. Since we could consider MZI as an external phase modulator, posing the limit of high chip-rate (e.g. 100 Gchips/s) and intending to keep the bit rate sufficiently high will be feasible (i.e. hundreds of Mbps).

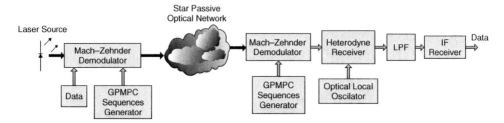

**Figure 6.8** Transceiver structure of heterodyne BPSK-OCDMA (© 2007 IEEE. Reprinted with permission from Performance analysis of heterodyne detected coherent optical CDMA using a novel prime code family, M.M. Karbassian and H. Ghafouri-Shiraz, *J. Lightw. Technol.*, **25** (10), 2007.)

We will then be limited to spreading sequences having lengths in the order of hundreds (less than a thousand). Obviously, a synchronous network (i.e. one in which all the transceivers are bit synchronized) shows very good results in terms of the number of allowed simultaneous users.

The spreading is taken to be the form of a pseudo-random rotation of the modulating signal's phase during each chip interval $T_c$ as depicted in Figure 6.9 according to the pseudo-random bit sequence (i.e. GPMPC) associated with the intended receiver (user). The receiver de-spreads the received signal using the exact pseudo-noise (PN) sequence by subtracting the same phase pattern used in the spreading process. After de-spreading, it is as if no spreading had been done. As mentioned, spreading and de-spreading can be done directly on the optical signal by using lithium-niobate crystal phase modulators. For proper de-spreading, the PN sequence used in the receiver must be synchronized with the one used in the transmitter as in the detail illustrated in Figure 6.7 for *user j → user k* transmission process. This synchronization is one of the tasks to be performed by the

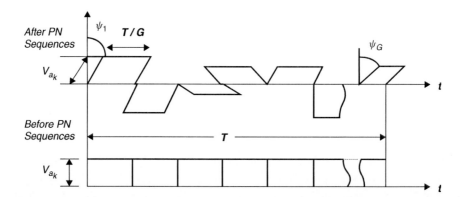

**Figure 6.9** PN-sequence applied to a unit energy pulse for direct phase modulation (© 2007 IEEE. Reprinted with permission from Performance analysis of heterodyne detected coherent optical CDMA using a novel prime code family, M.M. Karbassian and H. Ghafouri-Shiraz, *J. Lightw. Technol.*, **25** (10), 2007.)

call-start-up procedure, together with carrier-frequency acquisition. These are important functionsr, but we will not discuss them in this chapter, where the focus is on the key issue of the performance of an ongoing call under the (conservative) assumption that other users are actively engaged on a call.

### 6.3.1 Analysis of Phase Modulation with MZI

Here we take a more detailed look at the way in which a nominally transparent channel is provided between a generic pair of users $j \to k$ given a background of communications between other users. We use the complex representation for signals, so let $v_j$ denote the laser output at the $j^{\text{th}}$ transmitter destined for the $k^{\text{th}}$ receiver before the GPMPC sequence multiplier shown in Figure 6.7. $v_j(t)$ can be expressed as:

$$v_j(t) = u_j(t).e^{i\omega_j t} \tag{6.17}$$

where $u_j(t)$ is the modulating signal (i.e. laser driver) and $\omega_j$ is the optical angular frequency (i.e. $\omega_j = 2\pi f_j$).

Let $\{a_k(t)\}_{k=1}^{J}$ denote the set of addressing sequences of the $J$ receivers which is the same as the set of GPMPC sequences. Note that the chip duration $T_c$ divides the symbol duration $T$ equally. As the quotient $G (G = T/T_c)$ is called the *spread spectrum processing gain* and since the chip duration is denoted by $T_c$, we may represent each $a_k(t)$ mathematically as:

$$a_k(t) = \sum_{l=1}^{N} e^{i\phi_{lk}}.h(t - lT_c) \tag{6.18}$$

where

$$h(t) = \begin{cases} 1, & 0 \le t \le T_c \\ 0, & \text{otherwise} \end{cases} \tag{6.19}$$

with a Fourier transform of:

$$H(w) = 2T_c e^{-j\omega T_c/2} \frac{\sin(\omega T_c/2)}{\omega T_c/2} \tag{6.20}$$

Although the phase variables are known by the communicators, for analytical purposes they can be treated as random variables. The phase variables in the array $\{\phi_{lk}\}$ where $1 < l < N$ and $1 \le k \le J$ are assumed to have the properties of the independent and identically distributed random variables on the interval $[-\pi, \pi)$. It should be noted that $a_k(t) \cdot a_k^*(t) = 1$ for $\forall k : 1 \le k \le J$ and that $a_k(t) \cdot a_{k'}^*(t)(k \ne k')$ is statistically the same as $a_k(t)\forall k : 1 \le k \le J$. The accumulation of $J$ all signals, $\gamma$ is given by:

$$\gamma = \sum_{m=1}^{J} u_m(t) \cdot e^{i\omega_m t} \cdot a^{(m,k)}(t) \tag{6.21}$$

where $\{a^{(m,k)}\}_{m,k=1}^{J}$ means any permutation of the integer 1 through $J$, and $a^{(m,k)}$ conveys who is communicating to whom, i.e. the $m^{\text{th}}$ transmitter is communicating to the $k^{\text{th}}$

receiver; also $\gamma$ is assumed to be received at each receiver. At the $k^{th}$ receiver, we have upon acquisition of the transmitter angular frequency $\omega_j$:

$$\sum_{m=1}^{J} u_m(t).e^{i(\omega_m - \omega_j)t} \cdot a^{(m,k)}(t) \cdot a_k^*(t) = u_j(t) + \sum_{m=1}^{J} {}^j e^{i(\omega_m - \omega_j)t} \cdot u_m(t) \cdot b_m(t) \quad (6.22)$$

where $\sum^j$ means that the $j^{th}$ term is removed and the set of $b_m(t), \forall m : 1 \leq m \leq J$ is a statistically independent copy of the set of $a_m(t)$. The signal $u_j(t)$ has been received unaltered except for the additive noise of:

$$\sum_{m=1}^{J} {}^j e^{i(\omega_m - \omega_j)t} \cdot u_m(t) \cdot b_m(t) \quad (6.23)$$

Next, we look at quantifying the background noise level. Let the consecutive random phases of the desired pulse during an arbitrary symbol time be $\psi_n, n \in \{1, 2, \ldots, G\}$ and the corresponding phases of a generic interferer be $\phi_n, n \in \{1, 2, \ldots, G\}$. Let us assume that the interferer is displaced by a frequency $f_l$ ($\omega_l = 2\pi f_l$) from the desired transmission. At the end of a symbol period during which the $l^{th}$ transmitter has sent a 1, the unwanted contribution from the $l$ interferer, after the matched filter, is:

$$i_l = \frac{2}{T} \sum_{n=1}^{G} \int_{(n-1)T/G}^{nT/G} e^{i(\phi_n - \psi_n)}.e^{i\omega_l t} dt = \frac{2}{G} \sum_{n=1}^{G} e^{i\zeta_n} \frac{\sin(\omega_l T/2G)}{\omega_l T/2G} \quad (6.24)$$

where $\zeta_n = \phi_n - \psi_n$ is distributed in the same way as $\phi_n$ or $\psi_n$ and based on the central limit theorem [14], and for large values of $G$ the right-hand side of Equation (6.24) follows Gaussian distribution. So, if $g_l$ denotes a complex Gaussian variable of unit variance, we can say that the right-hand side of Equation (6.24) is approximately distributed as following variable $x_l$:

$$x_l = 2G^{-1/2} \left[ \sin(\omega_l T/2G)/(\omega_l T/2G) \right].g_l \quad (6.25)$$

The main guideline for the BPSK modulation is explained below. Let us assume that the $l^{th}$ signal is from the intended user. According to Figure 6.8, the signal is heterodyne detected, thus the receiver output after multiplication contains both the unwanted optical signal and the required intermediate frequency (IF) signal which is selected through the filter. The IF signal level is proportional to the phase shift of the incoming signal; the higher the difference in phase between the two states, the higher the difference between the two output voltage levels from the receiver will be. Decisions are made by the IF receiver (see Figure 6.8) on the basis of the IF signal level achieved from the following variable $Z(T)$ [13–16]:

$$Z(T) = \frac{\Re}{2N} \left[ N.d_i + \sum_{i=1, i \neq l}^{K} d_i.X_{li} \right] + n_B(T) \quad (6.26)$$

where $K$ is the number of simultaneous transmissions, $N$ is the length of the PN sequences (GPMPC), $d_i$ is the $i^{th}$ transmitter data bit, $n_B(t)$ is the sampled baseband Gaussian noise process and $X_{li}$ is a random variable representing the cross-correlation between the GPMPC sequences used by the $i^{th}$ and $l^{th}$ transmitters. If we define the new random variable, $W$, as:

$$W = \sum_{i=1, i \neq l}^{K} d_i . X_{li} \tag{6.27}$$

then its probability density function (PDF) can be obtained from the PDF of the random variable $X_{li}$. According to GPMPC correlation properties (see Section 2.4.5) in-phase cross-correlation values are $\lambda_c \leq 2$, in that the intended user interferes with other users in either the same or different groups, respectively. The cross-correlation values depending on whether the codes are in the same group or different groups will cause the interferences. The zero cross-correlation value means that the codes are orthogonal so there is no interference. The one or two cross-correlation values will cause interferences between the intended user and $P^2 - P$ users, referring to $P^2$ total number of sequences and $P - 1$ users sharing the same group with intended user, (one user itself) leading to $P^2 - P - 1 + 1 = P^2 - P$. Therefore cross-correlation values are uniformly distributed among interfering users, and thus the PDF of $W$ becomes:

$$P(W = i) = \frac{i}{P^2 - P} \tag{6.28}$$

where $P(W = i)$ is the probability that $W$ assumes the value $i$ (the number of actively involved users in the transmission). Based on Equation (6.28) following expression for the BER can be obtained as a function of a number of the simultaneous users, $K$ [14, 15]:

$$BER_K = \frac{1}{2} \sum_{i=0}^{W_m} erfc \left[ \frac{N - i}{N} \cdot \sqrt{r} \right] \cdot P(W = i) \tag{6.29}$$

where $W_m$, is the largest value assumed by the random variable which depends on the number of active users, $W$ denoting the interference and $r$ is the SNR, i.e.:

$$r = \frac{E_b}{N_0} = \frac{\eta P_r}{2hf \ B_{IF}} \tag{6.30}$$

where $\eta$ is the photodetector's quantum efficiency ($\eta = 0.9$), $P_r$ is the received signal power, $h$ is Planck's constant, $f$ is the employed optical frequency $\lambda = 1.55\,\mu m$ and $B_{IF}$ is the IF bandwidth.

In order to minimize the laser phase noise or chirp, the bandwidth of the intermediate frequency (IF) receiver in Figure 6.8 should be practically wider than a matched filter, i.e. a low-pass filter (LPF) in Figure 6.8 whose noise bandwidth is generally equal to the bit rate. In case of widening the matched filter bandwidth itself, to avoid the phase noise, the IF bandwidth should still be at least equal to the LPF bandwidth. However this inherently increases the noise levels since they are back-to-back filters with additional noise figures. Therefore in practice, the 10–25 per cent wider bandwidth with respect to the bit rate (i.e. hundreds of MHz or GHz) is usually considered for the IF receiver bandwidth, ignoring the phase noise impairment [12, 15, 17, 18].

Based on the above analysis, the performance (i.e. BER) of the BPSK-OCDMA transceivers is evaluated. Figure 6.10 illustrates the receiver BER variation against the number of simultaneous active users $K$ when $P = 23$ and 511-chip long Gold sequence. The system performs under different received signal power of $P_r = -30$ dBm and $P_r = -25$ dBm. Also in Figure 6.10, the desired BER threshold level (i.e. $10^{-9}$) is shown. In general, the system performs better (i.e. lower BER) when the received signal power $P_r$ (or SNR) increases. When $P_r = -30$ dBm and $P = 23$ (i.e. SNR = 13 dB), the maximum number of accommodated simultaneous users where the BER threshold can be achieved is $K_c = 240$ (i.e. $\approx$ 45 pre cent of 529 total users). This percentage increases to 59 per cent when $P_r = -25$ dBm (i.e. SNR = 16 dB and $K_c = 310$). The critical accommodated number of users ($K_c$) for the systems with 511-chip long Gold codes, when $P_r$ is $-30$ or $-25$ dBm are 220 ($\approx$ 43 per cent of 511 total users) and 300 ($\approx$ 57 per cent), respectively. This critical value depends on the system received power, $P_r$ (or SNR) that can be obtained by setting $BER = 10^{-9}$. For further analysis, let us consider the BER of a certain number of users, e.g. $K = 240$ and $P_r = -30$ dBm. The BER at the receiver with GPMPC is $1.4 \times 10^{-9}$, whereas that with Gold sequence is $3.2 \times 10^{-8}$ which implies BER performance of the receiver with GPMPC is significant improved.

Figure 6.11 shows the transceivers' BER variations against the received signal power $P_r$ under the different traffics when 30 per cent and 50 per cent of total number of users are present. As expected, the higher the $P_r$ (or SNR), the better the receivers perform. In this analysis, 30 per cent and 50 per cent of total number of users are assumed in the

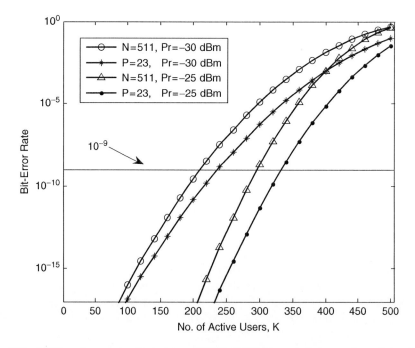

**Figure 6.10**   BER performance of heterodyne BPSK-OCDMA against the number of simultaneous active users $K$

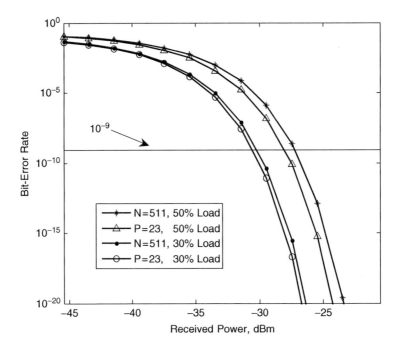

**Figure 6.11** BER performance of heterodyne BPSK-OCDMA against the received signal power $P_r$

communication when $P = 23$ and 511-chip long Gold sequence. As Figure 6.11 indicates, to maintain $BER = 10^{-9}$ the system $P_r$ should be $-30$ dBm and $-27$ dBm for 30 per cent and 50 per cent load, respectively. In other words, when the traffic is low, less SNR is required to maintain low BER. It is also apparent that GPMPC outperforms Gold sequence. For example, at the received power of $-27.45$ dBm, BER at the receiver with $P = 23$ is $7.8 \times 10^{-11}$ whereas that at the receiver with 511-chip long Gold sequence is $2.3 \times 10^{-9}$, which implies significant improvement.

## 6.4 Summary

We have investigated the performance of coherent homodyne and heterodyne OCDMA systems by employing a prime code family (GPMPC) and coherent dual-balanced detection. It should be noted that the analysis is still valid for similar code families with similar correlation properties. The homodyne scheme utilized either external phase modulation by the use of MZI or injection-locking of driving current of DFB laser diodes. The results indicate that the analyses are also valid for codes with the same correlation properties. Accordingly, employing unipolar codes outperforms the conventional bipolar codes regarding flexible code-lengths and accommodating more simultaneous active users. The limited phase excursion, generated by the injection-locking method, caused several complications. First, separate phase tracking is required as it can no longer be accomplished simultaneously with phase modulation; second, there is a dc bias level in the detector output requiring estimation and removal; finally, there is the degradation in BER equivalent to 1.2 dB signal loss. The overall performance of the transceivers shows that, by employing unipolar codes the system can be more power efficient and have enhanced network capacity as compared with commonly used bipolar Gold sequences. Above all, the coherent scheme can be a promising scheme for long-haul high-speed transmissions over the OCDMA networks.

# References

1. Wang, X. *et al*. (2005) 10-user, truly asynchronous OCDMA experiment with 511-chip SSFBG en/decoder and SC-based optical thresholder. In: *OFC*, Anaheim, CA, USA.
2. Wang, X. *et al*. (2005) Demonstration of 12-user, 10.71 Gbps truly asynchronous OCDMA using FEC and a pair of multiport optical-encoder/encoders. In: *ECOC*, Glasgow, UK.
3. Jiang, Z. *et al*. (2005) Four-User, 2.5-Gb/s, spectrally coded OCDMA system demonstration using low-power nonlinear processing. *J. Lightw. Technol.*, **23** (1), 143–158.
4. Ayadi, F. and Rusch, L.A. (1997) Coherent optical CDMA with limited phase excursion. *IEEE Comm. Letters*, **1** (1), 28–30.
5. Gnauck, A.H. (2003) 40-Gb/s RZ-differential phase shift keyed transmission. In: *OFC*, San Diego, USA.
6. Karbassian, M.M. and Ghafouri-Shiraz, H. (2008) Study of phase modulations with dual-balanced detection in coherent homodyne optical CDMA network. *J. Lightw. Technol.*, **26** (16), 2840–2847.
7. Wang, X. *et al*. (2006) Coherent OCDMA system using DPSK data format with balanced detection IEEE Photonics Tech. *Letters*, **18** (7), 826–828.
8. Liu, X. *et al*. (2004) Tolerance in-band coherent crosstalk of differential phase-shift-keyed signal with balanced detection and FEC. *IEEE Photonics Tech. Letters*, **16** (4), 1209–1211.
9. Karbassian, M.M. and Ghafouri-Shiraz, H. (2007) Phase-modulations analyses in coherent homodyne optical CDMA network using a novel prime code family. In: *WCE (ICEEE)*, London, UK.
10. Karbassian, M.M. and Ghafouri-Shiraz, H. (2008) Performance analysis of unipolar code in different phase modulations in coherent homodyne optical CDMA. *IAENG Engineering Letters*, **16** (1), 50–55.
11. Proakis, J.G. (1995) *Digital communications*. McGraw-Hill, New York, USA.
12. Wang, X. *et al*. (2006) Demonstration of DPSK-OCDMA with balanced detection to improve MAI and beat noise tolerance in OCDMA systems. In: *OFC*, Anaheim, CA, USA.
13. Karbassian, M.M. and Ghafouri-Shiraz, H. (2007) Performance analysis of heterodyne detected coherent optical CDMA using a novel prime code family. *J. Lightw. Technol.*, **25** (10), 3028–3034.
14. Benedetto, S. and Olmo, G. (1991) Performance evaluation of coherent code division multiple access. *Electronics Letters*, **27** (22), 2000–2002.
15. Huang, W., Andonovic, I. and Tur, M. (1998) Decision-directed PLL for coherent optical pulse CDMA system in the presence of multiuser interference, laser phase noise, and shot noise. *J. Lightw. Technol.*, **16** (10), 1786–1794.
16. Karbassian, M.M. and Ghafouri-Shiraz, H. (2008) Evaluation of coherent homodyne and heterodyne optical CDMA structures. *J. Optical and Quantum Electronics*, **40** (7), 513–524.
17. Foschini, G.J. and Vannucci, G. (1988) Noncoherent detection of coherent lightwave signals corrupted by phase noise. *IEEE Trans on Comm.*, **36** (3), 306–314.
18. Wang, X. and Kitayama, K. (2004) Analysis of beat noise in coherent and incoherent time-spreading OCDMA. *J. Lightw. Technol.*, **22** (10), 2226–2235.

# 7

# Hybrid Temporal Coherent and Incoherent OCDMA Networks

## 7.1 Introduction

In a conventional OCDMA, each time-slot is divided into chips consisting of (0,1)-sequences (i.e. the spreading code-length) addressed to each user. The data is modulated and assigned through optical pulses at certain chips of each allocated time-slot in either on-off keying (OOK) [1] or pulse position modulation (PPM) formats [2].

The modulated signal is then transmitted after being multiplied by the spreading code in the OCDMA encoder via optical tapped-delay lines (OTDL), i.e. the output optical pulse in the first chip of a time-slot is spread in the time domain to several chips corresponding to 1s of the spreading codes. The optical pulse sequences transmitted from users are multiplexed in the star passive optical network (SPON) couplers as an infrastructure reference and then transmitted over fibre to the destination (FTTx). At the receiver, in order to obtain the intended signal from the received signal, de-spreading is performed in a de-correlator which consists of an OTDL with inverse tap coefficients. The optical pulses are merged at the last chip in a slot, and the desired data is extracted in the demodulator based on the modulation scheme.

As mentioned, when the number of simultaneous active users increases, the effect of channel interference also inherently increases in incoherent OCDMA. On the other hand, multiple access interference (MAI) cancellation techniques at the receivers have been widely studied with different modulation schemes such as OOK and PPM. When OOK is used, the interference canceller is unable to completely eliminate the channel interference, since the reference signal has the components of the desired user [1]. In this chapter we analyse an interference cancellation technique in which the reference signal does not contain the component of the desired signal.

*Optical CDMA Networks: Principles, Analysis and Applications*, First Edition. Hooshang Ghafouri-Shiraz and M. Massoud Karbassian.
© 2012 John Wiley & Sons, Ltd. Published 2012 by John Wiley & Sons, Ltd.

Y. Gamachi *et al.* [3] have proposed the $M$-ary frequency shift keying (FSK)-OCDMA and indicated that in $M$-ary PPM the probability of overlapping pulses can be reduced by increasing the number of symbols ($M$). This is because each symbol is allocated to the corresponding 1 slot out of the $M$ existing slots. On the other hand, in WDMA transmission two wavelengths are allocated to each user, one for the upload link and the other one for the download link. Therefore, the more users allocated to the network, the more wavelengths must be assigned. In $M$-ary FSK-OCDMA, $\log_2 M$ encoded bits of data (symbol) are assigned to $M$ frequencies (wavelengths) for all users as a result of $M$-ary source coding. This results in a higher spectral efficiency, no wavelength-assignment and fewer wavelengths being used in the network [4].

Since the number of slots in a frame is independent of the number of symbols, the bit rate of FSK does not decrease as the number of symbols increases. When the number of slots in a frame is $\gamma$ (which corresponds to the repetition index of the tuneable laser), the bit rate of $M$-ary FSK becomes $M/\gamma$ ($\gamma < M$) times higher than that of $M$-ary PPM. Also, the probability of having interference for $M$-ary FSK is $1/M\gamma$, while the corresponding probability for $M$-ary PPM is $1/M$, so the probability of having interference can be mitigated by using FSK. Nevertheless, K. Iversen *et al.* [4] have studied an incoherent FSK-OCDMA and used OOC as spreading codes and found that the number of users was limited. Since systems without cancellation schemes have already been analysed and studied [4–7], we have focused on the transceivers with MAI cancellation as they have been less investigated, and this MAI cancellation technique is part of this book's research contribution. In the following, the $M$-ary FSK-OCDMA system with the proposed MAI cancellation technique taking advantage of group-padded modified prime codes (GPMPC) (see Section 2.4.5) is introduced and also the overall system performance in terms of bit error rate (BER) is analysed.

## 7.2   Coherent Transmitter with Incoherent Receiver

In this section, we examine and derive the BER of the FSK-OCDMA system with an interference canceller containing a reference signal which carries no data component of the intended signal.

$M$-ary FSK allocates $M$ symbols to the corresponding $M$ wavelengths, whereas $M$-ary PPM allocates the symbols to the slot positions. Figure 7.1 shows an $M$-ary FSK pulse series with $M = 4$, where four wavelengths $\lambda_0, \lambda_1, \lambda_2$ and $\lambda_3$, are assigned to contain two-bit information such as 00, 01, 10 and 11, respectively. $\gamma$ is the repetition index which corresponds to the number of slots between two subsequent transmitted optical pulses produced by a tuneable laser as shown in Figure 7.1. For the optical source, we consider a step-tuneable mode-locked laser diode with 100 GHz repetition rate [8]. In other words, we have multiple wavelength steps switched at the speed of 100 GHz which makes the $M$-ary FSK modulation possible. On the other hand in the CDMA encoder block shown in Figure 7.2, a passive OTDL has been used which makes the CDMA encoding at the incoming 100 GHz feasible [9].

We have shown in Section 7.4 that when the number of symbols $M$ is constant and $\gamma$ is small i.e. ($M > \gamma$), the throughput increases, resulting in an increase in the channel interference.

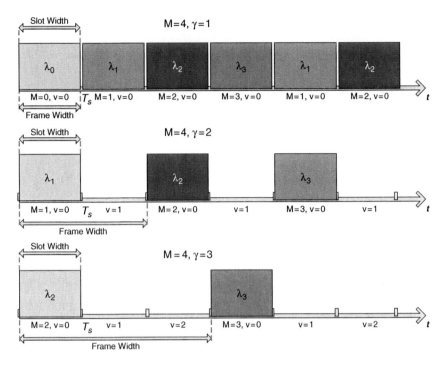

**Figure 7.1** $M$-ary FSK signalling format with $M = 4$ ($T_s$ is the slot time) (© 2007 IEEE. Reprinted with permission from Novel channel interference reduction in optical synchronous FSK-CDMA networks using a data-free reference, M.M. Karbassian and H. Ghafouri-Shiraz, *J. Lightw. Technol.*, **26**(8), 2007.)

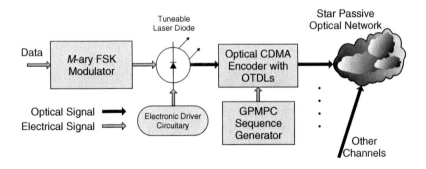

**Figure 7.2** Structure of FSK-OCDMA transmitter

On the other hand, as $\gamma$ increases, the bit rate reduces even though the channel interferences are reduced because the interrupted slots at each frame prevent interferences of in-phase correlation functions.

Figure 7.2 shows the transmitter block diagram of this FSK-OCDMA scheme. A frame consists of $\gamma$ slots, and each slot consists of $P^2 + 2P$ (code-length) chips. When $\gamma > 1$, one slot is selected for data transmission among $\gamma$ slots in a frame. The subsequent data is placed on the same slot positions in the following frames. As shown in Figure 7.1, the data is placed on the desired slot in a frame, and then it is inputted to the tuneable laser diode encoder as in Figure 7.2. The tuneable laser emits the optical pulse with a certain wavelength corresponding to the data at the particular chip positions in a slot coming from coherent FSK modulator. Note that optical pulses with other wavelengths are not emitted. The optical pulses are time-spread in the multiple chip positions corresponding to 1s of the spreading code by the encoder consisting of OTDLs in CDMA encoder. Then this desired user's signal is transmitted together with the signals of other users in the star passive optical network (SPON).

## 7.2.1   Interference Cancellation Technique

The FSK-OCDMA with an MAI canceller is similar to the PPM one with an interference canceller which was discussed in Section 5.3.2.2. However, the main difference between the two systems is that in FSK-OCDMA only one of the GPMPC sequences which contains the data-free component of the desired signal is used as a reference signal, and the cancellation is performed by subtracting the reference signal from the received signal of the desired user at each wavelength. The one reference signal is used to cancel the interference for all users. It is noted that the total number of subscribers increases to $P^2 - 1$ as compared with the previously mentioned PPM scheme in Chapter 5, which supports $P^2 - P$. This makes the system operate faster as it has to compare the received signal with only one reference rather than $P$ references and it makes receiver structure simpler as only one OTDL configuration in the reference branch is required instead of $P$ configurations.

Here, we allocate the first spreading code sequence in the first group to the intended user ($U_1$) and use the $P^{th}$ spreading code as a reference signal. The corresponding optical correlators are $OC_1$ and $OC_P$ for the main and reference branches, respectively. Figure 7.3 shows the receiver block diagram of the FSK-OCDMA system including the interference canceller. The transmitted signal with $M$ wavelengths on the SPON is separated incoherently into $M$ signals with different wavelengths by an arrayed waveguide grating (AWG) demultiplexer [10]. Here, we assume that the ideal AWG is used and that no interference between adjacent wavelengths occurs (i.e. no crosstalk). The received signals are separated by $M$ wavelengths and each is split into two paths passing through $OC_1$ and $OC_P$ as shown in Figure 7.3. Here, the data of $U_1$ with wavelength $\lambda_0$ is assumed to be in slot 0. The wavelengths in slot 0 only interfere with other users' data in the same wavelength with respect to their assigned spreading codes' cross-correlation value. Thus, other slots with $\lambda_0$ assigned to $U_1$ have the data component of $U_1$ and interference components. Now, since the main branch produces 'data, interference and noise' and the reference branch produces 'interference and noise', the interference and noise can be cancelled out by subtracting the reference signal from the components of $U_1$ at all slots. Ideally, only the signal component of $U_1$ with wavelength $\lambda_0$ is kept in slot 0 and there are no components in the other slots.

For the other wavelengths, there are no components in all slots, since $U_1$ has only components with $\lambda_0$. In practice, since the output of the photodetectors follows the Poisson process, and the photodetectors used in the branches have the same characteristics, they add similar amounts of thermal and shot noise to the signals; accordingly, it can be assumed that the noises are also cancelled out after subtraction. The outputs

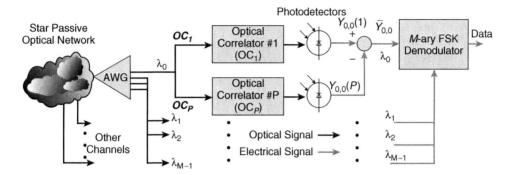

**Figure 7.3**  Structure of FSK-OCDMA receiver with MAI canceller (© 2007 IEEE. Reprinted with permission from Novel channel interference reduction in optical synchronous FSK-CDMA networks using a data-free reference, M.M. Karbassian and H. Ghafouri-Shiraz, *J. Lightw. Technol.*, **26**(8), 2007.)

of $OC_1$ and $OC_P$ for wavelength $\lambda_m$ and slot $v$ are converted from optical signal to electrical signal through the photodetectors, and the outputs of $OC_1$ and $OC_p$ are denoted by $Y_{m,v}(1), Y_{m,v}(P), m \in \{0, \ldots, M-1\}, v \in \{0, \ldots, \gamma-1\}$, respectively in Figure 7.3. Since we have focused on slot 0 and assigned $\lambda_0$ to $U_1$, hence $m = 0$ and $v = 0$. After the interference cancellation per wavelength, the signal with the highest power is selected by the maximum likelihood (ML) detector and then the corresponding data is obtained from the $M$-ary FSK detector unit.

## 7.3   Analysis of Transceivers with MAI Cancellation

In this section we derive an expression for the BER for this FSK-OCDMA with MAI canceller using GPMPC as the spreading codes. The analysis is based on unipolar spreading codes and it is assumed that the I/O characteristic of the photodetector follows the Poisson process, i.e. the number of photons has Poisson distribution. The signal power of the desired user with $\lambda_m$ in slot $v$ and the reference signal, $Y_{m,v}(i), i \in \{1, P\}$ after photodetector (see Figure 7.3) is given as:

$$Y_{m,v}(i) = Z_{m,v}(i) + I_{m,v}(i) \tag{7.1}$$

$$m \in \{0, 1, \ldots, M-1\}, v \in \{0, 1, \ldots, \gamma-1\}$$

where $Z_{m,v}(i)$ is the user (data) signal power and $I_{m,v}(i)$ is the interference power. Since the reference signal has only the reference $(P^{th})$ sequence multiplied by the received signal, the interference components of $Z_{m,v}(P)$ becomes 0 (i.e. data-free). Also, since all users in the same group receive an equal amount of interference from the users of other groups and no interference from the users from the same group, $I_{m,v}(1)$ equals $I_{m,v}(P)$. Here we assume that $U_1$ transmits the optical pulse of $\lambda_0$ $(m = 0)$ at the first slot $(v = 0)$ in a signal frame as shown in Figure 7.1.

Since the GPMPC sequences are employed as signature codes, the cross-correlation between the first and $x^{th}$ user, $C_{1,x}$, of the same group is equal to *zero*. Thus, $R$ users

$(0 \le R \le P - 2)$ in the first group do not affect the photon-count of the first user. The probability density function (PDF) of the random variable $R$ is given by [3]:

$$\Pr\{R = r\} = \frac{\binom{P^2 - 2P + 1}{K - r} \cdot \binom{P - 2}{r - 1}}{\binom{P^2 - P - 1}{K - 1}} \qquad (7.2)$$

where $r \in \{r_{\min}, \ldots, r_{\max}\}$, $r_{\max} = \min(K, P - 1)$ and $r_{\min} = \max(1, K - (P - 1)^2)$.

Here $K$ refers to the number of simultaneous active users. Let us define the random variable $\kappa_{m,v}$ to be the number of users who are in groups other than the first group, and have a pulse in the $v^{th}$ slot with wavelength $\lambda_m$. The probability that $l_{m,v}$ users have a pulse in the $v^{th}$ slot with wavelength $\lambda_m$ is $1/(M\gamma)$ and the probability of having no pulse is $1 - 1/(M\gamma)$ (recalling binomial distribution). Then, the PDF of $\kappa_{m,v}$ can be expressed as:

$$\Pr\{\kappa_{m,v} = l_{m,v} | R = r\} = \binom{K - r}{l_{m,v}} \cdot \left(\frac{1}{\gamma.M}\right)^{l_{m,v}} \cdot \left(1 - \frac{1}{\gamma.M}\right)^{K - r - l_{m,v}} \qquad (7.3)$$

When $U_1$ has an optical pulse with wavelength $\lambda_m$ in the $v^{th}$ slot $\left(b^1_{m,v} = \{m, v\}\right)$ the expected value of the random variable $Z_{m,v}(i), i \in \{1, P\}$ is expressed as:

$$\overline{Z}_{m,v}(i) = \begin{cases} \frac{(P+2) \cdot Q \cdot T_c}{2} & \text{if } b^1_{m,v} = \{m, v\} \\ 0 & \text{otherwise} \end{cases} \qquad (7.4)$$

where $T_c$ is the chip duration. Taking the fibre attenuation coefficient $\alpha$ into account, the average number of received photon-counts per pulse $Q$ can be expressed as [11]:

$$Q = \frac{\xi P_W}{hf} \cdot \frac{e^{-\alpha L}}{P + 2} \approx \mu \cdot \frac{\ln M}{P + 2} \qquad (7.5)$$

where $P_r = \xi \cdot P_W \cdot e^{-\alpha L}$ is the received signal power, $P_W$ is the transmitted peak power per symbol, $\xi$ is the quantum efficiency of the photodetectors, $h$ is Planck's constant, $f$ is the optical frequency, $L$ is the fibre-length and $\mu$ $(\mu = P_r/(h \cdot f \cdot \ln M))$ is the average number of photons per pulse (photons/nat) [2, 12]. The experimental values for these parameters are listed in Table 7.1. The expected value of the random variable $I_{m,v}(i), i \in \{1, P\}$ can be expressed as:

$$\overline{I}_{m,v}(i) = l_{m,v} \cdot Q \cdot T_c \qquad (7.6)$$

From 7.4 and 7.6, the expected value of $Y_{m,v}(i), i \in \{1, P\}$ is expressed as:

$$\overline{Y}_{m,v}(i) = \begin{cases} \frac{(P+2+l_{m,v}) \cdot Q \cdot T_c}{2} & b^1_{m,v} = \{m, v\} \\ \frac{l_{m,v} \cdot Q \cdot T_c}{2} & \text{otherwise} \end{cases} \qquad (7.7)$$

In order to cancel the interference, the reference signal is subtracted from the $U_1$ signal and also as mentioned, since both photodetectors have the same characteristics, the signals

**Table 7.1** Link parameters

| Descriptions | Symbols | Values |
|---|---|---|
| Optical wavelength | $\lambda_0$ | 1550 nm |
| PD quantum efficiency | $\xi$ | 0.8 |
| Linear fiber-loss coefficient | $\alpha$ | 0.2 dB/km |
| Chip-rate | $1/T_c$ | 100 Gchips/s |
| Fibre length | $L$ | 10 km |

suffer from the same shot noise which cancels out after subtraction. The subtracted signal $\tilde{Y}_{m,v}$ is then:

$$\tilde{Y}_{m,v} = \overline{Y}_{m,v}(1) - \overline{Y}_{m,v}(P) \tag{7.8}$$

Since $U_1$ has an optical pulse with wavelength $\lambda_0$ at slot 0, the symbol-error occurs when $\tilde{Y}_{j,0}(1) \geq \tilde{Y}_{0,0}(1), (j \neq 0)$. Suppose $P_E$ denotes the symbol-error rate, and then the BER ($P_b$) is expressed as [3, 13]:

$$P_b \leq \frac{M}{2(M-1)} \cdot \sum_{r=r_{\min}}^{r_{\max}} P_E \cdot \Pr\{R = r\} \tag{7.9}$$

where:

$$P_E = \sum_{j=0}^{M-1} \Pr\left\{\tilde{Y}_{j,0} \geq \tilde{Y}_{0,0} \middle| R = r, \kappa_{0,0} = l_{0,0}, \ldots, \kappa_{M-1,0} = l_{M-1,0}, b_{m,v}^1 = \{0,0\}\right\}$$

$$\times \Pr\left\{\kappa_{0,0} = l_{0,0}, \kappa_{1,0} = l_{1,0}, b_{m,v}^1 = \{0,0\} \middle| R = r\right\} \tag{7.10}$$

$$\leq (M-1) \sum_{l_{0,0}=0}^{K-r} \sum_{l_{1,0}=0}^{K-r-l_{0,0}} \Pr\left\{\tilde{Y}_{1,0} \geq \tilde{Y}_{0,0} \middle| R = r, \kappa_{0,0} = l_{0,0}, \kappa_{1,0} = l_{1,0}, b_{m,v}^1 = \{0,0\}\right\}$$

$$\times \Pr\left\{\kappa_{0,0} = l_{0,0}, \kappa_{1,0} = l_{1,0}, b_{m,v}^1 = \{0,0\} \middle| R = r\right\}$$

$$\leq \sum_{l_{0,0}=0}^{K-r} \sum_{l_{l,0}=0}^{K-r-l_{0,0}} \Pr\left\{\kappa_{0,0} = l_{0,0}, \kappa_{1,0} = l_{1,0} \middle| R = r\right\} \times \Phi(r, l_{m,0})$$

and:

$$\Phi(r, l_{m,0}) = (M-1) \cdot \Pr\left\{\tilde{Y}_{1,0} \geq \tilde{Y}_{0,0} \middle| R = r, \kappa_{0,0} = l_{0,0}, \kappa_{1,0}, = l_{1,0}, b_{m,v}^1 = \{0,0\}\right\} \tag{7.11}$$

The upper-bounded $\Phi(r, l_{m,0})$ is then given by [3]:

$$\Phi(r, l_{m,0}) \leq (M-1) \cdot E\left[z^{\overline{Y}_{1,0}(1)-\overline{Y}_{1,0}(P)\geq\overline{Y}_{0,0}(1)-\overline{Y}_{0,0}(P)} \middle| R = r,\right.$$

$$\left. \kappa_{0,0} = l_{0,0}, \kappa_{1,0} = l_{1,0}, b_{m,v}^1 = \{0,0\}\right] \tag{7.12}$$

where $z$ ($z > 1$ and integer) denotes the number of optical interference pulses in slot 0 and $E[\bullet]$ refers to the expected value, and by using the Chernoff bound $\Phi(r, l_{m,0})$ can be expressed as [3]:

$$\ln \Phi(r, l_{m,0}) \leq -\ln(M-1) - \overline{Y}_{1,0}(1)(1-z) - \overline{Y}_{1,0}(P)(1-z^{-1}) - \overline{Y}_{0,0}(1)(1-z^{-1})$$

$$-\overline{Y}_{0,0}(P)(1-z) + \overline{Z}_{0,0}(1)(1-z) \tag{7.13}$$

Now, by setting $z - 1 = \rho$ ($\rho > 0$ and integer) we have then $1 - z^{-1} \leq 1$ while $\rho - \rho^2 \leq 0$, thus by considering new boundaries, we obtain this lower-bounded equation [3]:

$$1 - z^{-1} \geq \rho - \rho^2 \tag{7.14}$$

By substituting Equations (7.4), (7.7) and (7.14) into Equation (7.13), we have:

$$\ln \Phi(r, l_{m,0}) \leq -\ln(M-1) - \left(\frac{(P+2+l_{1,0})QT_c}{2}\right)(-\rho) - \left(\frac{l_{0,0}QT_c}{2}\right)(\rho - \rho^2)$$

$$-\left(\frac{(P+2+l_{1,0})QT_c}{2}\right)(\rho - \rho^2) - \left(\frac{l_{0,0}QT_c}{2}\right)(-\rho) + \left(\frac{(P+2)QT_c}{2}\right)(-\rho) \tag{7.15}$$

Then,

$$\ln \Phi(r, l_{m,0}) \leq -\ln(M-1) + \left(\frac{(P+2+l_{1,0}+l_{0,0})QT_c}{2}\right)\rho^2 - \left(\frac{(P+2)QT_c}{2}\right)\rho \tag{7.16}$$

Now by minimizing the right-hand side of the new equation with respect to $\rho$, we will have:

$$\rho = \frac{P+2}{P+2+l_{0,0}+l_{1,0}} \tag{7.17}$$

By substituting $\rho$ back into Equation (7.16), we have the upper-bounded $\Phi(r, l_{m,0})$ as:

$$\Phi(r, l_{m,0}) \leq (M-1)\exp\left\{-\frac{Q \cdot (P+2)^2}{4 \cdot (P+2+l_{0,0}+l_{1,0})}\right\} \tag{7.18}$$

The upper-bound of symbol-error probability $P_E$ in Equation (7.10) is then rewritten as:

$$P_E \leq (M-1) \cdot \sum_{l_{0,0}=0}^{K-r} \sum_{l_{1,0}=0}^{K-r-l_{0,0}} \Pr\{K_{0,0} = l_{0,0}, \kappa_{1,0} = l_{1,0}, b_{m,v}^1 = \{0,0\}|R = r\}$$

$$\cdot \exp\left\{-\frac{Q \cdot (P+2)^2}{4 \cdot (P+2+l_{0,0}+l_{1,0})}\right\} \tag{7.19}$$

By substituting Equation (7.3) into Equation (7.19), and then substituting Equation (7.3) and new Equation (7.19) into Equation (7.9), the total upper-bounded probability of error $(P_b)$, depending on the number of active users $(K)$, can be derived as:

$$P_b \leq \frac{M}{2} \sum_{r=r_{\min}}^{r_{\max}} \sum_{l_{0,0}=0}^{K-r} \sum_{l_{1,0}=0}^{K-r-l_{0,0}} \binom{K-r-l_{0,0}}{l_{1,0}} \cdot \left(\frac{1}{\gamma.M}\right)^{l_{1,0}} \cdot \left(1 - \frac{1}{\gamma.M}\right)^{K-r-l_{0,0}-l_{1,0}}$$

$$\cdot \exp\left\{-\frac{\rho}{2} \cdot \frac{Q \cdot (P+2)}{2}\right\} \times \binom{K-r}{l_{0,0}} \cdot \left(\frac{1}{\gamma.M}\right)^{l_{0,0}} \cdot \left(1 - \frac{1}{\gamma.M}\right)^{K-r-l_{0,0}}$$

$$\times \frac{\binom{P^2-2P+1}{K-r} \cdot \binom{P-2}{r-1}}{\binom{P^2-P-1}{K-1}} \tag{7.20}$$

## 7.4 Results and Throughput Analysis

In this section, based on the theory presented earlier the BER of the FSK-OCDMA receiver with interference canceller is discussed. The link parameters used in simulations are listed in Table 7.1 and the numerical results are compared with those of PPM-OCDMA that uses GPMPC as the spreading codes (see Chapter 5).

Figure 7.4 shows the BER of FSK and PPM systems with the interference canceller against the average number of photons per pulse $\mu$. In this analysis we have assumed that all interfering users are present, i.e. full-load and $P = 13, M = 8$ and $\gamma = 1, 2, 3$ (i.e. in the graphs $\gamma$ is denoted by $j$). It is observable that the FSK-OCDMA is more power efficient than the PPM scheme, since the BER reaches to the reference level of $10^{-9}$ for a lower average number of photons per pulse, $\mu$. On the other hand, repetition index $\gamma$ can also enhance the FSK-OCDMA performance; for example when $\mu = 100$, the BER for FSK are $1.6 \times 10^{-7}, 2.1 \times 10^{-9}$ and $4.7 \times 10^{-11}$ for $\gamma = 1$, 2 and 3, respectively, whereas the error rate for PPM scheme is $2.2 \times 10^{-6}$.

Figure 7.5 also illustrates the bit-error probabilities against the number of simultaneous users $K$, for PPM and FSK receivers. It is clear that the FSK scheme outperforms the PPM one especially for higher repetition index $\gamma$. The results indicate that this hybrid FSK scheme (i.e. coherent modulation and incoherent demodulation) can mitigate the effect of multiuser interference much better. As mentioned, in the PPM system, the optical pulse of the desired signal can overlap with the pulse of any other user, having a probability of $1/M$ whereas, in the FSK, the corresponding probability is reduced by a factor of $\gamma$ (i.e. $1/(\gamma M)$).

When the repetition index $\gamma$ becomes large, the probability of pulse overlap decreases and the performance is improved, as indicated in Figures 7.4 and 7.5. The reason is that in PPM the probability of having interference is reduced by increasing the number of slots (corresponding to the number of symbols) in the time domain. On the contrary, in FSK the probability of having interference is reduced by both the number of wavelengths (corresponding to the number of symbols) and the repetition index. When the chip-rate is constant, in order to decrease the probability of having interference in the PPM scheme, the number of slots has to increase, which results in a larger frame-length and a lower bit rate.

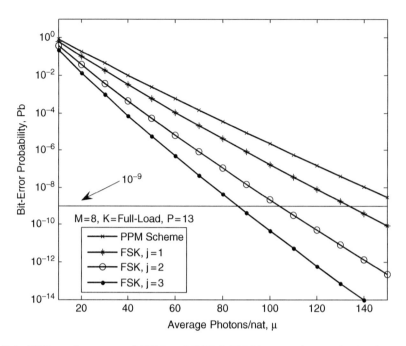

**Figure 7.4** BER performances of PPM and FSK-OCDMA transceivers with MAI cancellation against the average no. of photons per pulse $\mu$

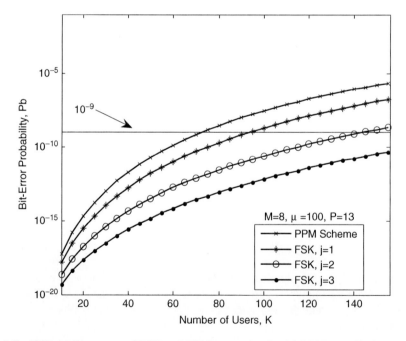

**Figure 7.5** BER performances of FSK and PPM transceivers with MAI cancellation against the no. of simultaneous users $K$

Alternatively, in the FSK scheme, the frame-length depends only on the repetition index, which is smaller than $M$, as shown in Figure 7.1. It is also shown in Figure 7.5 that the system capacity can be increased with the increase in repetition index.

By increasing the number of wavelengths $M$, the probability of having interference can therefore be further reduced in the FSK system with shorter frame-length than PPM; accordingly, FSK can attain higher bit rate transmission than PPM. The bit rates $R_b$ for FSK and PPM systems using GPMPC as spreading codes are expressed as [3, 12]:

$$R_b = \frac{\log_2^M}{\gamma \cdot T_c \cdot (P^2 + 2P)}, \quad M\text{-ary FSK-OCDMA} \qquad (7.21)$$

$$R_b = \frac{\log_2^M}{M \cdot T_c \cdot (P^2 + 2P)}, \quad M\text{-ary PPM-OCDMA} \qquad (7.22)$$

Additionally, in FSK the bit rate decreases as the repetition index increases. For example, the bit rates of FSK with $P = 7$, $M = 8$ and $\gamma = 2$ and 3 are 2.381 Gbps and 1.587 Gbps, respectively; whereas, the bit rate of PPM is 0.595 Gbps which is nearly four times slower. In other words, $R_{b-FSK}/R_{b-PPM} = M/\gamma$, where always $M > \gamma$.

Figures 7.6 and 7.7 illustrate the BER variations against $\mu$, for both FSK and PPM systems as functions of multiplicity $M$ and prime number $P$, respectively. All users are assumed to be active in the communications. The results reveal that in the FSK system lower BER as well as higher bit rate can be achieved as the number of symbols ($M$) increases, whereas in the PPM system the BER improves but the bit rate decreases. As the number of symbols increases, the probability of having interference decreases, which leads to the improvement of the performance.

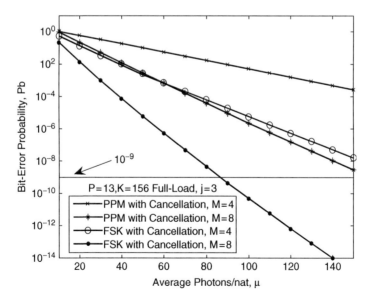

**Figure 7.6** BER performances of FSK and PPM transceivers with MAI cancellation against the average no. of photons per pulse $\mu$, with different multiplicities $M$

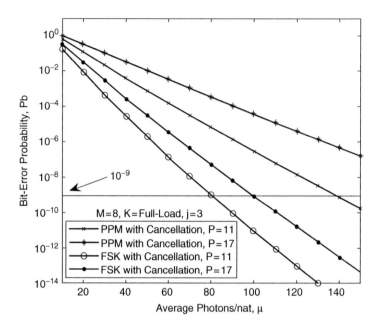

**Figure 7.7** BER performances of FSK and PPM transceivers with MAI cancellation against the average no. of photons per pulse $\mu$, with different prime numbers $P$

As for the bit rate, the number of FSK symbols is independent of the frame-length, while in the PPM the frame-length becomes larger as the number of symbols increases. As can be seen from Figure 7.7, when $P$ becomes larger, the performance is degraded due to the increase in the number of interfering users, and the bit rate decreases due to the increase in the code-length which corresponds to the increase of frame-length. However, the number of simultaneous active users can be increased in both FSK and PPM systems by utilizing a higher prime number $P$.

Figure 7.8 demonstrates the BER of different receivers with MAI canceller against the number of simultaneous active users $K$, with different multiplicities $M$. It can be seen from Figure 7.8 that the network capacity can be increased by raising the multiplicity in both PPM and FSK schemes.

It is clear that the FSK again outperforms; for example, assuming that 55 users (50 per cent of total number of users when $P = 11$) are present in the network, the error rates of the FSK receiver when $M = 4$ and 8 are $1.2 \times 10^{-7}$ and $4.2 \times 10^{-13}$, respectively, whereas the error rates for PPM receiver with $M = 4$ and 8 are $2 \times 10^{-5}$ and $2.7 \times 10^{-10}$, respectively. Figure 7.9 also displays the BER of different receivers with MAI canceller against the number of simultaneous active users $K$, with different average number of photons per pulse $\mu$. As can be seen, the higher received power makes the detection process more reliable due to elevating sufficient SNR that results in a low error rate and enhanced network capacity. In short, when $\mu = 100$ the PPM-OCDMA network can support 65 users at $BER = 10^{-9}$ whereas, under the same condition an FSK-OCDMA network can accommodate the total number of users (i.e. 110 users when $P = 11$).

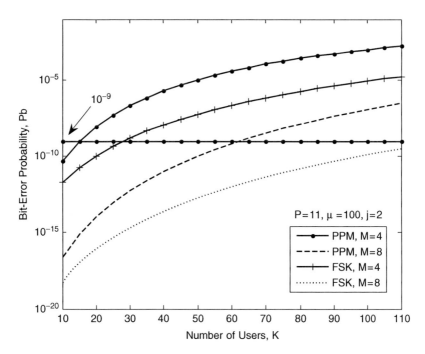

**Figure 7.8**  BER performances of FSK and PPM transceivers with MAI cancellation against the no. of simultaneous users $K$, with different multiplicities $M$

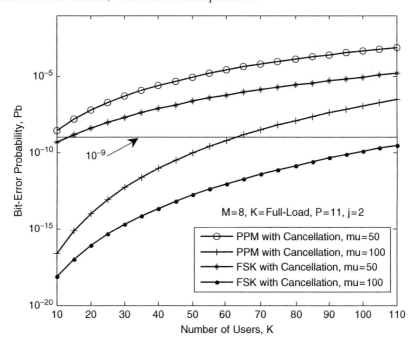

**Figure 7.9**  BER performances of FSK and PPM transceivers with MAI cancellation against the no. of simultaneous users $K$, with different average no. of photons per pulse $\mu$

## 7.5  Summary

A new interference cancellation technique taking advantage of code correlation properties, which simplifies the receiver structure in the synchronous $M$-ary FSK-OCDMA network, has been analysed in detail. Coherent FSK modulation along with incoherent demodulation using arrayed waveguide grating has been examined in the transceivers' structures. For the spreading codes, GPMPC has been considered. A reference signal is constructed by using one of the GPMPC sequences, and cancellation is performed by subtracting the reference signal from the received signal of the desired user. As a result, the FSK-OCDMA system with cancellation technique has a better performance (lower BER, higher bit rate) than the existing PPM-OCDMA with cancellation technique. Additionally, the results indicated that the FSK scheme is very power efficient and, when the bit rate is constant, its network capacity can be expanded to accommodate a large number of simultaneous active users with low error rate. It should be noted that the FSK-OCDMA architecture is more complex than the incoherent PPM-OCDMA scheme since driving the tuneable laser diode is challenging.

## References

1. Shalaby, H.M.H. (2002) Complexities, error probabilities and capacities of optical OOK-CDMA communication systems. *IEEE Trans Comm.*, **50** (12), 2009–2017.
2. Karbassian, M.M. and Ghafouri-Shiraz, H. (2007) Fresh prime codes evaluation for synchronous PPM and OPPM signaling for optical CDMA networks. *J. Lightw. Technol.*, **25** (6), 1422–1430.
3. Gamachi, Y. *et al.* (2000) An optical synchronous $M$-ary FSK/CDMA system using interference canceller. *J. Electro. Comm. in Japan*, **83** (9), 20–32.
4. Iversen, K. *et al.* (1996) $M$-ary FSK signalling for incoherent all-optical CDMA networks. In: *IEEE GlobeCom*, London, UK.
5. Iversen, K., Kuhwald, T. and Jugl, E. (1997) Ulm-Germany. D2-ary signalling for incoherent all-optical CDMA systems. In: *IEEE ISIT Conf*, Ulm, Germany.
6. Shalaby, H.M.H. (1995) Performance analysis of optical synchronous CDMA communication systems with PPM signaling. *IEEE Trans. Comm.*, **43** (2/3/4), 624–634.
7. Karbassian, M.M. and Ghafouri-Shiraz, H. (2008) Novel channel interference reduction in optical synchronous FSK-CDMA networks using a data-free reference. *J. Lightw. Technol.*, **26** (8), 977–985.
8. Lemieux, J.F. *et al.* (1999) Step-tunable (100GHz) hybrid laser based on Vernier effect between Fabry-Perot cavity and sampled fibre Bragg grating. *Electronics Letters*, **35** (11), 904–906.
9. Schröder, J. *et al.* (2006) Passively mode-locked Raman fiber laser with 100 GHz repetition rate. *Optics Letters*, **31** (23), 3489–3491.
10. Yang, C.C. (2006) Optical CDMA-based passive optical network using arrayed-waveguide-grating. In: *IEEE ICC, Circuits and Systems*, Istanbul, Turkey.
11. Shalaby, H.M.H. (1998) Chip-level detection in optical code division multiple access. *J. Lightw. Technol.*, **16** (6), 1077–1087.
12. Shalaby, H.M.H. (1998) Cochannel interference reduction in optical PPM-CDMA systems. *IEEE Trans. Comm.*, **46** (6), 799–805.
13. Karbassian, M.M. and Ghafouri-Shiraz, H. (2008) Frequency-shift keying optical code-division multiple-access system with novel interference cancellation. *J. Microw. Opt. Technol. Letters*, **50** (4), 883–885.

# 8

# Optical CDMA with Polarization Modulations

## 8.1 Introduction

Polarization shift keying (PolSK) is the only modulation scheme that makes use of the vector nature of the lightwave. Like frequency shift keying (FSK), it is a multiple-array ($M$-ary) signalling transmission scheme, and unlike FSK the spectrum of a PolSK signal corresponds to an equivalent amplitude shift keying (ASK) signal at the same bit rate [1].

When a polarized lightwave is transmitted through a single-mode fibre-optic (SMF), its state of polarization (SOP) changes due to the presence of waveguide birefringence. However the fibre birefringence only causes a rigid rotation of the lightwave's polarization constellation over the Poincaré sphere [2]. In other words, each signal point is displaced although the Stokes parameters illustrating the spatial relations among signal points in the Poincaré sphere are preserved. Thus, the information is not degraded. To compensate for the constellation rotation, some form of processing is required either optically by a polarization controlling and tracking or electronically by a digital signal processing (DSP) at the decision processor [3]. To perform PolSK detection avoiding optical birefringence compensation, it is necessary to use a receiver that extracts the Stokes parameters of the incoming lightwaves [4]. The receiver which extracts the Stokes parameters can generally be divided into two sections: (i) the optical front-end that probes the input electrical field and produces electric current at the output of its photodetector and (ii) the following electrical front-end that generates the Stokes parameters which are proportional to the produced photocurrent. In [5, 6], the polarized spreading codes have been considered to increase the number of accommodated users, since the polarization has been utilized to double the communication channel. However, here we employ the signal constellations in the polarization modulation as the information bearing to encode the data taking advantage of vector characteristic of lightwaves.

In this chapter, we introduce a new incoherent PolSK-OCDMA receiver which employs the optical correlators, i.e. optical tapped-delay lines (OTDL) to simplify the architecture

*Optical CDMA Networks: Principles, Analysis and Applications*, First Edition. Hooshang Ghafouri-Shiraz and M. Massoud Karbassian.
© 2012 John Wiley & Sons, Ltd. Published 2012 by John Wiley & Sons, Ltd.

with more enhanced performance than those mentioned in [6–8] for direct-detection (DD) PolSK. Thus, this PolSK-OCDMA transceiver does not require interferometer stability and the complicated DSP, although the polarization controller is needed here and it is very suitable for binary transmission as will be discussed. A comprehensive set of results in [9, 10] has shown that the performance of the binary system is approximately 3 dB better than intensity modulation/DD (on the peak optical power) of other PolSK or equivalent phase shift keying (PSK) modulations. Therefore, we have also considered binary PolSK for this transceiver architecture.

Originally, PolSK was theoretically analysed [10, 11] mostly in conjunction with coherent detection (CD). The early work on CD-PolSK has revealed that (i) the fibre birefringence does not corrupt polarization-encoded information and in particular the bit-error probability is relatively unaffected [10], (ii) birefringence compensation at the receiver is necessary, but it can be performed at the decision stage after photodetection [11], (iii) binary PolSK has a 40 photons/bit quantum-limited sensitivity [6], whereas coherent ASK requires 80 photons/bit (peak) [12], (iv) PolSK systems are largely insensitive to phase noise [3, 5, 13, 14, 15] and (v) for multilevel ($M$-ary) PolSK, quantum-limited (i.e. shot-noise limited) performance is even better than differential-PSK when 3 bits/symbol or more are transmitted [9].

DD-OCDMA technique inherently suffers from MAI, which requires estimation and removal through cancellation techniques or higher order modulations. In the past few years, advances in photonics technology (e.g. ultra-high-speed lasers, PDs and erbium-doped fibre amplifiers) have made it possible for DD systems to approach the sensitivity performance of CD systems. Hence, there has been a considerable shift of interest from CD scheme to constant-power-envelope DD schemes on the part of both research and industry, due to its (i) simple architecture, (ii) cost effectiveness, (iii) high immunity to laser phase-noise and (iv) insensitivity to self- and cross-phase modulations caused by the fibre nonlinearity [16, 17, 18].

On the other hand, FSK modulation has high phase noise insensitivity and high power efficiency. Therefore, we need to develop an advanced modulation scheme to take advantage of both modulations: FSK and PolSK. The advantages of FSK-OCDMA system has already been discussed in Chapter 7. Also the merits of hybrid frequency-polarization shift keying (F-PolSK) modulation has been introduced in [12, 19, 20] for the optical transmission.

Also in this chapter, we have proposed and analysed a novel design structure for the OCDMA transceiver, based on the two-dimensional array ($2D$-ary) hybrid F-PolSK modulation, which is based on the combination of both frequency and polarization modulations. This technique enhances distances between modulated signal points on the Poincaré sphere and hence reduces the required transmission power [10]. However, this advantage causes an increase in the transmission bandwidth requirement as a result of using multiple FSK tones. In the analysis we have assumed that the signal is degraded by (i) fibre amplified spontaneous emission (ASE) noise, (ii) electronic receiver noise (i.e. thermal noise), (iii) photodetector shot-noise and mainly (iv) multiple access interference (MAI). In addition, the intensity-modulated systems are vulnerable in terms of interception, which could easily be broken by simple power detection, even without any knowledge of the code, whereas the security can improve significantly by using 2D advanced modulation (i.e. F-PolSK).

## 8.2   Optical Polarization Shift Keying (PolSK)

### 8.2.1   Theory of Polarization Modulation

Polarization shift keying (PolSK) transmission encodes information on a constellation of signal points in the space of the Stokes parameters. In general, each signal point corresponds to a given state of polarization (SOP) and a given optical power. If only the polarization of the lightwave is modulated, and not its power, all the signal points lie on the Poincaré sphere. Examples of such signal constellations are shown in Figure 8.1. In this chapter we will restrict to constellations of equal power signal points.

The SOP of a fully polarized lightwave can be described through the Stokes parameters. Consider a reference plane $(\bar{x}, \bar{y})$ which is normal to the propagation axis $z$. An electromagnetic field $\bar{E}$ in this plane can be expressed as [2]:

$$\bar{E} = E_x\bar{x} + E_y\bar{y} \tag{8.1a}$$

where

$$E_x = a_x(x, y)e^{j(\omega t + \phi_x(t))}$$
$$E_y = a_y(x, y)e^{j(\omega t + \phi_y(t))} \tag{8.1b}$$

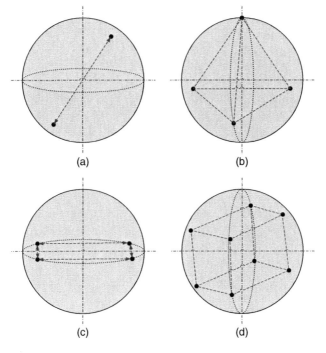

(a)                        (b)

(c)                        (d)

**Figure 8.1**   Signal-points constellation for multiarray PolSK inscribed into Poincaré sphere: (a) binary-PolSK; (b) quad-PolSK at the vertices of a tetrahedron; (c) quad-PolSK on a maximum circle of the Poincaré sphere; and (d) 8-PolSK at the vertices of a cube (© 1992 IEEE. Reprinted with permission from Theory of polarization shift keying modulation, S. Benedetto and P. Poggiolini, *IEEE Trans. on Comm.*, **40** (4), 1992.)

where $\omega$ is the optical angular frequency, $a_x$ and $a_y$ are the amplitudes of $x$ and $y$ components, $\Phi_x$ and $\Phi_y$ are the phases of $x$ and $y$ components before the transformation, respectively. The Stokes parameters can be calculated as:

$$S_1 = a_x^2 - a_y^2$$
$$S_2 = 2a_x a_y \cos(\delta) \tag{8.2}$$
$$S_3 = 2a_x a_y \sin(\delta)$$

where $\delta = \phi_x - \phi_y$. It should be noted that for notation simplicity in the following analysis we have omitted the time dependency of the above parameters.

In classical optics, an average is generally taken from the quantities appearing in the right-hand side of Equation (8.1) which have been utilized to define instantaneous values for the Stokes parameters. The fourth Stokes parameter is also defined as:

$$S_0 = a_x^2 + a_y^2 = S_1^2 + S_2^2 + S_3^2 \tag{8.3}$$

This represents the total electromagnetic power density travelling in the $z$-axis direction.

The $S_i$ can also be demonstrated in three-dimensional space with vectors $(\overline{S}_1, \overline{S}_2, \overline{S}_3)$, called the *Stokes space*. The waves having the same power density $\overline{S}$ lie on a sphere of radius $S_0$, called *Poincaré sphere*, as shown in Figure 8.1 with different signal constellations.

A fundamental feature of this demonstration is that the orthogonal SOPs, according to the Hermitian scalar product of:

$$\langle \overline{E}_1 \cdot \overline{E}_2^* \rangle = a_{1x}(x,y) \cdot a_{2x}^*(x,y) \cdot a_{1y}(x,y) \cdot a_{2y}^*(x,y) \cdot e^{j(\omega t + \phi_1)} \cdot e^{-j(\omega t + \phi_2)} \tag{8.4}$$

will map onto the points which are antipodal on the Poincaré sphere [2].

A generic transformation of the SOP of a fully polarized lightwave, propagating along the $z$-axis which preserves the degree of polarization, is explained below. Let $\overline{E}$ and $\overline{E}'$ be the electromagnetic field vectors before and after the transformation (i.e. modulation), respectively. Their associated decompositions can be expressed as:

$$E_x(t) = a_x(t)e^{j(\omega t + \Phi_x)} \qquad E_x'(t) = a_x'(t)e^{j(\omega t + \Phi_x')}$$
$$E_y(t) = a_y(t)e^{j(\omega t + \Phi_y)} \qquad E_y'(t) = a_y'(t)e^{j(\omega t + \Phi_y')} \tag{8.5}$$

where $a_x, a_x'$ and $a_y, a_y'$ are the amplitudes of $x$ and $y$ components, $\Phi_x, \Phi_x'$ and $\Phi_y, \Phi_y'$ are the phase of $x$ and $y$ components before and after the transformation, respectively. The electric field vectors are then given by:

$$\overline{E}(t) = E_x(t)\overline{x} + E_y(t)\overline{y}, \quad \overline{E}'(t) = E_x'(t)\overline{x}' + E_y'(t)\overline{y}' \tag{8.6}$$

In Equation (8.6) $\overline{x}, \overline{y}$ and $\overline{x}', \overline{y}'$ are the transverse reference axis sets before and after the transformation, respectively (i.e. normal to the direction of propagation). Thus, we have:

$$\begin{bmatrix} E_x'(t) \\ E_y'(t) \end{bmatrix} = \mathbf{Q} \begin{bmatrix} E_x(t) \\ E_y(t) \end{bmatrix} \tag{8.7}$$

where $\mathbf{Q}$ is a complex Jones matrix with unit determinant. A subset of Jones matrices, called the set of matrices of birefringence or optical activity, not only preserves the degree of polarization but also has the additional feature of preserving two orthogonal fields (according to the Hermitian scalar product [2]) which were orthogonal before the transformation [3]. Matrices of this kind are complex unitary matrices with unit determinant. Throughout this chapter we strictly refer to the $\mathbf{J}_j$ as the subset of $\mathbf{Q} = [\mathbf{J}_0 \ \ldots \ \mathbf{J}_j \ \ldots \ \mathbf{J}_{k-1}]$.

By using the Jones notation, the field can be represented by the vector, $\mathbf{J} = [E_x \ E_y]^T$ and the intensity of the beam can be normalized so that $|E_x|^2 + |E_y|^2 = 1$. Two SOPs represented by $\mathbf{J}_1$ and $\mathbf{J}_2$ are orthogonal if their inner product is zero, i.e. $\mathbf{J}_1^H \cdot \mathbf{J}_2 = E_{1x}^* \cdot E_{2x} + E_{1y}^* \cdot E_{2y} = 0$, where $H$ is the Hermitian. Any SOP can be transformed into another by multiplying it by Mueller matrices. A list of Mueller matrices that is required for SOP processing can be found in [2, 10].

## 8.2.2 Laser Phase Noise

The transmission capacity of optical communication systems, especially coherent systems, depends on the spectral emission capacity of the lasers (transmitter and local oscillator). It is known that the smaller the bandwidth, the higher the possible bit rate and the larger the possible repeater-less transmission span. In an ideal laser, the transmitted optical wave is composed of stimulated emission only, i.e. no spontaneous emission process occurs. Therefore, an ideal single-mode laser transmits real monochromatic light with a single wavelength. The power spectral density or emission spectrum of the ideal single laser is shown in Figure 8.2(a). Laser noise arises from spontaneous emission processes, which are unavoidable in a laser beam.

In a gas laser, spontaneous emission is primarily due to the local fluctuation of the laser mirrors. These fluctuations are in turn caused by changes of temperature and external mechanical disturbances. It is known that spontaneous emission can yield a time-varying amplitude and a time-varying phase. As a result, the laser emission spectrum is substantially broadened, as shown in Figure 8.2(b). The laser phase noise is a major cause of degradation of the performance of optical coherent communication.

The influence of laser phase noise could be reduced by narrowing the linewidth of the laser. Some of the physical approaches to narrowing the linewidth can be named [21, 22] as:

1. utilization of distributed feedback (DFB) laser
2. distributed-Bragg reflection (DBR) laser
3. external cavity laser
4. laser with optical reflection
5. coupled cavity lasers.

In addition to these physical approaches concerning the laser source, various system design techniques can also be employed to reduce the influence of laser phase noise such as error correction coding [23].

## 8.2.3 Self-Phase Modulation

With high capacity, long-haul communication systems, optical signal-to-noise ratio (OSNR) and receiver sensitivity considerations lead to an increasing power requirement

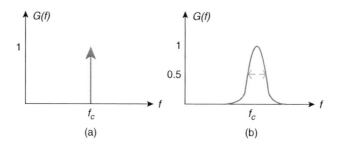

**Figure 8.2**  Emission spectrum of (a) ideal and (b) real single-mode laser

as the bit rate increases. As the fibre-input power for a very high bit rate system is increased beyond approximately 5 mW, the fibre cannot be assumed to behave as a linear transmission medium. The nonlinear Kerr effect also needs to be included for a more accurate model. The Kerr effect is due to the fact that the refractive index changes with optical power [24]. The Kerr effect can give rise to a phenomenon known as self-phase modulation (SPM) [1, 4, 12]. It can be described by the nonlinear dependence of the effective fibre refractive index $n$ on the intensity $I$ as:

$$n(\omega, I) = n_0(\omega) + n_2 \cdot I \tag{8.8}$$

where $n_0(\omega)$ is the linear index of refraction, and $n_2$ is the Kerr coefficient, which is approximately $3.2 \times 10^{-16}$ cm$^2$/W in silica fibres [2]. The propagation constant depends on $n(\omega, I)$ as:

$$\beta(\omega, I) = \frac{\omega \cdot n(\omega, I)}{c} \tag{8.9}$$

When the intensity $I(t)$ is modulated, the propagation constant changes with time. The *instantaneous* optical frequency is $d\omega/dt$ proportional to the time derivative of the propagation constant, so new frequency components are generated due to the intensity variations of the pulse. Hence, the SPM broadens the optical spectrum and this limits the maximum bit rate.

### 8.2.4  Polarization Fluctuation

Polarization fluctuations are the second major cause of degradation in the performance of coherent optical communications. It is demonstrated in [25] that when polarized light is sent through a single-mode optical fibre, it does not maintain the input state of polarization for more than a few metres due to the birefringence of the channel. Optical fibres exhibit a small difference in refractive index for a particular pair of orthogonal polarization states, a property called birefringence. The output state of polarization is not the same as the input state of the polarization. The polarization fluctuations in the received field produce fluctuations in the photodiode current, which may extinguish the detected signals completely.

The following techniques have been conceived to cope with the problem of polarization fluctuations in coherent systems which are briefly addressed in:

1. polarization control
2. polarization diversity
3. polarization scrambling
4. data induced polarization switching (DIPS)
5. use of polarization maintaining fibre.

### 8.2.4.1   Polarization Control

Polarization control requires an active component, which transforms the state of polarization [26, 27]. Figure 8.3 shows a polarization control lightwave communication system. It is noted that the speed of the device could be much slower than the bit rate, as long as it is fast enough to follow the polarization fluctuation. The oldest and most widely used type of polarization controller is based on fibre squeezers [28]. However, fibre squeezers can be bulky and therefore impractical. Other mechanical polarization controllers are highly birefringent fibre stretchers and rotating fibre cranks. More viable are the integrated optical polarization controllers due to their potential for lower cost and high reliability [28].

### 8.2.4.2   Polarization Diversity

A polarization diversity receiver, shown in Figure 8.4, splits the optical signal in two orthogonally polarized parts. The two signals are separately received and demodulated, and are combined after the demodulation. The advantages of using polarization diversity could relieve the following four problems:

1. limited receiver bandwidth
2. phase noise of the transmitter and local oscillator lasers
3. limited tuning range of local oscillators
4. polarization fluctuations of received signal

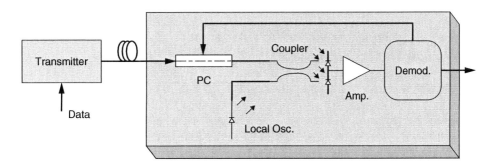

**Figure 8.3**   Polarization control

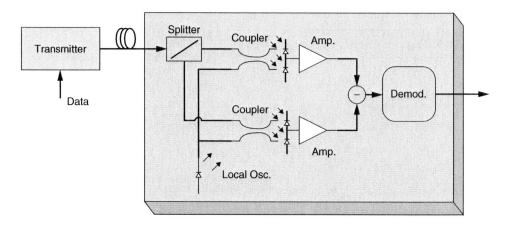

**Figure 8.4**  Polarization diversity method

It is reported in [26] that the polarization diversity method has a 0.39–0.7 dB sensitivity penalty over polarization control. Polarization diversity was introduced by Okoshi [29] and is currently the most commonly used method for polarization handling.

### 8.2.4.3  Polarization Scrambling

A polarization scrambling lightwave communication system is shown in Figure 8.5. This method involves scrambling the polarization at a rate faster than the bit rate. However, it has been reported that this technique achieved a reduction in polarization sensitivity compared to using polarization diversity approach [26, 28]. The main disadvantages of this technique are that it:

1. introduces a 3–5 dB penalty to the system
2. broadens the optical and intermediate frequency spectrum
3. needs a very fast polarization modulator.

### 8.2.4.4  Data-Induced Polarization Switching

A data-induced polarization switching optical communication system is shown in Figure 8.6. In this technique, polarization switching is combined with the wide deviation frequency shift keying (FSK) modulation. At the transmitter, a space (0) is sent as $f_o$ with some polarization and a mark (1) as $f_1$ with the orthogonal polarization. This is achieved by launching the FSK signal from the transmitter at $45°$ to the principal axes of a highly birefringence medium, usually a long segment of highly birefringence fibre. For a given frequency separation of $f_1 - f_{o'}$, the birefringence is chosen such that the polarization states at the output are orthogonal. The polarization orthogonality between the polarization states at the two frequencies remains essentially unaffected in transmission to the receiver. At the receiver, the signal is detected by a single balanced coherent receiver, and demodulated with a dual-filter FSK-demodulator or delay line

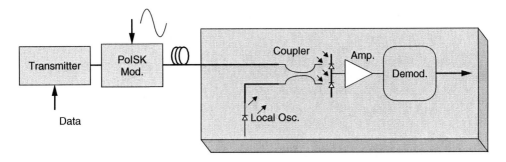

**Figure 8.5**   Polarization scrambling

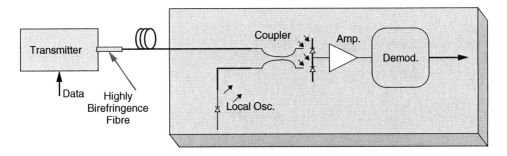

**Figure 8.6**   Data induced polarization switching

demodulator. The main advantage of this technique is relatively low cost and low system complexity, compared to the other polarization handling methods [28].

### 8.2.4.5   Polarization Maintaining Fibres

Polarization maintaining fibres are the simplest solution to polarization fluctuations. They are SMF in which a large amount of birefringence is intentionally introduced through design modifications so that small random birefringence fluctuations do not affect the light polarization significantly. In spite of the simplicity of this solution to the polarization matching problem, polarization maintaining fibres are difficult to produce and expensive, and they have a higher loss compared to standard SMFs. In addition, polarization maintaining fibres have difficulty in splicing, and most importantly almost all the already installed fibre is standard SMF. Normally, polarization maintaining fibres are only used for transporting the local oscillator light to a polarization controller or hybrid polarization diversity, or for data-induced polarization switching.

## 8.2.5   *Polarization Dependent Loss*

For a component, polarization-dependent loss (PDL) is defined as the maximum peak-to-peak insertion loss (or gain) in dB caused by the component when stimulated by all possible polarization states. PDL may also be referred to as polarization sensitivity, polarization-dependent gain (PDG) or extinction ratio (for optical polarizers) and is defined as:

$$PDL_{dB} = 10 \cdot \log \frac{P_{\text{Max}}}{P_{\text{Min}}} \qquad (8.10)$$

Some components are designed for maximum PDL. A linear optical polarizer, for example, must have high PDL in order to convert unpolarized lightwave into linearly polarized lightwave. Only one orientation of linearly polarized light passes through the polarizer unattenuated. Misaligned orientations of polarized light are attenuated by the polarizer's PDL. In other situations, any amount of PDL is a liability. Long-haul telecommunication systems, for example, are more cost effective as transmission distances between amplification stages become longer. Transmission-distance calculations are partly based on guaranteed power levels. Large variations in system power occur as the PDL of individual system components randomly combines. This makes power-budget calculations more difficult, expands design margins and reduces guaranteed performance. Long-haul considerations are discussed later in this chapter.

## 8.3    PolSK-OCDMA Transceiver Architecture

### 8.3.1    Signals and System Configuration

A binary PolSK constellation is made up of two points that are antipodal over the Poincaré sphere. If the diameter on which they lie is made to coincide with one of the Stokes axes, then the knowledge of only one of the Stokes parameters is sufficient to demodulate the signal without incurring any penalty [4]. The receiver model in Figure 8.7 extracts only one Stokes parameter and hence it is suitable only for binary systems. Its optical and electronic front-ends are remarkably simple but it does require an optical polarization controller to ensure that the axis of the binary constellation is aligned with the Stokes parameter axis at the input of the receiver. It can be seen from Figure 8.7 that an optical filter is placed at the input of the receiver to reduce the received ASE noise. For both technical and practical reasons (e.g. minimizing the laser phase-noise or chirp) the bandwidth of

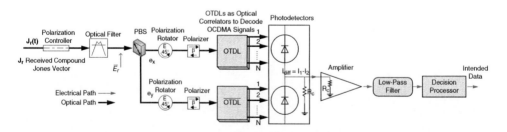

**Figure 8.7** Architecture of incoherent PolSK-OCDMA receiver with OTDL (© 2008 IEEE. Reprinted with permission from Transceiver architecture for incoherent optical CDMA networks based on polarization modulation, M.M. Karbassian and H. Ghafouri-Shiraz, *J. Lightw. Technol.*, **26** (24), 2008.)

this filter ($B_0$) should generally be wider than a matched filter whose noise bandwidth would be equal to the symbol-rate [1]. In order to reduce the penalty caused by using a wider filter bandwidth and by electrical noise, a tight low-pass filter (LPF) is placed after the polarization-demodulated signal as shown in Figure 8.7. This technique is also widely used in amplified intensity modulated systems [12]. The LPF is assumed to be an integrate-and-dump filter of integration time $T_s$ (i.e. symbol duration) whose bandwidth is $B_{el} = 1/T_s$. For binary systems the bit and the symbol duration coincide, but here every bit is also CDMA encoded at $T_c$ chip duration.

By assuming the complex envelope representation of band-pass signals, the electric field at the transmitter output (shown in Figure 8.8) can be written as:

$$\overline{E}_t = \sqrt{2P_t} \cdot \begin{pmatrix} e_x \\ e_y \end{pmatrix} \qquad (8.11)$$

where $P_t$ is the transmitted optical power launched into the fibre and $e_x$ and $e_y$ represent the $x$- and $y$-components of the signal with respect to two orthogonal linear fibre polarizations normalized in such a way that $|e_x|^2 + |e_y|^2 = 1$. They form a vector called a Jones vector shown in the transmitter (i.e. Figure 8.8) where the corresponding Stokes parameters $\overline{S} = (S_1, S_2, S_3)$ are derived from them and satisfy the equality $|\overline{S}|^2 = 1$. A vector in the Stokes space $\overline{S}$ is defined as the vector linking the origin to the point $(S_1, S_2, S_3)$ unless otherwise specified. The signal at the output of the optical filter at the receiver, in Figure 8.7, is given by:

$$\overline{E}_r = \sqrt{2P_r} \cdot \begin{pmatrix} e_x \\ e_y \end{pmatrix} + \begin{pmatrix} n_x \\ n_y \end{pmatrix} \qquad (8.12)$$

where $P_r$ is the received power and $n_x$ and $n_y$ are complex Gaussian random variables accounting for the total filtered ASE noise with the variance of:

$$\sigma^2{}_{ASE} = 2N_0B_0 \qquad (8.13)$$

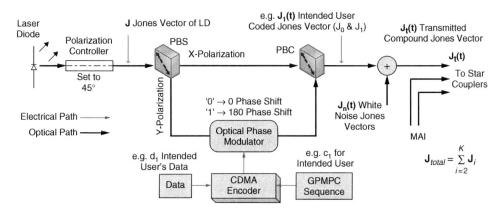

**Figure 8.8** Architecture of incoherent PolSK-OCDMA transmitter (© 2008 IEEE. Reprinted with permission from Transceiver architecture for incoherent optical CDMA networks based on polarization modulation, M.M. Karbassian and H. Ghafouri-Shiraz, *J. Lightw. Technol.*, **26** (24), 2008.)

where $N_0$ is the power spectral density (PSD) of the white ASE noise and $B_0$ is optical filter bandwidth. The apparent doubling of the overall ASE noise is due to the complex envelope notation adopted here. The power of the information signal is doubled as well, so that the OSNR is preserved. As shown in Figure 8.7, the signal $\overline{E}_r$ passes through a polarization beam splitter (PBS) that divides the two linear $\hat{x}$ and $\hat{y}$ polarizations. These signals are subsequently CDMA-decoded through OTDLs and then photodetected, and the following corresponding currents are generated at the outputs of the photodetectors (PD) for the decision rule:

$$I_x = \Re \cdot P_r \cdot |e_x|^2$$
$$I_y = \Re \cdot P_r \cdot |e_y|^2 \qquad (8.14)$$

where $\Re = e\eta/hf$ is the responsivity of PD, $\eta$ is the quantum efficiency of PD ($\eta = 0.9$), $h$ is the Planck's constant, $f$ is the optical frequency ($\lambda = 1.55\,\mu m$). Hence, the current $I_{diff}$ at the input of the LPF (after the amplifier) is:

$$I_{diff} = \Re \cdot P_r \cdot (|e_x|^2 - |e_y|^2) \qquad (8.15)$$

which is proportional to the Stokes parameter $S_1$ and, as mentioned, it is suitable for binary transmission. Since it extracts only one of the Stokes parameters, this remarkably simplifies the receiver structure. It is assumed, as shown in Figure 8.7, that the photodetector internal load is matched to the input resistance of the electronic amplifier ($R_c$). Hence, the output of the final integrate-and-dump LPF can be expressed as:

$$I = \Re \cdot P_r \cdot (|e_x|^2 - |e_y|^2) * h_{LP}(t) + n_{LP}(t) \qquad (8.16)$$

where $h_{LP}(t)$ is the impulse response of the electronic LPF and $n_{LP}(t)$ is a zero-mean Gaussian random variable that represents the receiver noise current with the following variance:

$$\sigma_{LP}^2 = \frac{2kT}{R_c} \cdot B_{el} \cdot F \qquad (8.17)$$

where $F$ is the noise figure of the electronic amplifier, $T$ and $k$ denote the absolute temperature and the Boltzmann constant, respectively. Apart from temperature, the fundamental parameter characterizing the electronic stage of the receiver is the ratio $F/R_c$. It is noticeable that the thermal noise power on the detected current is doubled if we place an amplifier after each photodetector before the subtraction node. Under the hypothesis that the LPF bandwidth is large enough to avoid phase-to-amplitude noise conversion, phase noise is cancelled out within the nonlinear receiver stage, extracting the Stokes parameters from currents (i.e. decision processor in Figure 8.7). The only side effect is the bandwidth enhancement of the noise passing through the LPF. This impairment can be almost completely recovered through the use of a post-detection filter, similar to the one used in ASK or FSK schemes [9, 10].

## 8.3.2 Decision Rule Analysis at PolSK-OCDMA Receiver

In PolSK, the angle of one polarization component is switched relative to the other between two angles; therefore, binary data bits are mapped into two Jones vectors. A block diagram of the proposed PolSK-OCDMA transmitter is illustrated in Figure 8.8. The light source is a highly coherent laser with a fully polarized SOP.

If a non-polarized source is used, then a polarizer can be inserted after the laser source. The light beam first passes through the polarization controller that sets the polarization to an angle of 45°; since the laser phase noise is minimal at this polarization and component's polarization matching (e.g. polarizer, polarization controller and rotator) at this angle is simpler [8, 14, 31]. Then, the lightwave is divided through PBS to become SOP-encoded in the PolSK modulator which switches the SOP of the input beam between two orthogonal states, that is referred to 0° and 180° at the phase modulator, $N$ times per bit according to an externally supplied code that spreads the optical signal into CDMA format. Thereafter, the PolSK-OCDMA modulated signals are combined through a polarization beam combiner (PBC) and transmitted. As shown in Figure 8.8, for a $K$-user system with the first user as the desired one (for example), the $i^{th}$ user SOP-encoded signal can be written as:

$$\mathbf{J}_i(t) = \begin{cases} \mathbf{J}_0 & \text{if } d_i(t) \oplus c_i(t) = 0 \\ \mathbf{J}_1 & \text{if } d_i(t) \oplus c_i(t) = 1 \end{cases} \qquad (8.18)$$

where $d_i(t) = \sum_{j=0}^{D-1} d_{i,j} \cdot u_T(t - jT_s)$ is the data signal with $D$ bits (i.e. the data-length with symbol duration of $T_s$), $c_i(t) = \sum_{j=0}^{N=1} c_{i,j} \cdot u_T(t - jT_c)$ is the code sequence signal with $N$ chips (i.e. the code-length with chip duration of $T_c$) and $d_i(t), c_i(t) \in \{0, 1\}$; also $u_T(t)$ denotes a unity rectangular pulse of width $T$, and $\oplus$ denotes the signal correlation. The emitted light is initially (linearly) polarized at an angle of 45°, and therefore $\mathbf{J}_0 = \frac{1}{\sqrt{2}}[1 \ 1]^T$ and $\mathbf{J}_1 = \frac{1}{\sqrt{2}}[-1 \ 1]^T$ [8]. In other words, we have:

$$\mathbf{Q} = [\mathbf{J}_0 \ \mathbf{J}_1] = \frac{1}{\sqrt{2}} \begin{bmatrix} 1 & -1 \\ 1 & 1 \end{bmatrix} \qquad (8.19)$$

So the polarization-modulated signal travels a distance of $L$ km through an SMF. Consequently, the SOP-encoded signal undergoes several impairments such as attenuation, dispersion, polarization rotation and fibre nonlinearity. The polarization rotation can be compensated for by a polarization tracking and control system, as discussed in [32], that also suppresses the polarization mode dispersion (PMD) as well as rotation. However, this will bring complexity to the structure and extra implementation cost. Instead, at the receiver end shown in Figure 8.7, the SOP rotation is compensated for by the polarization controller, rotators and polarizers whose functions are to ensure that the received signal and the optical components at the receiver have the same SOP reference axis. Hence, we have the received signal Jones vector $\mathbf{J}_r(t)$, containing the data from desired user $\mathbf{J}_1(t)$ generated at the transmitter and corrupted by other users' transmissions $\sum_{i=2}^{K} \mathbf{J}_i(t)$ (i.e. MAI) plus additive Gaussian white noise (AWGN) as follows:

$$\mathbf{J}_r(t) = \begin{bmatrix} E_{rx} \\ E_{ry} \end{bmatrix} = \mathbf{J}_1(t) + \sum_{i=2}^{K} \mathbf{J}_i(t) + \mathbf{J}_n(t) \qquad (8.20)$$

where $\mathbf{J}_n(t) = [E_{nx} \ E_{ny}]^T$ is the complex Jones vector of the AWGN. In Figure 8.7, it is assumed that the received composite signal undergoes a lossless split after PBS.

Thereafter, the signals in the both branches are rotated through the polarization rotators by 45° in order to align the output beam's polarization to the polarizer axis. The polarizer passes only the optical beam matched to its axis; it performs like a polarization filter. Then, the signals are optically correlated through OTDLs. The OTDLs are tuned to the intended user's assigned spreading code to de-spread the CDMA-encoded signals (i.e. OCDMA-decoder). The upper and lower branch signals are presented by the $(.)^0$ and $(.)^1$ notation respectively. Therefore, Jones vectors representing the upper and lower branches at the output of the PBS can be expressed as:

$$\mathbf{J}_{PBS}^z(t) = \begin{bmatrix} E_{rx} \\ |E_{ry}| \exp[j\,(\arg(E_{ry}) + x_z\pi)] \end{bmatrix} \tag{8.21}$$

where $z \in \{0, 1\}$ denotes upper or lower branch, thus $x_0 = c_1$, $x_1 = \bar{c}_1$, (complement of $c_1$) and arg(.) denotes the phase of $E_{ry}$. Then, the Jones vectors are applied into the rotators. The function of the rotator's Jones vectors is given as:

$$\mathbf{J}_R^z(t) = \frac{1}{\sqrt{2}} \begin{bmatrix} 1 & 1 \\ -1 & 1 \end{bmatrix} \cdot \mathbf{J}_{PBS}^z(t) \tag{8.22}$$

Then the polarizers (i.e. polarization filters) produce only the $x$-polarization at their output corresponding to the first elements of the Jones vector and only allow the signal with assigned polarization to pass. Hence, we have:

$$\mathbf{J}_P^z(t) = \begin{bmatrix} E_{Rx}^z & 0 \end{bmatrix}^T \tag{8.23}$$

The output of each polarizer is then:

$$\mathbf{J}_P^z(t) = E_{Rx}^z = \frac{1}{\sqrt{2}} \left\{ E_{rx} + |E_{ry}| \cdot \exp[j\,(\arg(E_{ry}) + x_z\pi)] \right\} \tag{8.24}$$

$\mathbf{D}^z$ is defined as the decision variable from which the polarizers' decision rules can be obtained by:

$$\mathbf{D}^z = |E_{Rx}^z|^2 = E_{Rx}^z \cdot E_{Rx}^{*z} \tag{8.25}$$

After substituting, it can be written as:

$$\begin{aligned} \mathbf{D}^z = \frac{1}{2} &\left\{ E_{rx} + |E_{ry}| \exp[j\,(\arg(E_{ry}) + x_z\pi)] \right\} \\ &\times \left\{ E_{rx}^* + |E_{ry}| \exp[-j\,(\arg(E_{ry}) + x_z\pi)] \right\} \end{aligned} \tag{8.26}$$

And it can be expanded to:

$$\mathbf{D}^z = \frac{|E_{rx}|^2 + |E_{ry}|^2}{2} + |E_{rx}| \cdot |E_{ry}| \cdot \cos(\phi + x_z\pi) \tag{8.27}$$

where $\phi = \arg(E_{ry}) - \arg(E_{rx})$. The decision variable $\mathbf{D}$ is the difference between the upper and lower branch outputs which is given by:

$$\mathbf{D} = \int_0^{T_s} |E_{Rx}^1|^2 \, dt - \int_0^{T_s} |E_{Rx}^0|^2 \, dt = \int_0^{T_s} (\mathbf{D}^1 - \mathbf{D}^0) \, dt \tag{8.28}$$

Finally, the decision rule $\tilde{d}$ is defined, in order to extract the final bit value from the modulated signal, as:

$$\tilde{d} = \begin{cases} 0 & if \ \mathbf{D} < 0 \\ 1 & if \ \mathbf{D} \geq 0 \end{cases} \tag{8.29}$$

### 8.3.3 PolSK-OCDMA Signal Processing

We previously discussed the configuration of the transceiver. Now we consider the alignment and analysis of the received optical signal. The electric field of the received polarization-modulated optical signal can be expressed [6] as:

$$\overline{E}'(t) = Re \left\{ \overline{E}(t) \cdot \sum_{i=1}^{K} \mathbf{Q} \cdot \begin{bmatrix} d_i(t) \\ 1 - d_i(t) \end{bmatrix} \cdot u_T(t - iT_s) \cdot c_i(t) \right\} \tag{8.30}$$

The channel is represented by the Jones matrix $\mathbf{Q}$ and $Re\{\bullet\}$ refers to the real part of complex $\overline{E}'(t)$. Equipower signal constellations have been considered throughout our analysis in this chapter. That is, we have assumed both orthogonal components have the same power and are equally attenuated and dispersed as they travel the same distance in the fibre and suffer from the same noise sources. Because the two polarizations are orthogonal (i.e. reciprocal), the electric field $\overline{E}(t)$ will have a constant amplitude. It has also been discussed in detail [3] that the loss of orthogonality in any linear optical medium is related to the transferred maximum and minimum power coefficients and a bound on this loss is calculated from these coefficients. In this analysis, as the power coefficients are clarified later in Equation (8.38), the orthogonality is therefore preserved with equality as a result of the $(x, y)$-components equipower constellations. While the switching time in SOP (i.e. bit rate) is much slower than the chip-rate, the elements of the Jones matrix can be considered as time-independent (i.e. $T_c << T_s$). The $x$-component of the received electric field vector based on $\mathbf{Q} = [\mathbf{J}_0 \ \mathbf{J}_1]$ in Equation (8.19) becomes:

$$E'_x(t) = Re \left\{ \overline{E}(t) \cdot \sum_{i=1}^{K} [\mathbf{J}_0 \cdot d_i(t) + \mathbf{J}_1(1 - d_i(t))] \cdot u_T(t - iT_s) \cdot c_i(t) \right\} \tag{8.31}$$

Thus, orthogonal components of the $i^{th}$ user are given as $E_{xi}(t) = \mathbf{J}_0 \cdot d_i(t) \cdot c_i(t) \cdot \overline{E}(t)$ and $E_{yi}(t) = \mathbf{J}_1 \cdot (1 - d_i(t)) \cdot c_i(t) \cdot \overline{E}(t)$ and the $(x, y)$-components of received modulated PolSK-OCDMA signal are [6]:

$$E'_{xi}(t) = \left( \frac{E_{xi}(t) + E_{yi}(t)}{2} + \sum_{i=1}^{K} c_i(t) \cdot d_i(t) \cdot \frac{E_{xi}(t) - E_{yi}(t)}{2} \cdot u_T(t - iT_s) \right) \cdot \cos(\varphi_{xi}) \tag{8.32}$$

$$E'_{yi}(t) = \left( \frac{E_{xi}(t) + E_{yi}(t)}{2} + \sum_{i=1}^{K} c_i(t) \cdot d_i(t) \cdot \frac{E_{xi}(t) + E_{yi}(t)}{2} \cdot u_T(t - iT_s) \right) \cdot \cos(\varphi_{yi})$$

where $\varphi_{xi} = \omega_{xi} t + \theta_{xi}$ and $\varphi_{yi} = \omega_{yi} t + \theta_{yi}$ describe the frequencies and phases of the transmitting lasers. Based on the concept of OCDMA, the field vectors of all $K$

transmitters are added up and transmitted (i.e. multiplexed) over the same channel. Thus, the overall channel field vector can be expressed as:

$$\overline{E}_{Channel} = \sum_{i=1}^{K} \overline{E}'_i(t) \tag{8.33}$$

Figure 8.7 illustrates the application of the OTDLs as the optical correlator in incoherent PolSK-OCDMA. They correlate the incoming signal with pre-reserved user assigned spreading code to de-spread the signal. In doing so, the time delays and coefficients in OTDLs are designed to perform as a correlator in both branches. Additionally, OTDL in the lower branch is set up with complement of code (i.e. 180° phase difference) used in the upper branch (i.e. shown by $\overline{\text{OTDL}}$ in Figure 8.7) to decode other symbols (i.e. 1). As can be observed from Figure 8.7, the OTDLs output contains $N$ chip pulses that can be assumed to be a parallel circuit of many single photodetectors so that their currents are added and no interference between the pulses is possible. The signals are photodetected in the balanced-detector format to generate the differential electrical current ($I_{diff} = I_1 - I_2$) ready for data-extraction in the decision processor. The individual branch current at a certain chip time $T_c$ in the upper branch (i.e. $x$-component) after photodetection is:

$$I_n^0 = \Re \int_{t=0}^{T_c} \left( \sum_{i=1}^{K} \left( \frac{E_{xi}(t) + E_{yi}(t)}{2} + d_i(t) \cdot c_i(t - nT_c) \right. \right.$$
$$\left. \left. \cdot \frac{E_{xi}(t) - E_{yi}(t)}{2} \right) \cdot \cos(\varphi_{xi}) \right)^2 dt \tag{8.34}$$

where $\Re$ is the responsivity of the photodetector, $c_i(t - nT_c)$ is the $n^{th}$ chip of assigned spreading code of the $i^{th}$ user. Hence, the total current in upper branch is:

$$I^0 = \Re \int_{t=0}^{T_s} \sum_{n=1}^{N} \left\{ \frac{c(nT_c) + 1}{2} \left( \sum_{i=1}^{K} \left( \frac{E_{xi}(t) + E_{yi}(t)}{2} + d_i(t) \cdot c_i(t - nT_c) \right. \right. \right.$$
$$\left. \left. \left. \cdot \frac{E_{xi}(t) - E_{yi}(t)}{2} \right) \cdot \cos(\varphi_{xi}) \right)^2 \right\} dt \tag{8.35}$$

which can be rewritten as:

$$I^0 = \frac{\Re}{4} \int_{t=0}^{T_s} \sum_{n=1}^{N} \left\{ \frac{c(nT_c) + 1}{2} \left( \sum_{i=1}^{K} \left( E_{xi}^2(t) + E_{yi}^2(t) + d_i(t) \cdot c_i(t - nT_c) \cdot (E_{xi}^2(t) \right. \right. \right.$$
$$\left. \left. \left. -E_{yi}^2(t) \right) \right) \cdot (1 - \cos 2\varphi_{xi}) \right) \right\} dt \tag{8.36}$$

$$+ \frac{\Re}{8} \int_{t=0}^{T_s} \sum_{n=1}^{N} \left\{ \frac{c(nT_c) + 1}{2} \left[ \sum_{i=1}^{K} \sum_{\substack{j=1 \\ j \neq i}}^{K} \left\{ (E_{xi}(t) + E_{yi}(t) + d_i(t) \cdot c_i(t - nT_c) \cdot (E_{xi}(t) \right. \right. \right.$$
$$\left. -E_{yi}(t))) \left( E_{xj}(t) + E_{yj}(t) + d_j(t) \cdot c_j(t - nT_c) \right. \right.$$
$$\left. \left. \left. \left. \cdot (E_{xj}(t) - E_{yj}(t)) \right) \cdot (\cos(\varphi_{xi} + \varphi_{xj}) + \cos(\phi_{xi} - \varphi_{xj})) \right\} \right] \right\} dt$$

Photodetector frequency response is similar to an LPF, so in Equation (8.36) the term $\cos 2\varphi_{xi}$ in the first bracket and $\cos(\varphi_{xi} + \varphi_{xj})$ in the second bracket are filtered out as they are outside of the photodetector frequency range. Furthermore by choosing suitable laser frequencies, the term $\cos(\varphi_{xi} + \varphi_{xj})$ can also be neglected, provided that $\omega_{xi} - \omega_{xj} \gg \omega_c$ where $\omega_c$ is the cut-off frequency of the photodetector. Hence, the total current of the upper branch can be expressed as:

$$I^0 = \frac{\Re}{4} \sum_{n=1}^{N} \left\{ \frac{c(nT_c) + 1}{2} \left( \sum_{i=1}^{K} \left( E_{xi}^2(t) + E_{yi}^2(t) + d_i(t) \cdot c_i(t - nT_c) \right. \right. \right.$$
$$\left. \left. \left. \cdot (E_{xi}^2(t) - E_{yi}^2(t)) \right) \right) \right\} \qquad (8.37)$$

The Stokes parameters are defined as:

$$S_i^0 = E_{xi}^2(t) + E_{yi}^2(t)$$
$$S_i^1 = E_{xi}^2(t) - E_{yi}^2(t) \qquad (8.38)$$

where $S_i^0$ refers to the signal intensity part, generated in the upper branch of the polarization modulator at the transmitter while $S_i^1$ refers to the linear polarized part, generated in the lower branch containing data (see Figure 8.8). These parameters also denote the maximum and minimum transmitted power, respectively, in order to have a bound on the negligible loss of orthogonality [3]. Thus, Equation (8.37) can be rewritten as:

$$I^0 = \frac{\Re}{4} \sum_{n=1}^{N} \left\{ \frac{c(nT_c) + 1}{2} \left( \sum_{i=1}^{K} \left( S_i^0 + d_i(t) \cdot c_i(t - nT_c) \cdot S_i^1 \right) \right) \right\} + n_1(t) \qquad (8.39)$$

Similarly the total current of the lower branch (i.e. the $y$-component) can be derived as:

$$I^1 = \frac{\Re}{4} \sum_{n=1}^{N} \left\{ \frac{1 - c(nT_c)}{2} \left( \sum_{i=1}^{K} \left( S_i^0 + d_i(t) \cdot c_i(t - nT_c) \cdot S_i^1 \right) \right) \right\} + n_2(t) \qquad (8.40)$$

where:

$$n_1(t) = n_{1x}(t) + jn_{1y}(t)$$
$$n_2(t) = n_{2x}(t) + jn_{2y}(t) \qquad (8.41)$$

In Equation (8.41) $n_1(t)$ and $n_2(t)$ represent the filtered AWGN with independent Gaussian processes of $n_{1x}(t)$, $n_{1y}(t)$, $n_{2x}(t)$, and $n_{2y}(t)$ with equal variances $\sigma^2$ which originate as $n_1(t)$ and $n_2(t)$. Thus, the balanced-detector output ($I = I^0 - I^1$) is derived as:

$$I = \frac{\Re}{4} \sum_{n=1}^{N} c(nT_c) \sum_{i=1}^{K} \left( S_i^0 + d_i(t) \cdot c_i(t - nT_c) \cdot S_i^1 \right) + n(t) \qquad (8.42)$$

The noise $n(t)$ includes noise contributions from both $I^0$ and $I^1$ and consists of optical ASE noise mentioned in Equation (8.13), photodetector shot-noise $\langle i_{av}^2 \rangle = 2eiB_0$, where

$i_{av}$ is the average photocurrent, and also electronic receiver noise at the output of LPF introduced in Equation (8.17). Thus, the total noise $n(t)$ variance can be represented as:

$$\sigma_{n(t)}^2 = \langle i^2 \rangle + \sigma_{ASE}^2 + \sigma_{LP}^2 \tag{8.43}$$

The differential output current, Equation (8.42), can be modified considering the first user (#1) as the intended user as:

$$I = \frac{\Re}{4} \sum_{n=1}^{N} S_1^0 \cdot c(nT_c) + \frac{\Re}{4} \sum_{n=1}^{N} c(nT_c) \cdot c_1(t - nT_c) \cdot d_1(t) \cdot S_1^1$$

$$+ \frac{\Re}{4} \sum_{i=2}^{K} \sum_{n=1}^{N} c(nT_c) \cdot c_i(t - nT_c) \cdot d_i(t) \cdot S_i^1 + n(t) \tag{8.44}$$

The first term in Equation (8.44) is a direct current (dc) that needs estimation and removal in the balanced detector. The second term represents the intended data mixed with its assigned spreading code auto-correlation and polarization while the third term assumes the interference (i.e. MAI) caused by other transmitters and the last term is the noise. Thus, the system SNR can be expressed as:

$$SNR = \frac{\left( \dfrac{\Re}{4} \displaystyle\sum_{n=1}^{N} c(nT_c) \cdot c_1(t - nT_c) \cdot d_1(t) \cdot S_1^1 \right)^2}{\left( \dfrac{\Re}{4} \displaystyle\sum_{i=2}^{K} \sum_{n=1}^{N} c(nT_c) \cdot c_i(t - nT_c) \cdot d_i(t) \cdot S_i^1 \right)^2 + \sigma_{n(t)}^2} \tag{8.45}$$

Now, according to the group-padded modified prime codes' (GPMPC) auto-correlation, introduced in Section 2.4.5, we have:

$$\sum_{n=1}^{N} c(nT_c) \cdot c_1(t - nT_c) = P + 2 \tag{8.46}$$

By defining the variable $X_{li}$ as the GPMPC cross-correlation value, we have:

$$X_{li} = \sum_{n=1}^{N} c_l(nT_c) \cdot c_i(t - nT_c) \tag{8.47}$$

Now, the probability density function (PDF) of $X_{li}$ can be obtained from the independent values of cross-correlation values as a random variable. The in-phase cross-correlation value is $\lambda_c \leq 2$ depending on whether the codes are in the same group or from different groups. Obviously, the 0 value does not cause the interference due to perfectly orthogonal sequences, but the $\lambda_c \in \{1, 2\}$ values cause interference which is only between the intended user and $(P^2 - P)$ users from the different groups (i.e. $P^2$ whole sequences and $P$ sequences from the same group of intended user which are orthogonal). As the cross-correlation values are uniformly distributed among interfering users, the PDF of $w$, the realization of $X_{li}$, is:

$$P(w - i) = \frac{i}{P^2 - P} \tag{8.48}$$

where $P(w = i)$ is the probability that $w$ assumes the value $i$ (the number of actively involved users in the transmission). Therefore, by substituting Equations (8.46) and (8.48) into Equation (8.45), the system SNR as a function of the number of active users $K$ can be further simplified as:

$$SNR(K) = \cfrac{1}{\left(\cfrac{(K + 2) \cdot (K - 1)}{2 \cdot (P^2 - P) \cdot (P + 2)}\right)^2 + \cfrac{16\sigma_{n(t)}^2}{\Re^2 \cdot d_1^2(t) \cdot S_1^{1^2} \cdot (P + 2)^2}} \qquad (8.49)$$

Note that the single-user SNR is given by $SNR(1) = \Re^2 \cdot d_1^2(t) \cdot S_1^{1^2} \cdot (P + 2)^2/16\sigma_{n(t)}^2 = E_b/N_0$, where $E_b$ is the energy of one bit and $N_0$ is the noise PSD, denotes. Equation (8.49) is one of the main results of this chapter as it represents the SNR of a polarization-modulated OCDMA system. It should be noted that the analysis is also valid for other unipolar spreading codes with similar correlation properties.

## 8.4 Evaluation of PolSK-OCDMA Transceiver Performance

In this section, the numerical results of the BER performance of the proposed incoherent PolSK-OCDMA transceiver based on the above detailed analysis of the system SNR are demonstrated and discussed.

Figure 8.9 shows the BER of the proposed structure against the single-user SNR (shown as *Sdb* on the graphs) for 10–25 per cent traffic when $P = 19$. *Traffic* here is defined as the percentage of number of simultaneous active users to the total number

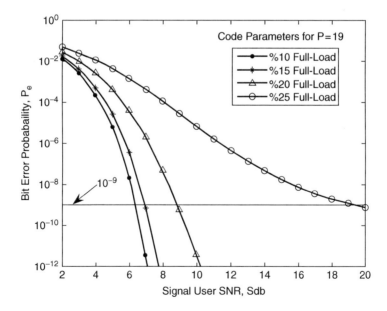

**Figure 8.9** BER performance of PolSK-OCDMA transceiver against the single-user SNR, *Sdb*

of users. Figure 8.9 shows clearly that as *Sdb* increases as the bit error probability decreases. The analysis shows that a system that can accommodate 20 per cent of all users (72 users) is able to meet $BER = 10^{-9}$ with $Sdb = 10$ dB; while at $Sdb = 7.2$ dB the system can support 15 per cent of all users (54 users) which is sufficiently superior to deliver the network services in this case.

However, the system is unable to guarantee a reliable communication for more than 25 per cent of all users unless there is a high single-user SNR of 20 dB. The system introduced in [6], employed Gold sequences with lengths of 511 and 1023 to accommodate 40 and 80 users which is 8 per cent of the full load in both cases). However, in the proposed system with $P = 19$ the code-length will be only 399. It should be noted that by using a longer code (greater $P$) the system performance improves and provides higher network capacity.

On the other hand, according to Figure 8.9, it is noted that this polarization-based OCDMA transceiver architecture is able to accommodate 10–20 per cent of all active users with a single-user SNR of as little as 10 dB. It is recommended to deploy codes with greater $P$ value to support a higher number of users and to apply higher single-user SNR for lower BER.

Figure 8.10 also displays the BER performance against the number of simultaneous active users $K$ for the proposed architecture. As it is apparent from Figure 8.10 that when a large number of users exist, the higher bit error rate occurs due to the growing interferences. The analysis indicates that a system that employs GPMPC with $P = 19$ and $Sdb = 14$ dB can tolerate 80 simultaneous users (i.e. 23 per cent of full load). This implies a cost-effective link-budget, $Sdb = 12$ dB, consuming less power as compared with the previous coding schemes and architectures introduced in [6, 8].

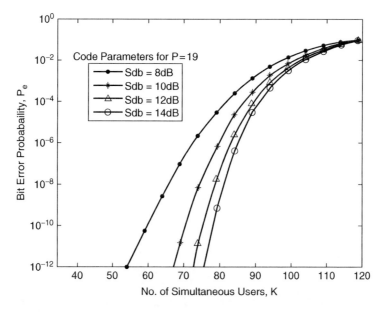

**Figure 8.10** BER performance of the PolSK-OCDMA transceiver against the number of simultaneous users, $K$

## 8.5    Transceiver Architecture for Hybrid F-PolSK-OCDMA

### *8.5.1    Transmitter Configuration*

Here we investigate the two-dimensional array (2D-ary) F-PolSK-OCDMA transmitter and its operation. This system structure includes sequentially connected optical $M_1$-ary FSK modulator and $M_2$-ary polarization modulator with the CDMA encoder. In the following, we illustrate the mathematical formulation of the hybrid modulated signals.

The 2D-ary F-PolSK transmitter for the signal generation is shown in Figure 8.11(a) which consists of a FSK modulator with a tuneable laser diode (TLD) that allocates $M_1$ symbols (i.e. $k_1 = \log_2^{M_1}$ bits per symbol) to the corresponding $M_1$ wavelengths emitted from the TLD corresponding to the data. As discussed in detail in Chapter 7, both the bit rate and OCDMA interferences will increase when the number of symbols ($M_1$) is constant, and the repetition ratio of TLD becomes smaller. In contrast, as the TLD repetition ratio becomes larger, the bit rate becomes lower and since the interrupted slots at each data frame prevent interference of in-phase correlation, the channel interference will decrease.

The PolSK-OCDMA modulator, similar to the previously discussed model, is shown in Figure 8.11(b). Having the signal FSK modulated, each wavelength from the TLD is

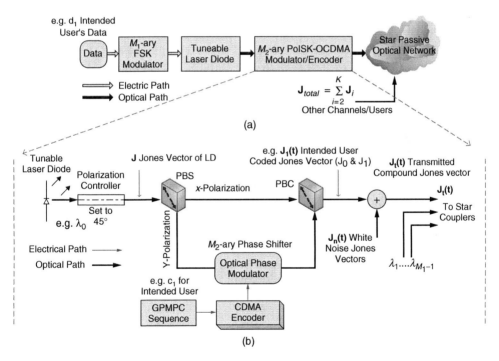

**Figure 8.11**    (a) Incoherent 2D-ary F-PolSK-OCDMA transmitter and (b) structure of $M_2$-ary PolSK-OCDMA modulator/encoder (© 2009 IEEE. Reprinted with permission from Incoherent two-dimensional array modulation transceiver for photonic CDMA, M.M. Karbassian and H. Ghafouri-Shiraz, *J. Lightw. Technol.*, **27** (8), 2009.)

initially linearly polarized and set to $45°$ for simplicity through the polarization controller at the beginning of the PolSK modulator. Then the wavelengths enter the PolSK-CDMA modulator. The signal is $M_2$-ary PolSK modulated where $M_2$ denotes number of the lightwave SOP (i.e. $k_2 = \log_2^{M_2}$ bits per SOP), and CDMA encoded by means of the GPMPC spreading sequences.

At the transmitter output and given a reference plane $(\hat{x}, \hat{y})$ normal to the direction of propagation $\hat{z}$, the transmitted lightwave during $M$ symbol intervals can be written [19], in the complex form, as:

$$\overline{v}(t) = \sum_{m=1}^{M} \overline{v}_{m_1 m_2}^{(m)}(t) = \sum_{m=1}^{M} \overline{p}_{m_2}^{M} \cdot \exp[j(2\pi \cdot (f_c + f_{m_1}^{(m)}) \cdot (t - mT_s)] \cdot u_T(t - mT_s)$$

$$0 \leq t \leq MT_s \tag{8.50}$$

where $T_s$ is the symbol interval, $m_1 = 1, 2 \ldots, M_1$, $m_2 = 1, 2 \ldots, M_2$, as signal multiplicities.

Also $\overline{v}_{m_1 m_2}^{(m)}(t)$ is the complex-valued transmitted signal during the $m^{th}$ signalling interval, $\overline{p}_{m_2}^{(m)}$ is the two-dimensional (2D) vector that gives the signal amplitudes and phases over $\hat{x}$ and $\hat{y}$ directions where $f_c$ is the carrier frequency with offsets of $f_{m_1}^{(m)}$ (increment or decrement) representing the assigned frequency from the TLD, and $u_T(t)$ is a rectangular pulse of unit amplitude. The complex-valued correlation coefficient between two different frequency-polarization modulated signals $\overline{v}_{m_1 m_2}^{(m)}(t)$ and $\overline{v}_{i_1 i_2}^{(m)}(t)$ is defined as [20]:

$$\rho \left[ \overline{v}_{m_1 m_2}^{(m)}(t), \overline{v}_{i_1 i_2}^{(m)}(t) \right] = \frac{1}{\sqrt[2]{E_{v_{m_2}} \cdot E_{v_{i_2}}}} \int_{(k-1)T_s}^{kT_s} LP \left[ \overline{v}_{m_1 m_2}^{(m)}(t) \cdot \overline{v}_{i_1 i_2}^{(m)*}(t) \right] dt \tag{8.51}$$

where $LP[.]$ denotes the low-pass component of the complex-valued quantity. $E_{v_{m_2}}$ and $E_{v_{i_2}}$, are the signal energies of the transmitted $\overline{v}_{m_1 m_2}^{(m)}(t)$ and $\overline{v}_{i_1 i_2}^{(m)}(t)$, respectively. The asterisk (*) denotes the complex conjugate. The F-PolSK signals satisfy the orthogonality principle if the magnitude of the correlation coefficient $\left| \rho[\overline{v}_{m_1 m_2}^{(m)}(t), \overline{v}_{i_1 i_2}^{(m)}(t)] \right|$ is zero [20]. It can be shown that the orthogonality can be satisfied when the minimum frequency separation between any adjacent frequency tones is $\Delta f = 1/T_s$ [19]. Thus, $f_{m_1}^{(m)}$ can take its value from the set $\{r_{m_1}^{(m)} \Delta f / 2\}_{m_1=1}^{M_1}$, where $r_{m_1}^{(m)} = 2m_1 - 1 - M_1$.

In Equation (8.50), $\overline{p}_{m_2}^{(m)}$ defines the transmitted signal amplitudes and phases over the orthogonal $\hat{x}$ and $\hat{y}$ channels during the $m^{th}$ signalling interval, given by:

$$\overline{p}_{m_2}^{(m)} = \begin{pmatrix} \overline{p}_{x_{m_2}}^{(m)} \\ \overline{p}_{y_{m_2}}^{(m)} \end{pmatrix} = \begin{pmatrix} a_{x_{m_2}}^{(m)} e^{j\theta_{x_{m_2}}^{(m)}} \\ a_{y_{m_2}}^{(m)} e^{j\theta_{y_{m_2}}^{(m)}} \end{pmatrix} \tag{8.52}$$

where $a_{x_{m_2}}^{(m)}$ and $a_{y_{m_2}}^{(m)}$ are the amplitudes of the $(\hat{x}, \hat{y})$-components of the lightwave. Also, $\theta_{x_{m_2}}^{(m)}$ and $\theta_{y_{m_2}}^{(m)}$ are their corresponding phases. The discrete random sequences $\{\overline{p}_{m_2}^{(m)}\}_{m_2=1}^{M_2}$ and $\{r_{m_1}^{(m)}\}_{m_1=1}^{M_1}$ are stationary and independent, and are referred to as source symbol sequences. Additionally, $\overline{p}_{m_2}^{(m)}$ determines the SOP of the fully polarized lightwave given in Equation (8.50) during the $m^{th}$ symbol interval, that is corresponding to symbol number $m_2$, where $m_2 = 1, 2, \ldots, M_2$.

The electromagnetic wave $\overline{v}_{m_1 m_2}^{(m)}(t)$ consists of two sets of 2D-ary signals ($M_1 = 2^{k_1}, M_2 = 2^{k_2}$). The signals at each set are uncorrelated and orthogonal (i.e. orthogonal frequencies and SOP), and all the $M = M_1 \times M_2 \rightarrow k = k_1 + k_2$ signals have the same time duration. Now, the data source emits a data symbol from a set of $M$ symbols every $T_s = k \cdot T_b$ seconds, where $T_b$ is the bit duration.

The SOP is described using $(S_1, S_2, S_3)$ as Stokes parameters, each of which is determined by the amplitudes $a_{x_{m_2}}^{(m)}$ and $a_{y_{m_2}}^{(m)}$ plus the phase difference $\Delta\theta^{(m)} = \theta_{y_{m_2}}^{(m)} - \theta_{x_{m_2}}^{(m)}$ of $m^{th}$ symbol. These parameters are given as follow regarding the symbol number $m_2$ [9]:

$$
\overline{S}_{m_2}^{(m)} = \begin{pmatrix} S_{1_{m_2}}^{(m)} \\ S_{2_{m_2}}^{(m)} \\ S_{3_{m_2}}^{(m)} \end{pmatrix} = \begin{pmatrix} \left|p_{x_{m_2}}^{(m)}\right|^2 - \left|p_{y_{m_2}}^{(m)}\right|^2 \\ p_{x_{m_2}}^{(m)} \cdot p_{y_{m_2}}^{(m)*} + p_{x_{m_2}}^{(m)*} \cdot p_{y_{m_2}}^{(m)} \\ \left[p_{x_{m_2}}^{(m)*} \cdot p_{y_{m_2}}^{(m)} - p_{x_{m_2}}^{(m)} \cdot p_{y_{m_2}}^{(m)*}\right] e^{j\pi/2} \end{pmatrix} = \begin{pmatrix} \left(a_{x_{m_2}}^{(m)}\right)^2 - \left(a_{y_{m_2}}^{(m)}\right)^2 \\ 2a_{x_{m_2}}^{(m)} \cdot a_{y_{m_2}}^{(m)} \cdot \cos\left(\Delta\theta^{(m)}\right) \\ 2a_{x_{m_2}}^{(m)} \cdot a_{y_{m_2}}^{(m)} \cdot \sin\left(\Delta\theta^{(m)}\right) \end{pmatrix}
$$
(8.53)

The average number of photons representing the energy of the transmitted multi-SOP lightwave is directly proportional to:

$$
S_{0_{m_2}}^{(m)} = \sqrt{\sum_{i=1}^{3} \left(S_{i_{m_2}}^{(m)}\right)^2} = \left(a_{x_{m_2}}^{(m)}\right)^2 + \left(a_{y_{m_2}}^{(m)}\right)^2
$$
(8.54)

where $S_{0_{m_2}}^{(m)} = E_{s_{m_2}}/T_s$ is the fourth Stokes parameter, representing the transmitted optical waveform power with energy of $E_{s_{m_2}}$. Only three out of four Stokes parameters are mutually independent since $S_{0_{m_2}}^{(m)}$ is expressed in terms of $S_{1_{m_2}}^{(m)}$, $S_{2_{m_2}}^{(m)}$, and $S_{3_{m_2}}^{(m)}$.

Because of nonlinear effects in the optical fibre, it is very convenient to generate lightwaves with a constant power envelope [9], because the constant-envelope lightwaves are immune from relative intensity and phase noises [9]. Also, when all electromagnetic signals are transmitted on the same power level (i.e. $S_{0_{m_2}}^{(m)} = S_0$ or equivalently, $E_{s_{m_2}} = E_s$ for $m_2 = 1, 2 \ldots, M_2$) and having the same carrier frequency, then they can be assumed as the vectors in the form of $\overline{S}_{m_2}^{(m)} = S_{1_{m_2}}^{(m)}\hat{s}_1 + S_{2_{m_2}}^{(m)}\hat{s}_2 + S_{3_{m_2}}^{(m)}\hat{s}_3$. These vectors are allocated on the surface of the Poincaré sphere with a constant radius of $S_0$ [12]. For any signal constellation, the upper half of the sphere corresponds to right-oriented polarizations, and the lower half corresponds to left orientations. Poles of the sphere correspond to circular polarization with two opposite orientations. The right-hand circular polarization is presented by the points $S_{1_{m_2}} = S_{2_{m_2}} = 0$ and $S_{3_{m_2}} = S_0$, while the left-circular polarization corresponds to the points $S_{1_{m_2}} = S_{2_{m_2}} = 0$ and $S_{3_{m_2}} = -S_0$. Linearly polarized signal points are located on the equator of the Poincaré sphere.

### 8.5.2 Receiver Configuration and Signal Processing

By assuming negligible nonlinear effects in the fibre and low polarization mode dispersion (PMD), the received optical power $\overline{r}_{m_1 m_2}(t)$ at the receiver input can be expressed as [19]:

$$
\overline{r}_{m_1 m_2}(t) = \mathbf{Q} \cdot e^{-(\alpha + j\phi(w))} \cdot \overline{v}_{m_1 m_2}^{(m)}(t) \qquad 0 \leq t \leq T_s
$$
(8.55)

where $\overline{v}_{m_1 m_2}^{(m)}(t)$ is the transmitted modulated signal, $\alpha$ and $\phi(w)$ are the fibre attenuation coefficient and phase shift, respectively, and $\mathbf{Q}$ is the complex Jones matrix. $\mathbf{Q}$ is a unitary operator that takes the polarization variation into account along the fibre due to coupling between the SOPs. Additionally, due to the low chromatic dispersion over the frequency range of the transmitted field, the value of $\phi(w)$ remains constant. Moreover, $\alpha$ has a negligible effect on the analysis as all transmitted signals have the same power level. In fact, these are some of the PolSK advantages being employed in the CDMA-based optical transmission links.

The front-end of the 2D-ary F-PolSK-OCDMA receiver is initially with frequency selective arrayed waveguide grating (AWG) as illustrated in Figure 8.12(a). In the analysis we have assumed an ideal AWG so that no interference between adjacent wavelengths occurs. The wavelengths assigned to symbols (i.e. $M_1$-ary) at the transmitter are divided into $\lambda_0, \lambda_1, \dots \lambda_{M_1-1}$ through AWG. Each wavelength enters the PolSK-OCDMA demodulator as displayed separately in Figure 8.12(b) and then the symbols are extracted from each wavelength. Thereafter, SOPs are obtained from the SOP extractor block based on generated Stoke parameters in the PolSK-OCDMA decoder. In the data processor block, the FSK part of the signal is demodulated, where the FSK detector determines the part of the symbols used to generate the frequency tones $\left\{f_{m_1}^{(m)}\right\}_{m_1=1}^{M_1}$ in every symbol

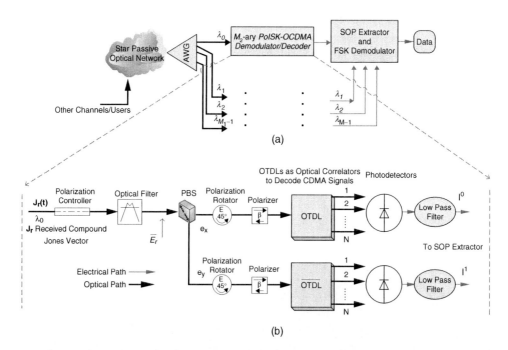

(a)

(b)

**Figure 8.12**    (a) Incoherent 2D-ary F-PolSK-OCDMA receiver and (b) structure of $M_2$-ary PolSK-OCDMA demodulator/decoder with OTDL (© 2009 IEEE. Reprinted with permission from Incoherent two-dimensional array modulation transceiver for photonic CDMA, M.M. Karbassian and H. Ghafouri-Shiraz, *J. Lightw. Technol.*, **27** (8), 2009.)

interval $m$. Also, the SOP of the transmitted optical field is estimated similar to the one described previously.

The signal alignment and analysis of the optical signals after passing the optical channel are investigated when reaching to the receiver end and ready for CDMA decoding in the OTDL and photodetection. The delay coefficients in OTDLs are designed in such a way as to make them perform as a CDMA chip decoder in both branches at each wavelength came from AWG similar to the PolSK-OCDMA receiver shown in Figure 8.7. However, the only difference is that this receiver works for $M_2$-ary PolSK with variable polarizers' output function according to the Jones vectors defining the symbols' SOPs. Consequently, Equation (8.23) is modified to a $1 \times M_2$ matrix instead, while the rest of the calculation is untouched and the same expression for SNR can be expected throughout the analysis for the PolSK-OCDMA receiver in Figure 8.12(b) [9].

### 8.5.3  Analysis of Receivers Error Probability

It is assumed that (i) $\overline{v}_{m_1 m_2}^{(m)}(t)$ is transmitted, (ii) the correct symbols are carried by the correct wavelength and (iii) the decision variables $\lambda_m$ are calculated at the demodulator for $m = 1, 2 \ldots, M_1$. Hence, the correct decision rule at the FSK demodulator can be expressed as:

$$|\lambda_m|^2 = \begin{cases} |E_s + N_{m_1}| & m = m_1 \\ |N_m|^2 & m \neq m_1 \end{cases} \tag{8.56}$$

where $E_s$ is the symbol energy and $\{N_m\}_{m=1}^{M_1}$ is an independent Gaussian noise with zero mean and PSDs of $\sigma_n^2$.

The correct decision is made if and only if $|\lambda_{m_1}|$ in the decision rule of $M_1$-ary FSK, for $m_1 = 1, 2 \ldots, M_1$ satisfies:

$$|\lambda_{m_1}|^2 = \max\{|\lambda_m|\}_{m=1}^{M_1} \tag{8.57}$$

Also, it is assumed that the received estimated noisy parameters $\overline{R}_{m_1} = (R_{1m_1}, R_{2m_1}, R_{3m_1})$ are in the decision region of the un-noisy transmitted parameters $\overline{S}_{m_2} = (S_{1m_2}, S_{2m_2}, S_{3m_2})$, as shown in Figure 8.13. Based on the maximum likelihood (ML) decision rule and assuming that all possible transmitted vectors $\{\overline{S}_l\}_{l=1}^{M_2}$ are equipower and equiprobable, the decision metric is implemented for multiple-array signalling as follows:

$$f_{\overline{R}_{m_1}|\overline{S}_{m_2}}(\overline{R}|\overline{S}_{m_2}) = \max\left\{f_{\overline{R}_{m_1}|\overline{S}_l}(\overline{R}|\overline{S}_l)\right\}_{l=1}^{M_2} \tag{8.58}$$

where $f_{\overline{R}_{m_1}|\overline{S}_l}(\overline{R}|\overline{S}_l)$, for $l = 1, 2 \ldots, M_2$, is the conditional PDF of the estimated noisy Stokes vector $\overline{R}_{m_1}$, given that $\overline{S}_l$ is transmitted. This PDF has already been given in spherical coordinates $(\rho_m, \theta_m, \phi_m)$ [12] in the form of:

$$f_{(\rho_m, \theta_m, \phi_m)}(\alpha_m, \beta_m, \delta_m) = \frac{\alpha_m}{16\pi\sigma_n^4} \cdot \sin(\beta_m) \cdot e^{\frac{E_s^2 + \alpha_m}{2\sigma_n^2}} \cdot I_0\left(\frac{E_s}{\sigma_n^2}\sqrt{\alpha_m}\cos\left(\frac{\beta_m}{2}\right)\right) \tag{8.59}$$

$$\alpha_m > 0, \quad \beta_m \in [0, \pi], \quad \delta_m \in [0, 2\pi]$$

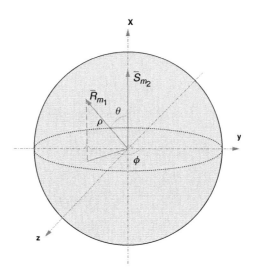

**Figure 8.13** Representation of noisy received signal in polar coordinates according to the un-noisy transmitted signals inscribed into Poincaré sphere

where $m = 1, 2 \ldots, M_2$ and $I_0(.)$ is the $0^{th}$ order modified Bessel function of the first kind. The random variables $\rho_m$ and $\theta_m$ are statistically independent of $\phi_m$, which is uniformly distributed over $[0, 2\pi][12]$. Based on the ML rules, particularly $\overline{S}_{m_2}$ was chosen as the transmitted vector when it satisfies:

$$\overline{R}_{m_1} \cdot \overline{S}_{m_2} = \max\{\overline{R}_{m_1} \cdot \overline{S}_l\}_{l=1}^{M_2} \equiv \max\{\cos(\beta_m)\}_{m=1}^{M_2} \tag{8.60}$$

Since birefringence polarization transformation only causes a rigid rotation of the signal constellation, the decision metric in Equation (8.60) is insensitive to this perturbing effect [10]. In the absence of noise, the index $m$ in Equation (8.57) and $l$ in Equation (8.58) should be equal to $m_1$ and $m_2$ of $\overline{v}_{m_1 m_2}^{(m)}(t)$, respectively.

The probability of the correct decision for the system equals the probability that satisfies Equation (8.57) times the probability that satisfies Equation (8.60) conditional on Equation (8.57). That is given by:

$$P_c = \Pr\left(|\lambda_{m_1}|^2 = \max\{|\lambda_m|\}_{m=1}^{M_1}\right) \tag{8.61}$$

$$\times \Pr\left(\overline{R}_{m_1} \cdot \overline{S}_{m_2} = \max\{\overline{R}_{m_1} \cdot \overline{S}_l\}_{l=1}^{M_2} \Big| |\lambda_{m_1}|^2 = \max\{|\lambda_m|\}_{m=1}^{M_1}\right)$$

The first probability term in Equation (8.61), for $m = 1, 2 \ldots, M_1$, is the probability that a correct decision is made on the transmitted frequency at the FSK demodulator shown in Figure 8.12(b). It is noted that the normalized decision variables $\chi_m = |\lambda_m|^2/2\sigma_n^2$ for $m = 1, 2 \ldots, M_1$ are mutually and statistically independent where they follow the chi-square distributed random variables with two degrees of freedom. It was also shown that their PDFs, conditioned on the transmitted signal $\overline{v}_{m_1 m_2}^{(m)}(t)$, are given by [19]:

$$f_{\chi_m | \overline{v}_{m_1 m_2}}(\mu | \overline{v}_{m_1 m_2}) = \begin{cases} e^{-(\mu + \gamma(K))} I_0(\sqrt{4\mu \cdot \gamma(K)}) & m = m_1, \mu \geq 0 \\ e^{-\gamma(K)} & m \neq m_2, \mu \geq 0 \end{cases} \tag{8.62}$$

where $\gamma(K)$ represents the system SNR per transmitted symbol, which is directly proportional to the number of photons representing a transmitted symbol and the number of users derived as Equation (8.49). Based on Equation (8.62), the probability of the correct decision on the transmitted frequency tone is obtained [34] as:

$$
P_c^{FSK} = \int_0^\infty e^{-(z+\gamma(K))} I_0(\sqrt{4\gamma(K) \cdot z}) \left[ \prod_{m=1}^{M_1} P(\chi_{m_1} \geq \chi_m | \chi_{m_1} = z) \right] dz
$$

$$
= \int_0^\infty e^{-(z+\gamma(K))} I_0(\sqrt{4\gamma(K) \cdot z})(1 - e^{-z})^{M_1-1} \qquad (8.63)
$$

$$
= \sum_{i=0}^{M_1-1} (-1)^i \binom{M_1 - 1}{i} \frac{1}{i+1} e^{-\frac{i}{i+1}\gamma(K)}
$$

The second probability term in Equation (8.61) is the probability that the transmitted SOP is correctly chosen at the SOP extractor shown in Figure 8.12(b), and that the correct decision is made on the transmitted frequency at the frequency demodulator. The probability of correct detection was evaluated in [9] for some regular equipower $M_2$-PolSK modulations with different variables of $n, \theta_0$ and $\theta_1$ which are constant, referring to different modulation levels as shown in Table 8.1, reproduced from [9], and given by:

$$
P_c^{PolSK} = F_\theta(\theta_1) - \frac{n}{\pi} \int_{\theta_0}^{\theta_1} \cos\left(\frac{\tan\theta_0}{\tan\tau}\right) \cdot f_\theta(\tau) d\tau \qquad (8.64)
$$

$f_\theta(\tau)$ the marginal PDF and $F_\theta(\tau)$ is the cumulative distribution function (CDF) of $\theta$, which can be derived from the joint PDF, i.e. Equation (8.59), by integrating over $\rho_m$ and $\phi_m$, gives:

$$
f_\theta(\tau) = \frac{\sin\tau}{2} e^{\frac{\gamma(K)}{2} \cdot (1-\cos\tau) \cdot \left[1+\frac{\gamma(K)}{2}(1+\cos\tau)\right]} \qquad (8.65)
$$

$$
F_\theta(\tau) = 1 - \frac{1}{2} e^{\frac{-\gamma(K)}{2} \cdot (1-\cos\tau) \cdot (1+\cos\tau)}
$$

where $\tau \in [0, \pi]$.

In the binary signalling format (BPolSK), as shown in Figure 8.1(a), the signal set consists of two antipodal points on the Poincaré sphere. The un-noisy received signals

**Table 8.1** The values for $n, \theta_0$ and $\theta_1$ for $M_2$-PolSK (© 1994 IEEE. Reprinted with permission from Multilevel polarization shift keying: optimum receiver structure and performance evaluation, S. Benedetto and P. Poggiolini, *IEEE Trans. on Comm.*, **42** (2,3,4), 1994.)

| $M_2$-PolSK | $n$ | $\theta_0$ | $\theta_1$ |
|---|---|---|---|
| 4-PolSK Circular | 2 | $\pi/4$ | $\pi/2$ |
| 4-PolSK Tetrahedron | 3 | $[\pi - \tan^{-1}(2\sqrt{2})]/2$ | $\pi - 2\theta_0$ |
| 6-PolSK Octahedron | 4 | $\pi/4$ | $\pi/2 \tan^{-1}(1/\sqrt{2})$ |
| 8-PolSK Cube | 3 | $\tan^{-1}(1/\sqrt{2})$ | $\pi/2 - \theta_0$ |

are $S_{1_{m_2}}, S_{2_{m_2}}$ where $S_{1_{m_2}} = -S_{2_{m_2}}$. Given a transmitted SOP such that the un-noisy received SOP is $\overline{S}_{m_2}$, chosen within the decision region, and the received vector $\overline{R}_{m_1}$, an error occurs each time the scalar product $\overline{R}_{m_1} \cdot \overline{S}_{m_2}$ becomes negative. This is due to the fact that the maximum likelihood criterion in this binary case implies a decision based on the sign of the scalar product (see Equation (8.29)). Hence, the error event turns out to be:

$$\{E\} = \left\{\theta > \frac{\pi}{2}\right\} \tag{8.66}$$

where $\theta = \cos^{-1}\left(\overline{R}_{m_1} \cdot \overline{S}_{m_2} / \left(|\overline{R}_{m_1}| \cdot |\overline{S}_{m_2}|\right)\right)$.

It is noted that the error event is independent of both $\rho$ and $\phi$, with this signal set. Therefore, by using $F_\theta(\tau)$ in Equation (8.65), the error probability of BPolSK can be obtained as:

$$P_e^{BPolSK} = P\left(\theta > \frac{\pi}{2}\right) = \frac{1}{2}e^{-\gamma(K)} \tag{8.67}$$

In the higher-order signalling format, the analysis presents the upper-bounds on the actual probability of errors in that the error is a function of both $\theta$ and $\phi$. The system probability of error is now chosen based on the following upper-bound:

$$P_e^{MPolSK} \leq P_e | \max \phi \tag{8.68}$$

Now for circular-quad-PolSK (CQPolSK), as shown in Figure 8.1(b), the error condition is assumed based on Equation (8.68) and signal constellation as:

$$\{E\}_{upper\ bound} = \left\{\theta > \frac{\pi}{4}\right\} \tag{8.69}$$

Thus, the error probability for CQPolSK is obtained as:

$$P_e^{CQPolSK} = 1 - F_\theta\left(\frac{\pi}{4}\right) = \frac{1}{2}\left(1 + \frac{\sqrt{2}}{2}\right) e^{-\gamma(K)\left(1\frac{\sqrt{2}}{2}\right)} \tag{8.70}$$

This scheme has an SNR penalty of 2.5 dB with respect to BPolSK in a generic polarization-modulated system [9].

Similarly for tetrahedron-quad-PolSK (TQPolSK), as shown in Figure 8.1(c), the error condition can be derived by half the angle subtended between the centre of the sphere and two adjacent signal points as follows:

$$\{E\}_{upper\ bound} = \{\theta > 0.304\pi\} \tag{8.71}$$

Thus the probability of error can be calculated as:

$$P_e^{TQPolSK} = 1 - F_\theta(0.304\pi) = (0.7882) \cdot e^{-\gamma(K)(0.4226)} \tag{8.72}$$

It is again worth noting that this scheme has an SNR penalty of 0.8 dB with respect to BPolSK in a generic polarization-modulated system.

Also for the cubic-PolSK (8-PolSK), as shown in Figure 8.1(d), the error condition, derived as above, is:

$$\{E\}_{upper\ bound} = \{\theta > 0.196\pi\} \tag{8.73}$$

And the resulting error probability bound is thus:

$$P_e^{8-PolSK} = 1 - F_\theta(0.196\pi) = (0.9082) \cdot e^{-\gamma(K)(0.1835)} \tag{8.74}$$

This scheme also has an SNR penalty of 2.6 dB with respect to BPolSK in a generic polarization-modulated system.

These bounds indicate that the multilevel polarization modulation in transmission lines can be accomplished with high performance at the cost of relatively small SNR penalties.

Finally, having the probability of $M_1$-FSK, $M_2$-PolSK and CDMA encoded SNR $\gamma(K)$, the overall system error probability, which denotes the BER of the transceivers, is derived [33] as:

$$P_e^{F-PolSk} = 1 - P_c = 1 - \left[ \sum_{i=0}^{M_1-1} (-1)^i \binom{M_1-1}{i} \frac{1}{i+1} e^{\frac{-i}{i+1}\gamma(K)} \right] \tag{8.75}$$

$$\times \left[ F_\theta(\theta_1) - \frac{n}{\pi} \int_{\theta_0}^{\theta_1} \cos\left( \frac{\tan\theta_0}{\tan\tau} \right) f_\theta(\tau) d\tau \right]$$

## 8.6 Performance of F-PolSK-OCDMA Transceiver

In this section, the BER probability of hybrid F-PolSK-OCDMA transceivers as a function of single-user SNR $\gamma(1)$ (shown as *Sdb* on the graphs) has been evaluated based on the above analysis. The numerical results of the proposed architecture are demonstrated in Figures 8.14 and 8.15. Further studies are found in [33].

Figure 8.14 shows the performance of the hybrid transceiver under conditions of $P = 17$, $Sdb = 14$ dB and binary FSK modulation with different polarization constellations. Figure 8.14 clearly indicates that binary FSK with binary PolSK enhances the overall BER as compared with the other polarization constellations. The 2-FSK/2-PolSK configuration is able to accommodate 44 simultaneous users, for $P = 17$. It has been observed that by increasing the polarization constellation, although the system complexity grows, the performance degrades. This is because the decision region for higher degrees of polarization becomes smaller, and subsequently the demodulation process becomes more complicated and requires precisely designed components. The proposed coding scheme and architecture can support more throughput as the spreading code is much smaller than Gold sequences of 511 or 1023 employed in the literature [6, 8].

It has been found that the combination of two binary modulations has potential as a secure, efficient and accommodating OCDMA architecture. Therefore, the variation of the binary F-PolSK-OCDMA transceiver's BER against the number of active users with different single-user SNR is illustrated in Figure 8.15 when $P = 23$. As can be seen, the higher SNR values reduce the error rate as well as enhancing the network capacity. The employed SNR values are still adequate for given circumstances to make the proposed architecture very power efficient.

## 8.7 Long-Haul PolSK Transmission

Here we discuss the considerations required for long-haul transmission regarding the signal degradations. The discussion in this section considers the sensitivity results for

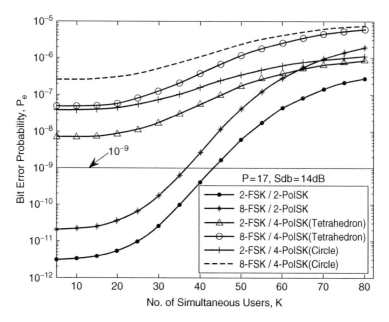

**Figure 8.14** BER performances of BFSK/$M_2$-PolSK-OCDMA receivers against the number of simultaneous active users, $K$

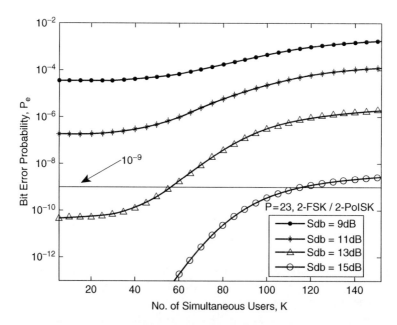

**Figure 8.15** BER performances of binary F-PolSK-OCDMA receivers with different single-user SNRs against the number of simultaneous active users, $K$

direct-detection (DD) and hybrid-polarization-based modulations and tries to point out the main PolSK advantages and drawbacks over other modulation formats. We stick to linear transmission regimes, while the nonlinear effects are seen as impairments.

## 8.7.1 Direct-Detection PolSK

The transmission format universally adopted so far for DD is intensity modulations (IM) such as on-off keying (OOK) or pulse position modulations (PPM), and obviously the simplicity is the major benefit. The principal drawback is inconstant signal power which makes the signal exposed to self-phase modulation. In long-haul systems, this phenomenon restricts the maximum amplifier output power. This implies a limit to the maximum amplifier spacing or in the other word the maximum bit rate. In contrast, PolSK produces a constant-envelope modulation format and in principle it is immune from self-phase modulation. In addition, fundamentally both theoretical [10] and experimental [4] results suggest that the impact on multicarrier PolSK (e.g. WDM) of four-wave mixing is substantially lower than on multicarrier IM/DD, again allowing for higher per-channel power levels. In multicarrier OCDMA systems the extra power margin could be used to increase either the number of code channels, the per-channel bit rate or the amplifier span.

Since the nonlinearity is proportional to the signal power, it is noted that the peak power required for binary PolSK transmission is half that needed for IM/DD. This implies that all nonlinear effects also benefit from the PolSK scheme. In addition to the nonlinear effects, dispersion is the other main limiting factor in long-haul high-capacity networks. To counter the effects of dispersion, the easiest and potentially most effective solution is still that of using a narrow-spectrum signal. The dispersion-shifted fibres potentially allow for dispersion-free transmission. However, it has been shown [3] that operating a thousands of kilometres link at the zero dispersion point causes a catastrophic build-up of beat ASE-signal four-wave-mixing contributions.

PolSK has an information-limited bandwidth which is the same as that of PSK or IM/DD. Waveform shaping can provide further narrowing, better than other methods including MSK, due to the two orthogonal polarizations (i.e. two channels) in the fibre. By appropriately forming the modulated waveform, the data bandwidth can be divided between the two channels, meaning ideally a halving of the spectral occupancy for a given bit rate. Since self-phase modulation is virtually absent in PolSK modulated signals, unlike externally modulated IM/DD, ideally spectrum re-broadening does not occur even at high average power in the fibre. Another way of substantially reducing the effect of dispersion is multilevel modulation. This technique offers an improvement that is quadratic with respect to the number of bits per symbol that are transmitted. If three bits/symbol are transmitted, the system tolerates a nine-time-higher dispersion than a binary system transmitting at the same bit rate due to the fact that the signal spectrum is then three times narrower and the duration of the symbol is three times longer. Moreover, multilevel transmission considerably relieves the requirements on the bandwidth expansion of the transmitter and receiver electronics and optoelectronics to support higher bit rates.

In this chapter, we have shown that multilevel polarization modulation alongside hybrid F-PolSK has a very good sensitivity performance. For realistic values of the optical filter bandwidth both 4-PolSK and 8-PolSK incur no substantial penalty with respect to binary PolSK. When three or more bits per symbol are sent, PolSK even outperforms DPSK.

The generation of 4-PolSK and 8-PolSK constellations has been experimentally obtained with a specifically designed modulator [1].

To summarize, as far as binary transmission is concerned, it seems that DD-PolSK could be a very interesting competitor for the long-haul high-performance system segment, because it is immune to self-phase modulation and substantially more tolerant to four-wave mixing. It is noted that PolSK transmitters and receivers may be more complex than IM/DD. However, in long-haul systems the head-end complexity is less of a factor than in short-haul distribution systems. In fact, in transoceanic links the cost of the head-end transmission and reception equipment may be less considered regarding the billion-dollar investment required in deploying the cables and transceivers. The state-of-the art IM/DD links are likely to require two external modulators at the transmitter (one for data modulation and the other for polarization scrambling) and numerous other tricks and technologies to counter the effects of dispersion and nonlinear phenomena. DD-PolSK still requires a substantial amount of both theoretical and experimental research to fully prove its potential and features, whereas the above considerations in the authors' opinion make it a technique valuable to investigate.

## 8.7.2 Noises in Polarization Modulated Systems

The necessity of polarization control, shot noise tolerance, phase noise tolerance and bandwidth occupancy are considered in the following. The homodyne and heterodyne detection schemes significantly require the states of polarization (SOP) control.

The basic methods of polarization tracking and control are reported in [32]. The use of rather sophisticated optics and electronics are common to all methods. The other option can be polarization diversity [15] where the received lightwave is divided into two orthogonal polarizations that are separately heterodyne-detected and then combined after phase adjustment. Polarization diversity usually features relatively simpler optics and no electronics, but more complicated receiver architecture. The full Stokes receiver binary-PolSK [7] does not require optical polarization control but an electronic feed-forward. As for the shot noise tolerance, PolSK shows 3.5 and 3 dB of penalty with respect to the best performing systems i.e. heterodyne PSK or phase-diversity DPSK [35].

Differential demodulation of continuous-phase (CP) FSK exhibits shot noise figures which depend on the modulation index $m$. An acceptable performance, as with DPSK, can be obtained for $m = 0.5$, i.e. with minimum shift keying (MSK) modulation [20]. However, it is shown that $m = 0.5$ is somewhat controversial. It can be shown that the delay-line receiver allows the use of the matched IF filtering only with DPSK, but not with MSK [21]. In short, the best performance of MSK cannot reach that of DPSK.

Dual filter envelope FSK shows shot noise performance equal to that of PolSK whereas single filter envelope FSK and envelope ASK have 3 dB degradations. SOP tracking is not required by the differential (D)PolSK scheme [1]. When implementation simplicity and receiver start-up time are a premium, this scheme could be considered a serious candidate, in spite of its slightly poorer shot noise performance, which shows 2.4 dB degradations compared with PolSK. Concerning bandwidth occupation, PolSK systems present a power spectral density (PSD) in two orthogonal channels whose magnitudes depend on the SOPs. It has been noted that the shape of the continuous part of the PSD, and thus its bandwidth, on both channels is the same as those of ASK, PSK and DPSK. Both CPFSK and envelope FSK present a wider spectral occupancy, whose bandwidth depends on the modulation index $m$.

We can guess from the digital radio systems that the best trade-offs between bandwidth and power efficiency obtained so far was by using two-dimensional, i.e. binary modulation, schemes such as PSK, DPSK and quadrature-amplitude modulation (QAM). In this chapter, we have proposed and preliminarily analysed multilevel PolSK modulations. The error probability showed that the third degree of freedom available in the Stokes space allows the shot noise tolerance to come closer to that of PSK and DPSK when increasing the number of signal constellations. As discussed in detail in [9], the 8-PolSK (cubic set) is approximately 0.6 dB better than the 8-DPSK and only 2.1 dB worse than the 8-PSK.

For a given bit rate, multilevel transmission reduces the system bandwidth and thus the required speed for the electronics. This can allow the use of digital signal processing applied to forward error correcting (FEC) codes and faster electronic decision rules to achieve even more enhanced BER or coding gains at digital coherent receivers.

### 8.7.2.1  Phase Noise in F-PolSK Modulation

As introduced and discussed earlier, laser phase noise is a major source of performance degradation at the receiver in optical communication systems. PolSK detection is highly insensitive to laser phase noise in a fashion similar to ASK and FSK, unlike PSK and DPSK. Since F-PolSK modulation can be considered as an extension to conventional PolSK modulation over orthogonal domains, it might be expected to show very low sensitivity to the laser phase noise effects [19]. In fact, F-PolSK, like PolSK, can be considered as a class of phase noise cancelling heterodyne receiving scheme. For the special case of BPolSK, it has been shown that this scheme has exactly the same phase noise insensitivity as the dual filter FSK receiver [36]. However, infrequent sudden fluctuations in the laser phase noise may affect this scheme [20]. As mentioned, this problem can be solved by choosing the demodulation IF filter bandwidth to be wider than that of the IF optimal matched filter, which has a comparable bandwidth to the transmitted signal.

### 8.7.2.2  Fibre Dispersion in F-PolSK Modulation

Another major source of performance degradation in high-capacity high-speed optical communication systems is dispersion, including chromatic dispersion (CD) and polarization mode dispersion (PMD). Although the effect of CD is deterministic, employing electrical or photonic equalization techniques may not completely compensate for the random nature of dispersions. The effect of dispersion is characterized by broadening the transmitted signals in time, resulting in producing inter-symbol interference (ISI) between the received signals at the receiver end. Accordingly, extensive work has been accomplished to propose optical or electrical equalizers in order to minimize power penalties in optical systems due to dispersion effects [1, 4, 9–12]. It is reported in [5, 9] that the simplest and potentially most effective solution for overcoming the effect of dispersion is multilevel (higher order) modulations. Multilevel modulations reduce the transmitted signal spectrum that can be achieved by the increase in the symbol duration. For example, if four bits per symbol are sent, the system tolerance against dispersion is 16 times higher than binary systems transmitting at the same bit rate [1]. Beside their tolerance to dispersion effects, multilevel modulations ease the requirements on the bandwidth of the transmitter and receiver optoelectronics. In fact, multilevel modulations do not need to realize data up to the bit rate, but only up to the symbol-rate. This fact, along with dispersion tolerance property, enables very high throughput as well.

## 8.8  Summary

The polarization-modulated OCDMA technique has been introduced in this chapter followed by a novel incoherent transceiver architecture which employed optical tapped-delay lines (OTDLs) to decode CDMA signals. From a detailed analysis, we obtained the system SNR and accordingly demonstrated the overall network BER performance.

Furthermore, the transceiver design of the proposed 2D-ary frequency-polarization modulated OCDMA has also been presented and analysed. The generated signals have the advantage of spreading over higher dimensional constellations which provides greater geometric distances between the transmitted signals. The results demonstrated that the binary combination of two modulations remarkably improves the transceivers' performance that can reliably and power-efficiently accommodate a greater number of simultaneous users that implies capacity enhancement of 10–15 per cent as compared with similar systems with existing coding schemes. It should be mentioned that the overall promising performance of PolSK modulation is a trade-off between complex architecture and physical implementation.

Summarizing the long-haul considerations, we can say that binary PolSK modulation schemes represent interesting alternatives to both envelope-based modulations and phase modulations. In most cases a direct comparison between BPolSK schemes and other systems highlights advantages and disadvantages, thus the ultimate choice should consider the technological challenges and requirements of specific applications in detail.

Moreover, the system security is also boosted by two-dimensional advanced modulation in the optical domain. The performance of OCDMA receivers in cooperation with the spreading code have been presented, taking into account the effects of optical ASE noise (i.e. optical filter), electronic receiver noise (i.e. LPF), photodetector shot-noise and mainly the multiuser interferences (i.e. MAI). The results indicated that the architectures can reliably and power-efficiently accommodate greater numbers of simultaneous users.

## References

1. Carena, A. *et al.* (1998) Polarization modulation in ultra-long haul transmission system: a promising alternative to intensity modulation. In: *ECOC*, Madrid, Spain.
2. Born, M., Wolf, E. and Bhatia, A.B. (1999) *Principles of optics*. 7th ed. Cambridge University Press, Cambridge, UK.
3. Cimini, L.J. *et al.* (1987) Preservation of polarization orthogonality through a linear optical system. *Electronics Letters*, **23** (25), 1365–1366.
4. Benedetto, S. *et al.* (1994) Coherent and direct-detection polarization modulation system experiment. In: *ECOC*, Firenze, Italy.
5. Huang, J. *et al.* (2005) Multilevel optical CDMA network coding with embedded orthogonal polarizations to reduce phase noises. In: *ICICS*, Bangkok, Thailand.

6. Iversen, K., Mueckenheim, J. and Junghanns, D. (1995) Performance evaluation of optical CDMA using PolSK-DD to improve bipolar capacity. In: *Proc. SPIE*, Amsterdam, The Netherlands.
7. Betti, S., Marchis, G.D. and Iannone, E. (1992) Polarization modulated direct detection optical transmission systems. *J. Lightw. Technol.*, **10** (12), 1985–1997.
8. Tarhuni, N., Korhonen, T.O. and Elmustrati, M. (2007) State of polarization encoding for optical code division multiple access networks. *J. Electromagnetic Waves Applications (JEMWA)*, **21** (10), 1313–1321.
9. Benedetto, S. and Poggiolini, P. (1994) Multilevel polarization shift keying: optimum receiver structure and performance evaluation. *IEEE Trans. on Comm.*, **42** (2/3/4), 1174–1186.
10. Benedetto, S. and Poggiolini, P. (1992) Theory of polarization shift keying modulation. *IEEE Trans. on Comm.*, **40** (4), 708–721.
11. Betti, S. *et al.* (1991) Homodyne optical coherent systems based on polarization modulation. *J. Lightw. Technol.*, **9** (10), 1314–1320.
12. Benedetto, S., Guadino, R. and Poggiolini, P. (1995) Direct detection of optical digital transmission based on polarization shift keying modulation. *IEEE J. Selected Areas Comm.*, **13** (3), 531–542.
13. Huang, W., Andonovic, I. and Tur, M. (1998) Decision-directed PLL for coherent optical pulse CDMA system in the presence of multiuser interference, laser phase noise, and shot noise. *J. Lightw. Technol.*, **16** (10), 1786–1794.
14. Foschini, G.J. and Vannucci, G. (1988) Noncoherent detection of coherent lightwave signals corrupted by phase noise. *IEEE Trans. on Comm.*, **36** (3), 306–314.
15. Gisini, N., Huttner, B. and Cyr, N. (2000) Influence of polarization dependent loss on birefringent optical fiber networks. In: *OFC*.
16. Liang, W. *et al.* (2008) A new family of 2D variable-weight optical orthogonal codes for OCDMA systems supporting multiple QoS and analysis of its performance. *Photonic Network Communications*, **16** (1), 53–60.
17. Batayneh, M. *et al.* (2008) Optical network design for a multiline-rate carrier-grade Ethernet under transmission-range constraints. *J. Lightw. Technol.*, **26** (1), 121–130.
18. Yang, C.C., Huang, J.F. and Hsu, T.C. (2008) Differentiated service provision in optical CDMA network using power control. *IEEE Photonics Tech. Letters*, **20** (20), 1664–1666.
19. Matalgah, M.M. and Radaydeh, R.M. (2005) Hybrid frequency-polarization shift keying modulation for optical transmission. *J. Lightw. Technol.*, **23** (3), 1152–1162.
20. Pun, S., Chan, C. and Chen, L. (2005) A novel optical frequency-shift keying transmitter based on polarization modulation. *IEEE Photonics Tech. Letters*, **17** (7), 1528–1530.
21. Bass, M. and Stryland, E.W.V. (2002) *Fiber Optics Handbook: Fiber, Devices, and Systems for Optical Communications*. McGraw-Hill, USA.
22. Ilyas, M. and Moftah, H.T. (2003) *Handbook of optical communication networks*. CRC Press.
23. Gho, G.H., Klak, L. and Kahn, J.M. (2011) Rate-adaptive coding for optical fiber transmission systems. *J. Lightw. Technol.*, **29** (2), 222–233.
24. Sabella, R. and Lugli, P. (1999) *High speed optical communications*. Kluwer Academic Publishers.
25. Kaminow, I.P. (1981) Polarization in Optical Fibres. *IEEE J. Quantum Electronics*, **QE-17** (1), 15–21.
26. Oskar, M. (1996) *Fundamental of bidirectional transmission over a single optical fibre*. Kluwer Academic Publishers, The Netherlands.
27. Kazovsky, L.G. (1989) Phase and polarization diversity coherent optical techniques. *J. Lightw. Technol.*, **7** (2), 279–292.
28. Noe, R. *et al.* (1991) Comparison of polarization handling methods in coherent optical systems. *J. Lightw. Technol.*, **9** (10), 1353–1365.
29. Okoshi, T., Ryu, S. and Kikuchi, K. (1983) Polarization diversity receiver for heterodyne/coherent optical fibre communications. In: *IOOC*, Tokyo, Japan.
30. Karbassian, M.M. and Ghafouri-Shiraz, H. (2008) Transceiver architecture for incoherent optical CDMA networks based on polarization modulation. *J. Lightw. Technol.*, **26** (24), 3820–3828.
31. Wang, X. and Kitayama, K. (2004) Analysis of beat noise in coherent and incoherent time-spreading OCDMA. *J. Lightw. Technol.*, **22** (10), 2226–2235.
32. Shin, S. *et al.* (2000) Real-time endless polarization tracking and control system for PMD compensation. In: *OFC*, Baltimore, USA, Paper TuP7-1.

33. Karbassian, M.M. and Ghafouri-Shiraz, H. (2009) Incoherent two-dimensional array modulation transceiver for photonic CDMA. *J. Lightw. Technol.*, **27** (8), 980–988.

34. Gamachi, Y. *et al.* (2000) An optical synchronous M-ary FSK/CDMA system using interference canceller. *J. Electro. Comm. Japan*, **83** (9), 20–32.

35. Cooper, A.B. *et al.* (2007) High spectral efficiency phase diversity coherent optical CDMA with low MAI. In: *Lasers and Electro-Optics (CLEO)*.

36. Idler, W. *et al.* (2004) Advantages of frequency shift keying in 10 Gb/s systems. In: *IEEE Workshop on Advanced Modulation Formats*, USA.

# 9

# Optical CDMA Networking

## 9.1 Introduction

In the past decade we have witnessed significant development in the area of optical networking. Such advanced technologies as dense wavelength-division multiplexing (DWDM), optical amplification, optical path routing, e.g. optical cross connect (OXC), reconfigurable optical add-drop multiplexer (ROADM) and high-speed switching have found their way into the wide-area networks (WANs), resulting in a substantial increase of the telecommunications backbone capacity and greatly improved reliability.

At the same time, enterprise networks converged on 100 Mbps fast Ethernet architecture [1]. Some LANs even moved to gigabit data rates [2], courtesy of a new gigabit Ethernet (GbE) standard adopted by the Institute of Electrical and Electronics Engineers (IEEE 802.3z and 802.3ab).

An increasing number of households have more than one computer. Home networks allow multiple computers to share a single printer or a single Internet connection. Most often, a home network is built using a low-cost switch or a hub that can interconnect 4 to 16 devices. Builders of new houses now offer an option of wiring a new house with a category-5 (CAT-5 or RJ45) cable to facilitate home networks or end-user demarcations [3]. Older houses have an option of using existing phone wiring, in-house power lines, or an ever more popular wireless network, based on the IEEE 802.11 standards. Different flavours of this standard can provide up to 150 Mbps bandwidth up to 250 m outdoors (IEEE 802.11n-2009). Whether it is a wireless or a wired solution, home networks are a sort of tiny LAN providing high-speed interconnection for multiple devices.

These advances in the backbone, enterprise and home networks, coupled with the tremendous growth of Internet traffic have accentuated the frustrating delay of access network capacity. The last-mile (from provider's point of view) still remains the bottleneck between high-capacity LANs and the backbone network.

In this chapter, we discuss the solutions for access networks from briefly recalled existing architectures to next generation architectures including various flavours of passive optical networks (PON) with an the emphasis on optical CDMA as an access technique

*Optical CDMA Networks: Principles, Analysis and Applications*, First Edition. Hooshang Ghafouri-Shiraz and M. Massoud Karbassian.
© 2012 John Wiley & Sons, Ltd. Published 2012 by John Wiley & Sons, Ltd.

and its compatibility with Internet protocol (IP) PON and multi-protocol label switching (MPLS) configurations. For complete use of efficient bandwidth and asynchronism with the concept of CDMA in optical domain, a few popular random access protocols for optical CDMA networks are also introduced.

### 9.1.1 Current Solutions

The most widely deployed broadband solutions today are digital subscriber line (DSL) and cable modem (CM) networks. Although they are improvements compared to old 56 Kbps dial-up lines, they are unable to provide enough bandwidth for emerging services such as full duplex video conferencing.

#### 9.1.1.1 Digital Subscriber Line (DSL)

Digital subscriber line (DSL) uses the same twisted pair as telephone lines and requires a DSL modem at the customer premises and a DSL access multiplexer (DSLAM) in the central office. The basic premise of the DSL technology is to divide the spectrum of the line into several regions with the lower 4 kHz used by plain old telephone service (POTS) equipment, while the higher frequencies are being assigned for higher-speed digital communications. There are a few basic types of DSL connections [4]. The basic DSL is designed with integrated services data network (ISDN) compatibility in mind. It has 160 Kbps symmetric capacity and provides the user with either 80 or 144 Kbps of bandwidth, depending on whether the voice circuit is supported or not. The high-speed digital subscriber line (HDSL) is made compatible with a T1 rate of 1.544 Mbps. The original specification required two twisted pairs. However, a single-line solution was standardized by International Telecommunication Union (ITU G.991.1).

#### 9.1.1.2 Asymmetric DSL

Asymmetric DSL (ADSL) is the most widely deployed flavour of DSL. It uses one POTS line and has an asymmetric line speed. In the downstream direction, the line rate could be in the range of 750 Kbps to 2 Mbps on loops of 5 km. On shorter loops, the rate can be as high as 10 Mbps. In the upstream direction, the rate could be in the range of 128 to 750 Kbps. The actual rate is chosen by the ADSL modem based on the link circumstances.

Finally, the very-high-speed digital subscriber line (VDSL) can have a symmetric or an asymmetric line speed. It achieves much higher speed than HDSL or ADSL, but operates over much shorter loops. The rates could range from 13 Mbps for 1500 m loops to 52 Mbps over 300 m loops [4].

While the maximum ADSL transmission capacity is 2 Mbps, in reality it could go much lower depending on the line conditions. Twisted-pair wires allow a number of impairments, most significant of which are crosstalk, induced noise, bridged taps and impulse noise. To cope with such impairments, ADSL employs a multicarrier modulation approach known as discrete multitone. In this approach, the system transmits data on multiple subcarriers in parallel. It adapts to line conditions by varying the bit rate on each subcarrier channel. A good channel may carry as many as 15 bits/symbol/s, while a really noisy channel may carry no data at all.

The asymmetric nature of ADSL is prompted by observation of user traffic at the time. The downstream traffic is a result of downloading large files and web pages, while the upstream traffic primarily consisted of short commands, uniform resource locator (URL) requests, and/or server login queries. Consequently, ADSL adopted a 10:1 ratio of the downstream bandwidth to the upstream bandwidth, with AT&T even advocating as high as a 100:1 ratio [5].

It is interesting to note that the highly asymmetric nature of the traffic is a feature of the past. New and emerging applications tend to skew the ratio toward greater symmetry. Such applications as video conferencing or storage-area networks (SAN) require symmetric bandwidth in both directions. A big impact on traffic symmetry can be attributed to peer-to-peer applications. It is reported that the recent ratio of downstream to upstream traffic is approximately 1.4:1 [3].

### 9.1.1.3   Community Access Television (CATV) Networks

The CATV networks were originally designed to deliver analogue broadcast TV signals to subscriber TV sets. Following this objective, the CATV networks adopted a tree topology and allocated most of its spectrum for downstream analogue channels. Typically, CATV is built as a hybrid fibre-coax (HFC) network with fibre running between either a head-end or a hub (optical node) and the final drop to the subscriber through coaxial cables as illustrated in Figure 9.1. The coaxial part of the network uses repeaters (amplifiers) to split the signal among many subscribers.

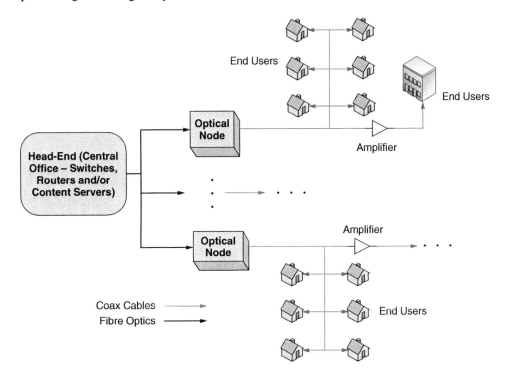

**Figure 9.1**   Hybrid fibre-coax architecture

The major limitation of CATV architecture for carrying modern data services is that this architecture was originally designed only for broadcast analogue services. Out of a total cable spectrum width of about 740 MHz, the 400 MHz band is allocated for downstream analogue signals, and the 300 MHz band is allocated for downstream digital signals. Upstream communications are left with about a 40 MHz band or about 36 Mbps of effective data throughput per optical node. This very modest upstream capacity is typically shared among 500 to 1000 subscribers, resulting in frustratingly low speed during peak hours [5].

## 9.1.2  Next Generation Networks (NGN)

Optical fibre is capable of delivering bandwidth-intensive, integrated voice, data and video services at distances beyond 20 km in the subscriber access network. A straightforward way to deploy optical fibre in the local access network is to use a point-to-point (P2P) topology, with dedicated fibre runs from the central office to each subscriber as shown in Figure 9.2(a). While this is a simple architecture, in most cases it is cost prohibitive because it requires significant outside fibre plant deployment as well as connector termination space in the local exchange.

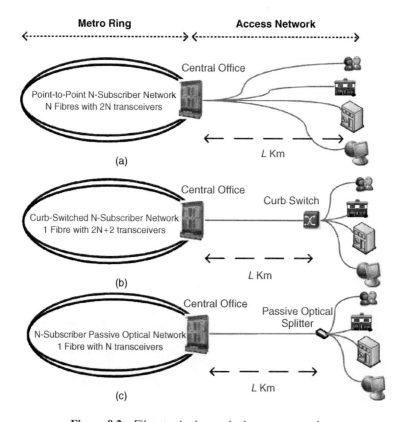

**Figure 9.2**  Fibre to the home deployment scenarios

As shown in Figure 9.2, considering $N$ subscribers at an average distance of $L$ km from the central office, a P2P architecture, Figure 9.2(a), requires $2N$ transceivers and $N \times L$ total fibre length, even assuming that a single fibre is used for bidirectional transmission [6].

To reduce fibre deployment, it is possible to deploy a remote concentrator (switch) close to the neighbourhood. This will reduce the fibre consumption to only $L$ km by assuming negligible distance between the switch and customers, whereas it will actually increase the number of transceivers to $2N + 2$, as there is one more link added to the network, as illustrated in Figure 9.2(b). In addition, curb-switched network architecture requires electric power as well as backup power at the curb switch. Currently, one of the most significant operational expenditures for local exchange carriers is that of providing and maintaining electric power in the local loop. Therefore, it is logical to replace the tough active curb-side switch with an inexpensive passive optical splitter [7].

### 9.1.2.1  Passive Optical Network (PON)

The passive optical network (PON) is a technology viewed by many as an attractive solution to the first-mile problem [1, 8]. A PON minimizes the number of optical transceivers, central office terminations and fibre deployment. It is a point-to-multipoint (P2MP) optical network with no active elements in the signal path from source to destination. The only interior elements used in PON are passive optical components, such as optical fibre, splices and splitters. An access network based on a single fibre PON only requires $N + 1$ transceivers and $L$ km of fibre.

An optical line terminal (OLT) at the central office is connected to many optical network units (ONU) at remote nodes through one or multiple 1:$N$ optical splitters. The network between the OLT and the ONU is passive, i.e. it does not need any power supply. An example of a PON using a single optical splitter is shown in Figure 9.2(c). The presence of only passive elements in the network makes it relatively more fault-tolerant and decreases its operational and maintenance costs once the infrastructure has been laid down. A typical PON uses a single wavelength for all downstream transmissions (from OLT to ONUs) and another wavelength for all upstream transmissions (from ONUs to OLT), multiplexed on a single fibre.

PON technology is being given more and more attention by the telecommunication industry as the first-mile solution. Advantages of using PON for local access networks are numerous [1, 5, 7, 8], including:

- A PON-based local loop can operate at distances of up to 20 km and even much longer with long-reach PON techniques which considerably exceed the maximum coverage afforded by various flavours of DSL.
- Only one strand of fibre is needed in the trunk (i.e. private exchange box), and only one port per PON is required in the central office. This allows for very dense central office equipment and low power consumption.
- PON provides higher bandwidth due to deeper fibre penetration. Although the fibre-to-the-building (FTTB), fibre-to-the-home (FTTH) or even fibre-to-the-PC (FTTPC) solutions have the ultimate goal of fibre reaching all the way to the customer premises, fibre-to-the-curb (FTTC) may be the most economical deployment today.

- PON eliminates the necessity of installing multiplexers and demultiplexers in the splitting locations, thus relieving network operators of the dreadful task of maintaining them and providing power to them. Instead of active devices in these locations, PON has passive components that can be buried in the ground at the time of deployment.
- PON allows easy upgrades to higher bit rates or additional wavelengths. Passive splitters and combiners provide complete path transparency.

### 9.1.2.2 Wavelength Division Multiple Access PON

One possible way of separating the ONUs' upstream channels is to use wavelength division multiple access (WDMA), in which each ONU operates on a different wavelength. From a theoretical perspective, it is a simple solution but it remains cost-prohibitive for an access network. A WDMA solution would need either a tuneable receiver or a receiver array at the OLT to receive multiple channels. An even more serious problem for network operators would be to have a wavelength specific ONU inventory instead of having just one type of ONU – there would be multiple types of ONUs differing in their laser wavelengths.

Each ONU would have to use a laser with narrow and controlled spectral width, and thus will become more expensive. It would also be more problematic for an unqualified user to replace a defective ONU because a unit with the wrong wavelength might interfere with some other ONU in the PON. Using tuneable lasers in ONUs may solve the inventory problem, but it is still expensive at the current state of technology. For these reasons, a WDMA-PON network is not an attractive solution in today's environment [2]. Several alternative solutions based on WDMA have been proposed, e.g. wavelength routed PON [9] where it uses an arrayed waveguide grating (AWG) instead of a wavelength-independent optical splitter/combiner. The AWG is also a passive device with a fixed routing matrix. An AWG provides fixed routing (i.e. there is no flexibility in routing and dealing with traffic) of an optical signal from a given input port to a given output port based only on the wavelength of the signal.

In one version of WDMA-PON, ONUs use external modulators to modulate the signal received from the OLT and send it back upstream. This solution is not cheap either, it requires additional amplifiers at or close to the ONUs to compensate for signal attenuation after the round-trip propagation, and it needs more expensive optics to limit the reflections, since both downstream and upstream channels use the same wavelength. Also, to allow independent (non-arbitrated) transmission from each of $N$ ONUs, the OLT must have $N$ receivers – one for each ONU [10].

In another version, ONUs contain inexpensive light-emitting diodes (LEDs) whose wide spectral band is sliced by the AWG on the upstream path. This approach still requires multiple receivers at the OLT. If a single tuneable receiver is used at the OLT, then a data stream from only one ONU can be received at a time, which would be very similar to time-division multiple-access (TDMA) PON. Both versions suffer heavily from beat noise generated through various wavelengths in the link [11].

### 9.1.2.3 TDMA-PON

In TDMA-PON, simultaneous transmissions from several ONUs will collide when they reach the combiner. To avoid data collisions, each ONU must transmit in its own

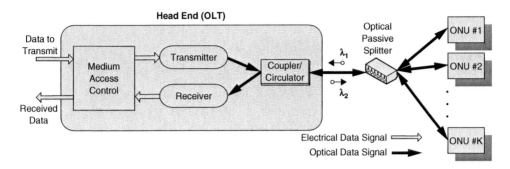

**Figure 9.3**  PON using a single fibre

transmission window (time-slot). One of the major advantages of a TDMA-PON is that all ONUs can operate on the same wavelength and are absolutely identical in design configuration. The OLT will also need a single receiver. A transceiver in an ONU must operate at the full line rate, even though the bandwidth available to the ONU may be lower. This property also allows the TDMA-PON to efficiently change the bandwidth allocated to each ONU by altering the assigned time-slot size, or even to employ statistical multiplexing to fully utilize the PON channel capacity [12]. In a subscriber access network, most of the traffic flows are downstream (from network to user) and upstream (from user to the network), not peer-to-peer (from user to user). Thus, it seems reasonable to separate the downstream and upstream channels. In doing so, two wavelengths are used: $\lambda_1$ for the upstream transmission and $\lambda_2$ for the downstream transmission as shown in Figure 9.3.

The channel capacity on each wavelength can be flexibly divided between the ONUs by using time-sharing techniques. Time-sharing appears to be the preferred method today for optical channel sharing in an access network, as it allows for a single upstream wavelength and a single transceiver in the OLT, resulting in a cost-effective solution. The medium access control (MAC) unit in Figure 9.3 will be discussed later.

### 9.1.2.4  Asynchronous Transfer Mode PON (APON)

In 1995, seven network operators formed the full service access network (FSAN) initiative with the goal of creating a unified specification for broadband access networks. FSAN members developed a specification for a PON-based optical access network that uses asynchronous transfer mode (ATM) as its layer-2 protocol, i.e. the medium access control (MAC) protocol. Such systems were called APON, an abbreviation for ATM-PON. The name APON was later replaced with BPON for broadband-PON. The name change was a reflection of the system's support for broadband services such as Ethernet access, video distribution and virtual private network (VPN) or leased-line services [13].

The downstream frame consists of 56 ATM cells (53 bytes each) for the basic rate of 155 Mbps, scaling up to 224 ATM cells for 622 Mbps. There are two dedicated cells referred to physical layer operation, administration and maintenance (PLOAM) cells; one at the beginning of the frame, and one in the middle [14]. The remaining 54 cells are

ATM data cells. The upstream transmission is in the form of bursts of ATM cells, with a three-byte physical overhead appended to each 53-byte ATM cell to allow for burst mode receivers. Burst mode receivers are required at the OLT to synchronize the different ONUs which may be located at different distances from the OLT. The ATM cell may be either an ATM data cell, or a PLOAM cell.

In the downstream, the PLOAM cells are used to carry grants from the OLT to the ONUs. Each grant is a onetime permission for an ONU to transmit payload data in an ATM cell – 53 grants for the 53 upstream frame cells are mapped into the PLOAM cells. The OLT sends a continuous stream of grants to the entire ONUs in the PON. Thus the OLT can moderate the portion of the upstream bandwidth assigned to each ONU. In the upstream direction, the PLOAM cells are used by the ONUs to transmit their queue sizes to the OLT. This information shall be used by the OLT for bandwidth allocation.

### 9.1.2.5 Gigabit PON (GPON)

GPON and BPON have been produced by the FSAN group and the ITU-T standards organization. As it was originally framed in 2002, GPON was designed to be an inclusive standard, covering the carriage of TDM, data and video services, using both ATM and packet payloads, at a wide variety of speeds and loss budgets. This strategy helped to promote quick adoption of GPON, since it could accommodate pretty much any service and fit into a wide range of network situations [7].

It proposes bit rates of up to 2.5 Gbps. It also aims towards providing higher efficiency while carrying multiple services over the PON. It proposes a protocol using the generic framing procedure (GFP) [15], which provides a generic mechanism to adapt traffic from higher layers (e.g. Ethernet, MAC/IP) over a transport layer such as synchronous optical network – synchronous digital hierarchy (SONET-SDH). Other functionalities such as dynamic bandwidth assignment, operation and maintenance etc., are borrowed from APONs. As discussed in [13], both APON and GPON have the disadvantage of complex protocols and implementations, because of which they have not gained much popularity among users and equipment vendors. Various service providers such as NTT and Bell-South have tested initial deployments and test beds of BPON [16].

### 9.1.2.6 Ethernet PON (EPON)

In January 2001, the IEEE formed a study group called Ethernet in the first-mile (EFM). This group was chartered with extending existing Ethernet technology into subscriber access area, focusing on both residential and business access networks [1, 8]. The group set the goal of providing a significant increase in the performance while minimizing equipment, operations and maintenance costs. Ethernet over PONs became one of the focus areas of EFM. EPON is a PON-based network that carries data traffic encapsulated in Ethernet frames as defined in the IEEE 802.3 standard. It uses a standard 8b/10b line coding (eight data bits encoded as ten line bits) and operates at standard Ethernet speed of 1 Gbps. EPON utilizes the existing 802.3 specification where possible, including the usage of existing 802.3 full-duplex MAC.

Ethernet has become a universally accepted standard, with hundreds of millions of ports deployed worldwide, offering staggering economies of scale [17]. High-speed gigabit Ethernet deployment is widely accelerating, and 10-gigabit Ethernet (10 GbE) products are becoming available, while 40/100 GbE is just around the corner. Ethernet, which is easy to scale and manage, is gaining new ground in wide area networks (WAN). Given that more than 95 per cent of enterprise LANs and home networks use Ethernet, it becomes obvious that ATM-PON may not be the best choice to interconnect two Ethernet networks [6, 14].

One of the drawbacks of the ATM is its high overhead for carrying variable-length IP packets, since they are the dominant Internet traffic. Below we compare the overheads imposed by Ethernet frame encapsulation and ATM cell encapsulation. It should be mentioned that a 1 Gbps Ethernet link has an actual line rate of 1.25 Gbps. The rate increase is performed by the physical coding sublayer (PCS) which preconditions the line signal by encoding each user byte with a 10-bit codeword [18]. The increased rate is only visible at the physical layer, whereas MAC, MAC-interface, and the MAC clients operate at the 1 GbE rate. Therefore, this encoding generally is not considered an overhead.

For a particular IP datagram size distribution obtained in an access network [19], the Ethernet frame encapsulation overhead equals 7.42 per cent, significantly lower than the ATM cell encapsulation overhead of 13.22 per cent [14]. The improved efficiency is just one of the advantages of using variable-size Ethernet frames to carry variable-size IP packets. ATM switches and network cards are significantly more expensive than Ethernet switches and network cards [18].

Newly adopted quality-of-service (QoS) techniques have made Ethernet networks capable of supporting voice, data and video. These techniques include full-duplex transmission mode, prioritization and virtual LAN (VLAN) tagging [1]. It is not surprising that Ethernet is poised to become the architecture of choice for next-generation subscriber access networks.

## 9.2   OCDMA-PON

Optical CDMA technique is a point-to-multipoint technology where each end-user picks up its own message from the broadcast signal. Similarly, PON architecture is also point-to-multipoint access technology with passive components, such as splitters, couplers, fibre-optics etc., where potentially the cost is reduced. The first-mile is a network with a central office (CO) which serves multiple users. There are several multipoint topologies suitable for the access network, including tree, ring or bus [5]. All transmissions in a PON are performed between an optical line terminal (OLT) and optical network units (ONU) which are premises or general end-users. The architecture of PON using a single fibre link is illustrated in Figure 9.4. The OLT may contain all encoder-decoder pairs required for communication with each ONU or a smaller number of tuneable encoders-decoders. The OLT resides in the CO and connects the optical access to the backbone or long-haul transport network.

TDMA-PON and WDM-PON have also been enabled to date. Even though TDMA-PON utilizes the bandwidth of fibre efficiently, it has limitations in its increased transmission speed, difficulty in burst synchronization and traffic control, low security

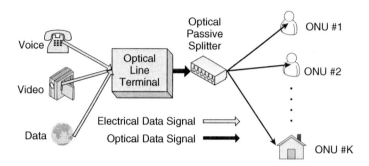

**Figure 9.4** PON architecture using single fibre link (© 2009 IEEE. Reprinted with permission from Analysis of scalability and performance in passive optical CDMA network, M.M. Karbassian and H. Ghafouri-Shiraz, *J. Lightw. Technol.*, **27** (17), 2009.)

and complex dynamic bandwidth allocation requirement [18]. The emerging WDM-PON overtook TDMA due to required bandwidth growth, whereas it came with extravagant cost from precise wavelength-dependent components. In addition, the effect of statistical multiplexing is insignificant in multimedia communications environments [20]. Although WDM-PON has several advantages over TDMA-PON, it has found its way to hardly any industries due to its high operation and maintenance costs.

OCDMA-PON, where each subscriber's channel is given its own code for spreading and de-spreading, is a good alternative in view of cost, simplicity and noise reduction [11]. The OCDMA link is transparent to the input channel's data protocol with security. It supports bursty traffic and random access protocols, which will be discussed later. Furthermore, the optical beat noise problem – which often arises in a system using several laser diodes, as in optical subcarrier multiplexing or WDM – does not have much effect on OCDMA-PON [22].

In this chapter, two of the previously introduced and analysed transceiver architectures are used in the context of networking. The coherent homodyne BPSK-OCDMA – introduced in Chapter 6 – is utilized in the proposed OCDMA-PON line terminal and network units. Also, a network node configuration featured with Internet protocol (IP) traffic transmission in hybrid $M$-ary FSK-OCDMA network – introduced in Chapter 7 – is studied in detail.

## 9.3   OCDMA-PON Architecture

The architectures of transmitter, receiver, optical network unit (ONU) and optical line terminal (OLT) as part of the OCDMA-PON are explained in detail here. The transmitter structure of the coherent homodyne BPSK-OCDMA with external Mach–Zehnder (MZ) phase modulator as an optical external modulator (OEM) is shown in Figure 9.5. The outgoing data is first BPSK encoded in order to generate the in-phase and quadrature-phase (IQ) signals electrically to drive the MZ modulator [23]. Then, the encoded BPSK-signal drives the MZ modulator to phase-modulate the lightwave. Finally, the lightwaves are CDMA encoded by means of the group-padded modified prime codes

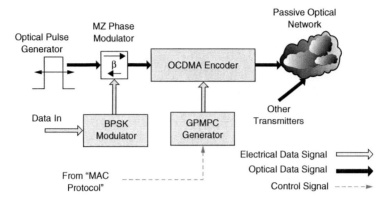

**Figure 9.5** Coherent OCDMA transmitter in PON architecture (© 2009 IEEE. Reprinted with permission from Analysis of scalability and performance in passive optical CDMA network, M.M. Karbassian and H. Ghafouri-Shiraz, *J. Lightw. Technol.*, **27** (17), 2009.)

(GPMPC), introduced in Section 2.4.5, and multiplexed via couplers and transmitted over the PON as a network infrastructure.

At the receiver, the local oscillator, which is modulated with the pre-reserved GPMPC address code (introduced in Chapter 2) as shown in Figure 9.6, is mixed and correlated with the received OCDMA signal. The polarization controller makes sure that all users have the same polarization to reduce any polarization-sensitive noise on the photodetectors (PD). In the CDMA decoding process, the portion of the received signal encoded with the same GPMPC spreading code sequence at the transmitter (i.e. intended data for the intended receiver) is de-spread, whereas signals encoded with other GPMPC spreading code sequences (i.e. MAI) are further spread and reduced. The coherently mixed optical signals are incident on a dual-balanced detector whose electrical output conserves the phase information. The generated bipolar electrical signal is integrated over a bit interval and the result is compared to a reference to form the final bit estimation.

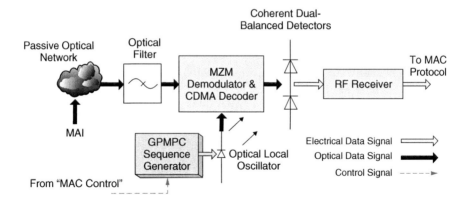

**Figure 9.6** Proposed coherent OCDMA receiver in PON architecture (© 2009 IEEE. Reprinted with permission from Analysis of scalability and performance in passive optical CDMA network, M.M. Karbassian and H. Ghafouri-Shiraz, *J. Lightw. Technol.*, **27** (17), 2009.)

Next, the passive optical network architecture in which coherent OCDMA is employed as a multiple-access technique is investigated. The optical line terminal configuration of this OCDMA-PON is shown in Figure 9.7. The multiple access is achieved by using GPMPC sequences as the address code to identify users in the all-optical domain. In the downstream from the optical line terminal to the optical network unit, at the 1550 nm wavelength, the optical pulses are encoded at the OLT by means of MZ external modulator driven by ASIC GPMPC generator at the transmitters and every user is assigned with one unique sequence code.

Since the OLT serves a number of ONUs, it contains multiple transceivers consisting of a reconfigurable GPMPC generator. The signal is then coupled and transmitted over fibre link to the receivers, i.e. ONUs, where each user is separated and identified by its optical address and medium access control (MAC). The ONU configuration of this OCDMA-PON is shown in Figure 9.8. In the upstream channel from ONU to OLT, at the 1310 nm wavelength, the signals are optically decoupled and divided to the decoder at the OLT where the information from each user is obtained, together with the MAC signal control from the ONUs to the OLT.

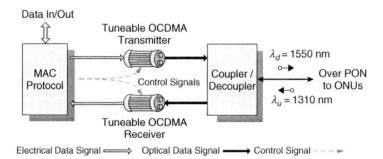

**Figure 9.7** Configuration of OLT in the OCDMA-PON architecture (© 2009 IEEE. Reprinted with permission from Analysis of scalability and performance in passive optical CDMA network, M.M. Karbassian and H. Ghafouri-Shiraz, *J. Lightw. Technol.*, **27** (17), 2009.)

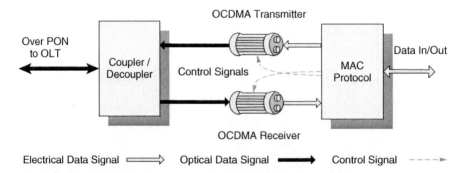

**Figure 9.8** Configuration of ONU in the OCDMA-PON architecture (© 2009 IEEE. Reprinted with permission from Analysis of scalability and performance in passive optical CDMA network, M.M. Karbassian and H. Ghafouri-Shiraz, *J. Lightw. Technol.*, **27** (17), 2009.)

The MAC signal is also fed back to the access protocol transmitter to manage the network operation, for example, for the allocation of the GPMPC to each user as shown in Figures 9.7 and 9.8. Usually, a stable upstream wavelength is required with a stabilized laser source at the transmitter of the ONU. The downstream signal passes through the decoupler and goes into the detector, then the data information for the user is separated by an optical correlation operation with their unique address sequence. The downstream control signal is also obtained and passed to the network control unit. For the upstream, the signal from the ONU to the OLT is encoded by the GPMPC for user identification by the optical encoder and is then transmitted towards the OLT through the fibre link. The MAC protocol can be carrier-sense multiple-access with collisions detection (CSMA/CD). However, a few common random access protocols will be discussed later in this chapter.

In this OCDMA-PON, the signal can be potentially modulated by frame information for data load switching as well as address code sequences for user identification. This brings the compatibility of the architecture to work also with IP and label switching techniques for routing and traffic management. By considering the ring topology with a number of nodes (i.e. OLT), the OLT can be treated as one of the nodes which generally link to the number of ONUs. The downstream and upstream traffic is at the different wavelengths and then can be broadcast on the same fibre link. For example, one node can add/drop data traffic by a 2 × 2 coupler. One port of the coupler is connected to a 2 × 2 optical cross connector (OXC) and the other port links to the fibre ring attached to the nodes based on the OCDMA-PON. The OXC control signals can be generated through an optical routing table or a label switch paradigm as discussed in [24]. The downstream traffic from a node and the upstream traffic from the ONU pass through the same optical coupler, where the former is directed to the OLT and the latter is directed to an ONU or a user. This architecture proposes a transparent protocol, flexible user allocation and mainly all-optical operations with cost effective solution since it reduces wavelength-sensitive devices with complex operations and control used in the other schemes.

## 9.3.1   OCDMA-PON Transmission Analysis

Before OCDMA-PON can be considered for use in large-scale networks, it must demonstrate scalability in terms of fibre transmission distance. Scaling is a concern because the encoding process broadens the spectra of individual OCDMA transceivers. This may result in higher sensitivity to wavelength-dependent impairment such as beat noise and dispersions in wideband multiwavelength transmission. Power-budget-based analysis of the network scalability relatively mitigates that impairment and is more practical to study. In the following we study the impact of coding parameters, number of nodes (i.e. number of ONUs per OLT), channel link length and optical components characteristics on the system biterror rate (BER).

If we assume that each node supports up to $N_u$ users (i.e. ONUs), then the number of nodes in the network ($N_n$) can be expressed as:

$$N_n = \frac{N_T}{N_u} \tag{9.1}$$

where $N_T$ is the total number of users in the network. The upstream signal power, from ONU to OLT, must satisfy the following power budget [21, 25]:

$$R_S \leq P_{UT} - \alpha_c \cdot N_n - \alpha_F \cdot L - \alpha_{IL} - \delta^2_{other} \tag{9.2}$$

where $P_{UT}$ is the upstream transmitter output power of ONU, $\alpha_c$ is coupler/decouplers' loss, $\alpha_F$ is the fibre attenuation coefficient, $L$ is the length of the fibre link, $\alpha_{IL}$ is the optical filter's insertion loss and $R_S$ is the photodetector (PD) sensitivity. Similarly, the downstream traffic power, from OLT to ONU, must satisfy the following power budget [25, 26],

$$R_S \leq P_{DT} - \alpha_c \cdot N_n - \alpha_F \cdot L - \alpha_{IL} - C \cdot \log_2^{S_P} - \delta^2_{other} \tag{9.3}$$

where $P_{DT}$ is the downstream transmitter power of OLT, $C$ is the filtering index and $S_P$ is the splitting ratio. In the two power budget equations above, the equivalent noise power budget term, shown as $\delta^2_{other}$, includes the noise contribution from the CDMA en/decoder $\delta^2_{coder}$ and the noise contribution from the MAI $\delta^2_{MAI}$. Thus:

$$\delta^2_{other} = \delta^2_{coder} + \delta^2_{MAI} \tag{9.4}$$

The en/decoder noise $\delta^2_{coder}$ comprises the MZ modulator voltage drift relevant to the number of chips and the chip duration that can be approximated as an average by 1 dB [27]. $\delta^2_{MAI}$ can be introduced as:

$$\delta^2_{MAI} = (K-1) \cdot \delta^2_{MAI-single} \tag{9.5}$$

where $K$ is the number of active users in the network (i.e. sending and receiving data). $\delta^2_{MAI-single}$ has been discussed and introduced in [25, 28] as follow:

$$\delta^2_{MAI-single} = \Re^2 \cdot P_{UT} \cdot P_{DT} \cdot Var[C_{mn} \cdot \cos(\theta_i - 2\pi f \tau_i)]/N^4 \tag{9.6}$$

where $\Re$ is the responsivity of the PDs, $Var[.]$ is the variance function, $\theta_i$ is the CDMA encoded phase angle of the $i^{th}$ user, $f$ is the optical carrier frequency, $\tau_i$ is the propagation delay between the $i^{th}$ user transmitter and the corresponding receiver, $N$ is the spreading code-length and $\delta^2_{MAI-single}$ is the variance of an interfering signal of a single user, i.e. its cross-correlation value. Based on the above analysis, the network scalability for this OCDMA-PON architecture will be discussed later.

To conserve the information contained in the phase of the optical carrier, coherent detection is deployed, whereby a local optical source is coherently combined with the received information-bearing signal. By following the procedure analysed in Chapter 6, the integration of the detector output, over a bit interval $T$, will result in (i.e. Equation (6.7)):

$$S_{out} = \Re \sum_{i=1}^{K} \int_0^T l(t)s(t)dt + \sqrt{N_0} \int_0^T n(t)dt \tag{9.7}$$

$$= \Re \hat{L} \hat{S} b_0^1 T + \sqrt{\Re T} \hat{L} \int_0^T n(t)dt + \Re \hat{L} \hat{S} \sum_{i=2}^{K} \left[ b_{-1}^i R_{i,1}(\tau_i) + b_0^i \hat{R}_{i,1}(\tau_i) \right] \cos \theta_i$$

where $l(t)$ is the local oscillator's signal with power of $\hat{L}$, $s(t)$ is the received signal with power of $\hat{S}$, $b_0^1$ represents the information bit being detected, $b_{-1}^i$ and $b_0^i$ are the overlapping of the previous and the following bits of the $i^{th}$ user, $N_0$ is the noise power spectral density (PSD) and $R_{i,j}(\tau)$ and $\hat{R}_{i,j}(\tau)$ are the continuous-time partial correlation functions. The noise $n(t)$ at the optical receiver includes mainly thermal and shot noise, relative intensity noise and the fibre attenuation, e.g. amplified spontaneous emission (ASE) noise. The thermal noise $\delta_{th}^2$ is given by:

$$\delta_{th}^2 = (2k_B^{B_r} T_r T)/(e^2 R_L) \tag{9.8}$$

where $k_B$ is the Boltzmann constant, $B_r$ is the ratio of the equivalent receiver bandwidth to the signal bandwidth, $T_r$ is the receiver noise temperature, $R_L$ is the receiver load resistance and $e$ is the fundamental electron charge. When a relatively high-power local oscillator is employed, the receiver operates under the shot-noise-limited regime (i.e. only shot noise remains). Then the noise has a one-sided power spectral density of $N_0 = \Re T \hat{L}^2$. Nevertheless, the shot noise $\delta_{sh}^2$ is introduced by [28]:

$$\delta_{sh}^2 = \hat{S}^2(2+m^2)/(8G_{PD}B_S^2) \tag{9.9}$$

where $m$ is the modulation index, $G_{PD}$ is the PD processing gain ratio ($G_{PD} = 60$) and $B_S$ is the baseband signal bandwidth.

The relative intensity noise $\delta_{RIN}^2$ is also introduced [26] as:

$$\delta_{RIN}^2 = 2P_{RIN} \cdot \hat{S}^2 \cdot R_b \tag{9.10}$$

where $P_{RIN}$ is the intensity PSD and $R_b$ is the data bit rate.

The fibre link noise $\delta_{link}^2$ such as ASE from the optical amplifiers is also defined as:

$$\delta_{link}^2 = \sum_{i=1}^{K} \frac{1}{2} i.e.R_L \cdot P_P \cdot \Re \cdot B_w + 2B_w \Re^2 [(\eta_{sp}(G_{amp} - 1)hv)/(\eta G_{amp})]^2 \tag{9.11}$$

where $P_P$ is the optical power per pulse, $B_w$ is the optical components bandwidth, $\eta_{sp}$ is the spontaneous emission factor, $hv$ is the photon energy, $\eta$ is the PD quantum efficiency and $G_{amp}$ is the gain of optical amplifiers.

Now with all the main contributing noise sources, the total noise $n(t)$ can be expressed as:

$$\delta_{n(t)}^2 = \delta_{th}^2 + \delta_{sh}^2 + \delta_{RIN}^2 + \delta_{link}^2 \tag{9.12}$$

The noise $n(t)$ is assumed to be a Gaussian random variable with zero mean and unit variance; all data bits are independent and equiprobable and the delays are independent and uniformly distributed over one bit interval. By following the analysis for a coherent homodyne system, the OCDMA-PON transmission signal-to-noise ratio (SNR), with respect to the number of active users $K$, is derived as:

$$SNR(K) = \frac{\Re^2 \hat{L}^2 \hat{S}^2 T^2}{\Re^2 \hat{L}^2 \hat{S}^2 T^2 \dfrac{K-1}{3N} + N_0 \cdot \delta_{n(t)}^2} = \frac{1}{\dfrac{K-1}{3N} + \dfrac{\delta_{n(t)}^2}{\hat{S}^2 \Re T}} \tag{9.13}$$

It should be noted that the signal-to-noise ratio for a single-user is: $SNR(1) = \hat{S}^2 \Re T/\delta_{n(t)}^2$.

## 9.3.2 Performance Discussion of OCDMA-PON

In this section some numerical results are presented based on the above analysis. The parameters used for the simulation are listed in Table 9.1 [25]. For the spreading code, GPMPC with $P = 23$ is employed that makes the code-length and total number of users 575 and 529, respectively.

We can obtain the maximum fibre length for acceptable receiver sensitivity. The number of nodes in this network architecture and the number of tolerable ONUs per node are shown in Figures 9.9 and 9.10, respectively, with different downstream transmitter power. As the results in Figure 9.9 show, the maximum fibre length decreases as the number of nodes in the network increases. As the results in Figures 9.9 and 9.10 show, the maximum fibre length decreases as the number of nodes, while the number of ONUs per node in the network increases. The maximum accessible length of fibre link also increases with higher transmitter output optical power. The results in Figure 9.9 show that by decreasing the number of nodes in the network, the distant between the central office (CO) and ONUs increases markedably. For example, when the downstream transmitter output power is 5 dBm and the network is able to support 10 nodes (i.e. $N_T/N_u = 10$), then the OLT and ONUs can be 25 km (maximum) apart, which indicates an enhanced power efficiency [22, 25, 26].

On the other hand, the investigation of the scalability with respect to the number of users is illustrated in Figure 9.10. It is apparent that the accessibility reduces by growth in the number of ONUs (i.e. users).

**Table 9.1**  OCDMA-PON link parameters

| Descriptions | Symbols | Values |
|---|---|---|
| Downstream transmitter output power | $P_{DT}$ | 5–6 dBm |
| Upstream transmitter output power | $P_{UT}$ | −4 dBm |
| Photodetector sensitivity | $R_S$ | −35 dBm |
| Coupler coefficient loss | $\alpha_c$ | 1 dB |
| Fibre attenuation coefficient | $\alpha_F$ | 0.2 dB/km |
| Optical filter insertion loss | $\alpha_{IL}$ | 1 dB |
| Filtering index | $C$ | 3 |
| Splitting ratio | $S_P$ | 16–64 |
| Receiver load resistor | $R_L$ | 1030 $\Omega$ |
| Amplifier gain | $G_{amp}$ | 20 dB |
| Photodetector processing gain | $G_{PD}$ | 60 |
| Photodetector quantum efficiency | $\eta$ | 0.8 |
| Ratio of the equivalent receiver's bandwidth | $B_r$ | 100 MHz |
| Chip duration | $T_c$ | 0.1 ns |
| Receiver baseband bandwidth | $B_s$ | 1 GHz |
| Receiver noise temperature | $T_r$ | 600 K |
| Modulation index | $m$ | 100 |
| Fibre length | $L$ | 0–45 km |

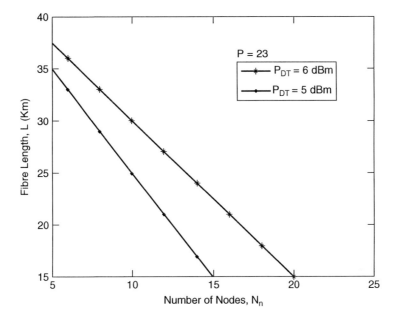

**Figure 9.9**   Fibre length against the tolerable number of nodes, $N_n$

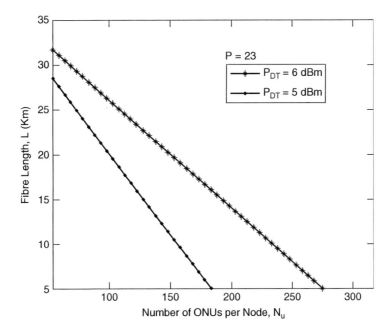

**Figure 9.10**   Fibre length against the tolerable number of ONUs per node, $N_u$

A reasonable number of 100 active users (i.e. 19 per cent total of the number of users when $P = 23$) can be accommodated up to 20 or 25 km fibre link as shown in Figure 9.10 when transmitter power is 5 dBm or 6 dBm, respectively.

The overall network performance in terms of BER against the number of active users and received signal power (i.e. $\hat{S}^2$) is investigated in Figures 9.11 and 9.12. Figure 9.11 shows that the BER degrades as the number of users increases due to increasing multiple user interferences from which CDMA inherently suffers, and obviously the higher the received power, the lower BER obtained. Figure 9.11 illustrates that 162 users can be reliably accommodated when $P_r = -20$ dBm at $BER = 10^{-9}$.

To support a greater number of users, higher $P$ value and higher received power should be considered; however, there will then be a balance in the network throughput and the number of users since the greater $P$ means longer code-length. The results are comparable with the CDMA-PON and WDM-PON studied in [20, 22, 26], since they indicate that this coding scheme and architecture enhanced the network capacity and also decreased the power consumption.

Figure 9.12 shows the BER performance of the network against the received signal power under the presence of various numbers of active users to share the channel from 15 to 45 per cent of the total number of users. Results in Figure 9.12 indicate that a lower received power is required for fewer users (e.g. 15 and 25 per cent) to maintain $BER = 10^{-9}$, since fewer users means less interference and accordingly higher SNR.

The BER reaches $10^{-9}$ by accommodating 35 per cent of the total number of users with the received power of $-17$ dBm which is still very power-efficient. To further improve the performance of the OCDMA-PON, we need to consider the degradation problems from the MAI, and improvement in the optical encoder/decoders.

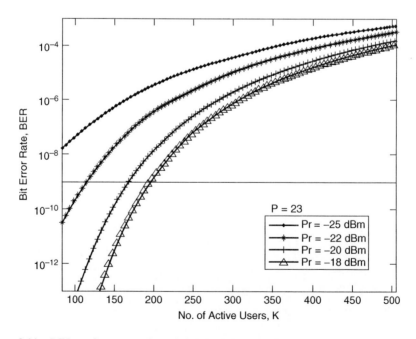

**Figure 9.11**    BER performance of the OCDMA-PON against the number of active users, $K$

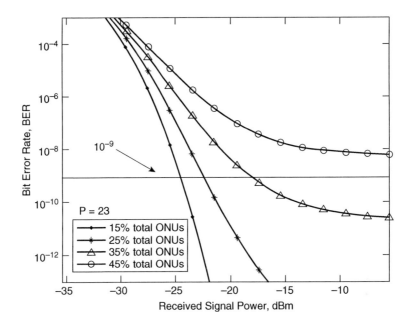

**Figure 9.12**    BER performance of the proposed OCDMA-PON against the received signal power

## 9.4   IP Traffic over OCDMA Networks

The IP routing operates electrically in the network layer hence it cannot be processed at a speed matching the ultrafast data rate offered by the fibre-optic. Therefore this issue has become the main challenge in optical networking.

In the current WDM networks [29] the electronic IP router receives the selected wavelength channels at its input ports, converts the data from optical to electronic form, and finally routes the packets by forwarding them through the output ports. In wavelength routed networks [9] the direct wavelength path can be established by introducing OXC switches at each node. The opportunity to establish better routing increases as the number of wavelengths increases, which means that wider bandwidth, a larger set of wavelengths and wavelength-sensitive devices will be required. Therefore, it will be very advantageous in the future to execute as many tasks as possible in the optical domain such as routing/switching [30] and dynamic signal processing [31] although they are still under development.

Another solution to keep the optical signals in the optical domain for further processing is the label switching scheme. The labels (i.e. header-like tag attached to the data or packet frame) can be used to establish end-to-end paths that are called label switched paths (LSP). Multi-protocol label switching (MPLS) is a switching protocol between layers 2 and 3, adding labels in packet headers and forwarding labelled packets in corresponding paths using switching instead of routing [32]. Optical MPLS will be explained later in this chapter, at Section 9.6. This is what exactly is done in the OCDMA concept if it is utilized as a network access protocol and then has the potential to support label switching as well. Major applications of MPLS are traffic engineering. Generalized MPLS (GMPLS) extends

MPLS to add a signalling and routing control plane for devices in the packet domain, time domain or wavelength domain, providing end-to-end provisioning of connections, resources and quality of services (QoS).

Even though (G)MPLS forms a good and fast solution, it cannot on its own solve the mismatch of the router switching speed with the data speed in the fibre, because the look-up table processing is still slower than the data rate. In an attempt to overcome this, research started focusing on optical packet switching (OPS) [33] and optical label swapping (OLS) [34], where the packet header (label) is processed (all) optically. OLS implements the packet-by-packet routing and forwarding functions of (G)MPLS directly in the optical domain. Ideally, this approach can route packets independent of bit rate, packet format and packet length. Advantages of OPS are particularly evident in core networks, where OLS can be used to replace both OXCs and IP core routers. With regard to OXCs, OLS is a multiclient transport platform used by IP, SDH, gigabit and Ethernet (GbE) clients to manage the bandwidth more efficiently [35]. With regard to IP routers, OLS offers an aggregation layer; it implies using multiple network cables/ports, e.g. Ethernet, in parallel to increase link speed [36]. The IP network is simplified (avoiding core devices) through the transport infrastructure realized by OPS nodes. From a networking perspective, an all-optical node is defined as a high-throughput packet-switched node. However, processing capabilities are rather limited, and the node essentially limits itself to a forwarding function based on the label of the incoming packets. In metro-regional networks, transport functionality is currently realized by means of different solutions such as SDH or WDM rings, etc.

MPLS can be then a solution since at the intermediate nodes a packet is forwarded only according to its label [32]. Since network layer label analysis is avoided, significant processing time is saved at each hop. The end-to-end delay can also be significantly reduced because IP routing is only needed at the edge routers. Although MPLS partially relieves the IP routing, the electrical routing scheme will still become a bottleneck as IP traffic increases. OPS can be another solution by use of pure optical signal processing. However, there are many difficulties in contention resolution and optical buffers [33] that make OPS still an immature technology, although the state-of-the-art industry platform for optical packet networking is emerging [37].

### 9.4.1 IP Transmission over OCDMA Network

The architecture of IP transmission in the OCDMA network is shown in Figure 9.13. At every transmitter network node, the destination of each incoming IP packet is recognized, the packet recognition can be performed by an address correlation process and then the packet is saved into the buffer. The buffer is divided into $K$ first-in-first-out (FIFO) subparts where $K$ is the total number of users accommodated in the network. IP packets that are destined for different receivers are stored in different subparts accordingly. When IP packets are to be routed to the same receiver, they are saved in one FIFO subpart in order [38].

It is important to note that the purpose of storing IP packets separately according to their destination address is to transmit all the IP traffic to the same receiver at one time and at a high speed, once the total length reaches a predefined threshold. Thus, the optical encoder is adjusted for the number of packets belonging to the same user rather than

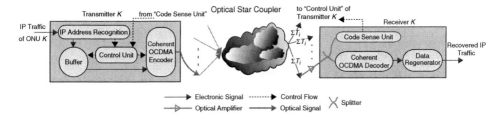

**Figure 9.13** IP routing and transmission over OCDMA network architecture

being tuned for every incoming IP packet individually. As a result, the encoder adjusting time requirement is significantly reduced. The control unit is responsible for recording the total traffic of each subpart. When the total traffic is greater than a certain value, i.e. a threshold, the control unit tries to send the packets to the assigned address. Before sending, the optical encoder has to be adjusted according to the desired address sequence, i.e. the GPMPC generator in Figure 9.14(a). It should be noted that with a higher threshold, each packet has to wait for a longer time in the buffer before transmission. When the threshold is large, the buffer delay becomes predominant. However, due to the higher transmission speed, proper selection of a threshold value will make this delay acceptable, even for real-time services. The star coupler mixes all the incoming optical signals and this superimposed signal is amplified and transmitted to the receiver of each user.

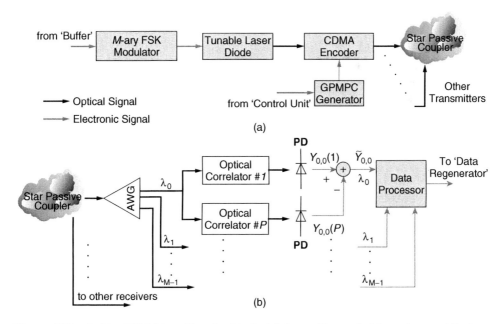

**Figure 9.14** Inside OCDMA en/decoder block (a) transmitter and (b) receiver architectures (© 2009 IEEE. Reprinted with permission from IP routing and transmission analysis over optical CDMA networks, M.M. Karbassian and H. Ghafouri-Shiraz, *J. Lightw. Technol.*, **27** (17), 2009.)

At the receiver network node, the optical decoder retrieves the right signal and regenerates the original data stream. When the GPMPC is employed, the number of users can be as large as the size of the code (i.e. $K = P^2$). It should be noted that when two (or more) transmitters send signals to the same receiver at the same time, a collision may occur. In order to prevent the collision, a code sense unit is used to sense whether others are sending data to the same address. In fact, the sense unit can be a correlator to recognize the intended user via the auto-correlation value, similar to the way that the IP recognition unit at the transmitter extracts the packet header. The sensing procedures can be similar to the CSMA/CD protocol; however, a modified one is required to fit in the timing and packet length specifications, which could be a subject for future work. Other functions of the code sense unit are to check whether the optical encoder is adjusted correctly to the desired address code and to prioritize the users, to avoid collision. It is noted that there must be one user that we can send data to, immediately after a collision, since $K$ different code sequences are assigned to $K$ users separately.

In addition, the probability of a collision is small because of the large number of available code sequences, although a few popular random access protocols will be studied later here. In this network, owing to the use of coherent OCDMA transmitter and incoherent receiver techniques, not only is fibre bandwidth utilized efficiently, but also IP traffic routing is automatically performed. This means that the OCDMA-encoded IP packets are broadcast through a star coupler and only the intended user recognizes the desired data by its assigned spreading code sequence. Since each IP packet is buffered only twice at the edge of the OCDMA network, the same as in an MPLS network, the buffer delay is significantly reduced compared with traditional routing schemes where IP packets are buffered at each hop.

While the FSK-OCDMA technique is considered, Figure 9.14 illustrates the inside of the optical encoder/decoder blocks as introduced in Figure 9.13 and analysed in detail in Chapter 7. As mentioned, a step-tuneable mode-locked laser diode with 100 GHz repetition rate can be utilized for the optical source to make the 100 Gchips/s feasible [39, 40].

## 9.4.2 Analysis of IP over OCDMA

The performance analysis for the FSK-OCDMA scheme with the MAI canceller is derived using GPMPC as studied in detail in Chapter 7. It is assumed that the input/output characteristic of the PDs follows the Poisson process. Since the reference signal has only the $P^{\text{th}}$ sequence (reserved at the receiver, i.e. there is no reference channel) multiplied by the received signal, the data components of the reference signal becomes 0 due to further spreading. Also, since all users in the same group receive an equal amount of MAI from the users of other groups and no interference from the users from the same group, i.e. GPMPC correlation properties, the interference signal of intended user $u_1$, equals the interference signal of $P$. It is assumed that $u_1$ transmits the optical pulse of $\lambda_0$ at the first slot in a data frame.

Since the GPMPC sequences are employed as signature codes and considering the number of interfering users in each group based on its correlation properties, and using various probability distribution functions based on interfering users and interference estimation, the final bit-error probability $(P_b)$ is derived as (i.e. Equation (7.20)):

$$
\begin{aligned}
P_b \leq \frac{M}{2} \sum_{r=r_{\min}}^{r_{\max}} \sum_{l_{0,0}=0}^{K-r} \sum_{l_{1,0}=0}^{k-r-l_{0,0}} \binom{K-r-l_{0,0}}{l_{1,0}} \times \left(\frac{1}{\gamma.M}\right)^{l_{1,0}} \times \left(1-\frac{1}{\gamma.M}\right)^{K-r-l_{0,0}-l_{1,0}} \\
\times \exp\left\{-\frac{\rho}{2}\cdot\frac{Q.(P+2)}{2}\right\} \times \binom{K-r}{l_{0,0}} \times \left(\frac{1}{\gamma.M}\right)^{l_{0,0}} \times \left(1-\frac{1}{\gamma.M}\right)^{K-r-l_{0,0}} \quad (9.14) \\
\times \binom{P^2-2P+1}{K-r} \times \binom{P-2}{r-1} \bigg/ \binom{P^2-P-1}{K-1}
\end{aligned}
$$

where $P$ is a prime number, $r$ is the number of interfering users in a same group in which $r \in \{r_{\min}, \ldots, r_{\max}\}$, $r_{\max} = \min(K, P-1)$ and $r_{\min} = \max(1, K - (P-1)^2)$. Here $K$ refers to the number of simultaneous active users and $l_{m,v}$ is the number of users who are in groups other than the first group and have a pulse in the $v^{\text{th}}$ slot with wavelength $\lambda_m$. Taking the fibre attenuation coefficient of $\alpha$ into account, the average received photon-count per pulse $(Q)$ can be expressed as:

$$
Q = \frac{\eta P_w}{hf} \cdot \frac{e^{-\alpha L}}{P+2} \approx \mu \cdot \frac{\ln M}{P+2} \quad (9.15)
$$

where $P_r = \eta \cdot P_w e^{-\alpha L}$ is the received power to the detector, $P_w$ is the transmitted peak power per symbol, $\eta$ is the quantum efficiency of the PDs, $h$ is Planck's constant, $f$ is the optical frequency, $L$ is the fibre length, and $\mu$ $(\mu = P_r/(h.f.\ln M))$ is the average number of photons per pulse (photons/nat). As introduced in Chapter 7, $\rho$ is the parameter minimizing the interference equal to:

$$
\rho = \frac{P+2}{P+2+l_{0,0}+l_{1,0}} \quad (9.16)
$$

On the other hand, when the bursty IP traffic is to be handled with the OCDMA concept, to obtain an acceptable performance without overload, the designed transmission rate for each user should be larger than the average traffic arrival rate. Hence each code channel cannot be fully utilized. It is easy to see that the average number of active users in the network changes when different channel utilizations are applied. Since the performance of an OCDMA network is a function of the number of active users, the channel utilization will have a significant effect on the network performance. For this impact analysis, all users (i.e. ONUs) are assumed to have the same channel utilization in the network as defined by:

$$
B = \frac{\textit{Average Output Bitrate}}{\textit{Maximum Transmission Bitrate}} \quad (9.17)
$$

Taking into account that both the 0 and 1 data bits are equiprobable, then the probability of each transmitted bit is 1/2. Since the ONUs are sending data independently, so the distribution of $K$ as a number of active users is $K/U$ where $U$ is the total number of users accommodated in the network. Consequently, the probability that $K$ users are active

$(P_{ac})$ equals the probability of a transmitted data bit times the probability of users involved in the transmission times the channel utilization. This can be expressed as:

$$P_{ac.} = \frac{1}{2} \times \frac{K}{U} \times B \tag{9.18}$$

Being active (i.e. sending IP packet) means that the number of users out of total number of users is selected randomly to be active. This behaviour can be treated as a binomial distribution. Thus, the PDF of $K$ active users out of $U$ users sending IP packet is obtained by:

$$P_{IP}(K) = \binom{U}{K} P_{ac.}^K (1 - P_{ac.})^{U-K} \tag{9.19}$$

Due to the independent events, the total probability of error function of the number of active users $K$, $P_T(K)$, denoting BER, can be expressed by the decoder probability of error $(P_b)$ times the probability of error stating the $K$ active users $(P_{IP})$. This is then derived by:

$$P_T(K) = \sum_{k=1}^{K} P_{IP}(k).P_b(k) \tag{9.20}$$

The packet-error rate (PER) of the IP traffic over this OCDMA (IP-over-OCDMA) network can be expressed [38] as:

$$PER = 1 - (1 - P_T(K))^w \tag{9.21}$$

where the average IP packet length is $w$ bits.

### 9.4.3 Performance of IP over OCDMA

The numerical results are presented based on the above analysis. The parameters used for the simulation are found in Table 9.2. For spreading code, GPMPC with $P = 11$ is employed that makes the code-length and total number of users 110 and 120 respectively. The repetition ratio $(\gamma)$ is shown by $j$ in the graphs and the BER threshold of $10^{-9}$ is also displayed in all graphs as a reference.

Figure 9.15 shows the performance (i.e. BER) comparison of PPM and FSK schemes against the number of users $K$ involved in the transmissions. The analysis for PPM-OCDMA employed GPMPC has also been analysed in Chapter 5. It is obvious that

**Table 9.2** IP-over-OCDMA link parameters

| Descriptions | Symbols | Values |
|---|---|---|
| Optical wavelength | $\lambda_0$ | 1550 nm |
| PD quantum efficiency | $\eta$ | 0.8 |
| Linear fibre-loss coefficient | $\alpha$ | 0.2 dB/km |
| Chip rate | $1/T_c$ | 100 Gchips/s |
| Fibre length | $L$ | 10 km |
| Packet length | $w$ | 12,000 bits |

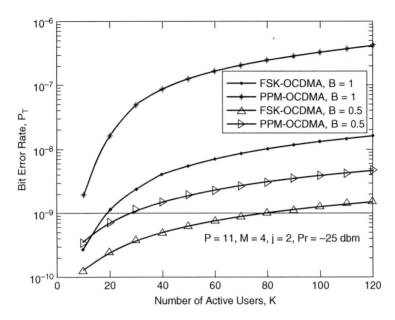

**Figure 9.15** BER performance of IP over different OCDMA against the number of active users, $K$

the performance degrades when the MAI increases by increasing the number of users as a result of the CDMA concept. The received power $(P_r)$ is set to $-25$ dBm in this analysis. It is apparent that the FSK outperforms PPM in that the repetition ratio $\gamma$ and $M$-ary frequency signal distribution mitigate the interference better than the signal position distribution. Figure 9.15 demonstrates two different cases when the channel utilization is fully and then 50 per cent occupied. It can be seen that when the channel utilization is moderate, i.e. $B = 0.5$, the FSK network is able to accommodate 80 active users while PPM supports only 30 users at $BER = 10^{-9}$. In the worst scenario, when $B = 1$, the IP-over-FSK-OCDMA network still tolerates 20 users at $BER = 10^{-9}$ while only 10 users are supported by PPM scheme close to $BER = 10^{-9}$.

Figure 9.16 illustrates the performance of IP-over-OCDMA against the number of active users in different conditions of signal multiplicity $M$, laser repetition ratio $\gamma$ and the channel utilization. The received power is again set to $-25$ dBm. It is shown that increasing the repetition ratio improves the performance markedably, though at the cost of throughput, see Section 7.4. It is indicated from Figure 9.16 that under the same conditions, the system with $\gamma = 3$ and $B = 1$ behaves very similarly to $\gamma = 2$ and $B = 0.5$, which presents the impact of repetition ratio on the performance. Increasing multiplicity means a larger number of positions to distribute the signal and more symbols to transmit; therefore, as seen in Figure 9.16, it can suppress the effect of co-channel interference (i.e. MAI). The number of users accommodated under $BER = 10^{-9}$ when $M = 8$ (42 users) is 100 per cent greater than that of $M = 4$ (20 users) in the worst case (i.e. $B = 1$). Also obviously, it is presented that the performance can be enhanced by reduction in the users' channel utilization.

In Figure 9.17, the variations of packet-error rate (PER) against the number of active users for different channel utilizations are presented. The received power and repetition

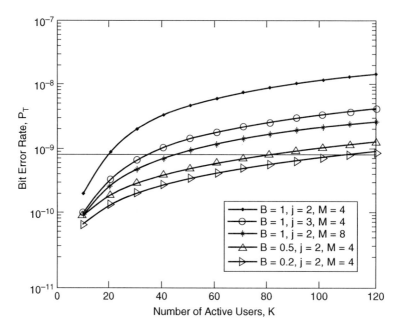

**Figure 9.16** BER performance of IP over OCDMA against the number of active users $K$, under different multiplicities $M$, channel utilizations $B$ and repetition ratios $\gamma$

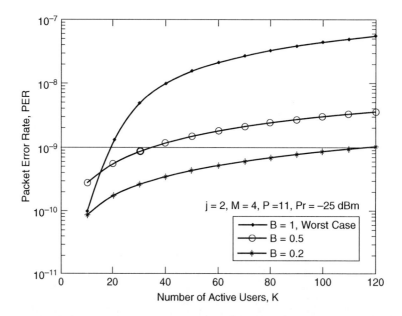

**Figure 9.17** PER performance of IP over OCDMA against the number of active users, $K$

ratio are set to $-25\,$dBm and 2, respectively. In this analysis, it has been assumed that the IP traffic has a packet length of 1500 bytes, i.e. the Ethernet local area network data frame length. Therefore, the calculated PER is estimated in the worst conditions. It is clearly shown that the performance of IP traffic becomes better with the reduction in the channel utilization. As observable from Figure 9.17, while $B = 1$ the performance degrades dramatically, but that 20 users are still accommodated at $BER = 10^{-9}$. When the probability that a user is active becomes relatively low, i.e. $B = 0.2$, the network is able to hold $BER = 10^{-9}$ when all users are active (i.e. 120 users). To compare with the scheme and conditions previously used, it should be noted that here $P = 11$ and received power is only $-25\,$dBm, whereas $P = 17$ and 19 (i.e. longer code-length) and effective power equals $-10\,$dBm (i.e. more power consumption) [38, 42].

When the channel utilization is 50 per cent the network is still able to provide a reliable communication link for 38 users (32 per cent of the total users). To achieve a consistent overall network performance when each user in the network has a fixed average bit rate, optimal channel utilization can be set for the network based on the network preferences and link-budgets at the design stage. To support a greater number of users, it is obviously recommended to employ higher $P$ and $P_r$ values.

The performance presented in Figure 9.18 is against the received signal power $(P_r)$ for different channel utilizations. In this analysis 100 users (83 per cent of the total users) are assumed to be involved in the transmission. It is obvious that by increasing the received power the detection is performed with assurance and BER becomes lower.

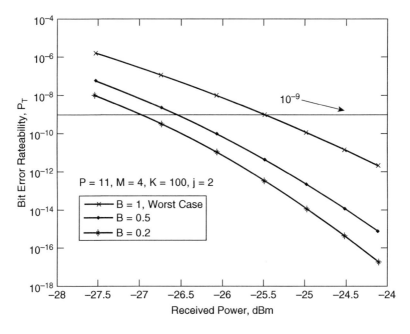

**Figure 9.18**   BER performance of IP over OCDMA against the received signal power, $P_r$

As Figure 9.18 noticeably shows, in order to mitigate the BER in the worst cases higher power consumption can be a solution. But the overall performance reveals that the system is very power efficient.

When $B$ is low, the probability that a user is sending traffic becomes relatively small. On the average, there are fewer active users sending data in each unit data frame, therefore the performance is enhanced. This means when the traffic burden is light, the network performance will be automatically improved. Further examination at $P_r = -25\,\text{dBm}$ in Figure 9.18 reveals that the error rates become $1.24 \times 10^{-14}, 2 \times 10^{-13}$, and $1.1 \times 10^{-10}$ when the channel utilization is 20 per cent, 50 per cent and fully occupied, respectively.

## 9.5 Random Access Protocols

The most significant noise in the optical CDMA network is multiple access interference (MAI). It is noted that MAI results from the chip overlaps between users in the system. MAI causes a dramatic drop in the normalized throughput of the system along with the increase in the normalized offered load which will be defined in Section 9.5.2. Therefore, in a high offered load situation, the performance of the system dramatically falls. It has been demonstrated in the literature that the normalized throughput of the system approaches zero when the normalized offered load is 100 per cent [43]. As mentioned, random access protocol or so-called transmission scheduling protocol is the solution to this problem.

Transmission scheduling is a higher layer protocol to minimize the chip overlaps between codewords so as to enhance the system normalized network throughput. The chip overlap relates to the situations where a 1 in a codeword encounters a 1 in another codeword. This will give a sum of 'double' power intensity in the signal on the line. The signal power will then be cut off and go back to normalized power in the hard-limiting operation.

As such, a result of the hard-limiting correlation decoder, many chip overlaps will lead to a bit error, which further creates a packet error. To avoid the chip overlap between codewords from different users, effort should be made to minimize the overlaps between 1 s in chips. This implies generating code families with very low or constant cross-correlation value. The overlaps cause intensity noise and/or beat noise in time-spreading or wavelength-spreading, respectively, or even both in two-dimensional coding system. Therefore, the specific transmission queuing mechanism becomes crucial. The idea of a transmission scheduling protocol is to equip every user with the ability to calculate the delay shifts with which the transmitted codeword can avoid interference in the resulting signal. In particular, this protocol decides the best time-slot to transmit a codeword so that its 1 chip has the best chance not to overlap with others as well as providing feasible asynchronous communications. P. Kamath et al. [44] have also shown that relying only on the orthogonal codewords with no scheduling or random access control leads the network to a sudden drop in the throughput in heavy traffic peaks. Thus there is a need for media access control in optical CDMA.

In this section, the principles of random access algorithms are discussed. Three recently proposed and popular algorithms (selfish, threshold and overlap section) based on these principles are explained in detail. Their advantages are described and their contributions to optical CDMA networks are also studied.

## 9.5.1  Random Access Protocol Algorithms

The principle of a transmission scheduling protocol algorithm is to preserve the codeword itself from overlapping with the signal on the line. In order to achieve this, it calculates the delay shifts for each transmission so that the codewords can survive from collision losses [44]. A basic algorithm like the 'pure selfish' algorithm only considers the survival of the particular codeword. But adding a codeword to the transmission line has direct impact on the future transmission, so an advanced algorithm has to consider the effect on the future transmission as well.

### 9.5.1.1  Pure Selfish Algorithm

The pure selfish algorithm purely preserves the codeword itself [44]. This algorithm considers each node as an independent individual and schedules a codeword whenever the signal on the line allows. It selfishly occupies the line as long as its own data can be transmitted without a loss under the current signal situation. The flow chart algorithm of the pure selfish algorithm is shown in Figure 9.19.

As observed from Figure 9.19, the initial shift is set to zero, which indicates that no shift delay applies to the current codeword or packet. Then the hard-limiting operation is executed to the signal on the line (i.e. the state of the line: SL). Then the algorithm compares the user codeword with the hard-limited signal on the line through an AND operation. The shift will be marked as a feasible shift if the result of the AND operation is not identical to the codeword. In other words, this actually means there is a 0 in

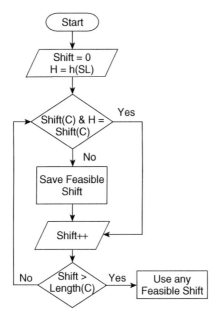

**Figure 9.19**  Flow chart of pure selfish algorithm for transmission scheduling in optical CDMA networks

the signal on the line where there is a 1 in the codeword (i.e. according to the AND operation, 0 AND 1 = 0 which makes the result not identical to the original codeword). This comparison is carefully demonstrated in Figure 9.20.

As shown in Figure 9.20, every possible shift is examined by the pure selfish algorithm. In the condition Shift = 2, there is a 0 in the hard-limited signal on the line where there is a 1 in the shifted codeword. Therefore, the AND operation gives a different result from the original codeword. The corresponding shift (equals to 2 in this case) is marked as a feasible shift. This process will be executed for all possible shifts. Normally, this will be executed $N$ times, where $N$ is the length of the codeword, since there are altogether

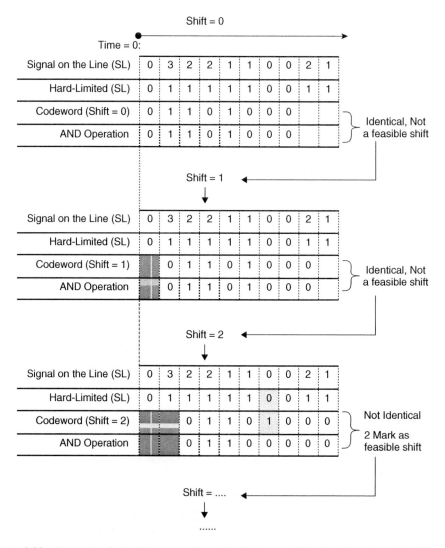

**Figure 9.20** Example of scheduling a codeword with pure selfish algorithm (possible shifts are examined one by one, shift = 2 is marked as a feasible shift)

$N$ possible shifts for a codeword of length $N$. After all the feasible shifts are found, this algorithm will randomly use one of them to schedule the data transmission.

### 9.5.1.2  Threshold Algorithm

The threshold algorithm preserves the data itself and allows the transmitted data to overlap with the signal on the line below a threshold level [43]. The threshold value indicates the maximum allowed percentage of 1 chips which are overlapped in the signal on the line. It is initially set to a fixed value in the system. This algorithm is stricter than the pure selfish algorithm. It does not allow the codeword to be scheduled if it overlaps with the signal on the line beyond a certain threshold. The threshold algorithm considers the impact of adding the codeword to future packet transmission and keeps the resulting signal on the line in a 'clean' status. The flow chart of this 'threshold algorithm' is shown in Figure 9.21.

Figure 9.21 shows whether, after a decision has been made, the shift is a feasible shift under the principles of pure selfish algorithm. This shift is not marked as feasible straightaway, but instead the shifted codeword is added to the signal on the line (SL) and the *overlap factor* of them, which is defined as the ratio of overlapped 1s in the signal to the length of the codeword is calculated. Then the overlap factor is compared with the threshold value. Finally, if the overlap factor is below the threshold value, the shift will be marked as a feasible. As with the pure selfish algorithm, this process is executed for all

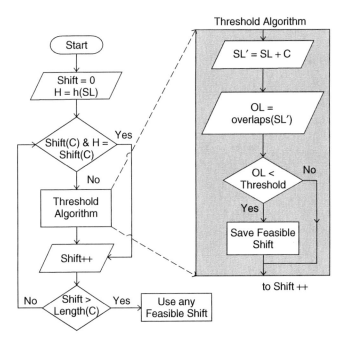

**Figure 9.21**  Flow chart of threshold algorithm for transmission scheduling in optical CDMA networks

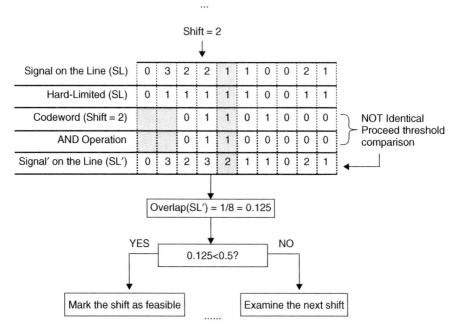

**Figure 9.22**  Scheduling a codeword with threshold algorithm (possible shifts are examined one by one, the threshold value is set to 0.5; overlap (SL′) is calculated as 0.125; shift = 2 is marked as a feasible shift)

possible shifts, and the system picks up the feasible shift on a random basis. Figure 9.22 demonstrates the process of comparing the overlap factor with the threshold value in the threshold algorithm:

Figure 9.22 illustrates the threshold algorithm process. First of all, the codeword is compared with the hard-limited signal on the line to see if there is a 0 chip where the codeword has a 1 chip. If so, further threshold consideration will be taken. An assumption is made that this codeword is transmitted with the particular shift. Then the algorithm observes how many 1 chips in the signal on the line will be overlapped by the newly added codeword. In Figure 9.22, there is only one 1 chip in the signal (grey box) that is overlapped by the codeword. Therefore, the overlap factor is 1/8, where eight is the length of the codeword. This value will be compared with the threshold value, which is set initially as 0.5. If it is smaller than the threshold, this shift will be marked as a feasible shift. If not, the next shift will be examined. In this case, 0.125 is smaller than 0.5, so Shift = 2 is marked as a feasible shift.

### 9.5.1.3  Overlap Section Algorithm

The overlap section algorithm, also called smart threshold algorithm, is an improved version of the threshold algorithm. It considers the number of 1 chips in the resulting signal on the line. Instead of initially setting a fixed threshold value, the threshold value is set as a function in the overlap section algorithm. This function calculates the number of 1s in the resulting signal after the codeword is transmitted. It is noted as *ones*() in the flow chart in Figure 9.23 at 'overlap section algorithm'. This algorithm compares the result of the *ones*() function with the number of 1 chips overlapped in the signal by the codeword. If the number of overlapped chips is fewer than the number of 1 chips in the resulting signal, this particular shift will be marked as feasible.

The overlap section algorithm considers the future tidiness of the signal on the line (SL). Furthermore, having a variable 'threshold value' which is the result of the *ones*() function in this algorithm, the system will be more flexible to deal with different offered load situations. The flow chart of the overlap section algorithm is shown in Figure 9.23.

Figure 9.23 shows that the overlap section algorithm is carried out after the overlaps in the signal on the line are calculated. Then the number of 1 chips in the resulting signal is set to be the threshold value. After that, this value is compared with the number of overlaps in the signal on the line. The codeword is transmitted if the number of overlaps is smaller than the number of 1 chips in the resulting signal Thus in overlap section algorithm, since the number of overlapped chips including the newly added codeword is less than the number of 1 chips in the resulting signal, the shift will be marked as feasible. As in the pure selfish algorithm, every possible shift will be examined and the algorithm will pick up a feasible shift to schedule the transmission randomly. The Figure 9.24 below further explains the process of overlap section algorithm by an example.

The number of overlapped chips in the signal on the line is 1, and there are three 1 chips in the resulting signal as illustrated in Figure 9.24. Thus the condition of overlap section

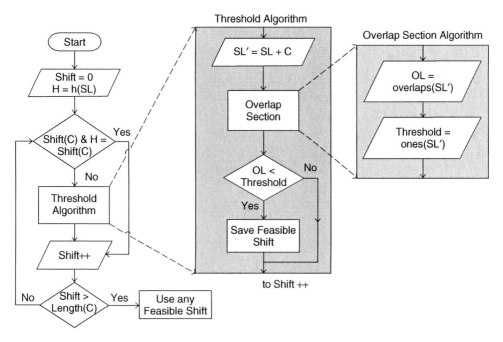

**Figure 9.23** Flow chart of overlap section algorithm for transmission scheduling in optical CDMA networks

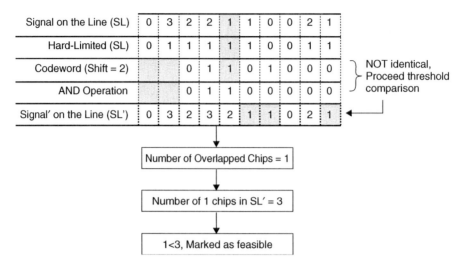

**Figure 9.24** Scheduling a codeword with overlap section algorithm (possible shifts are examined one by one, one overlapped chip in the signal on the line; three 1 chips in the resulting signal (SL′); shift = 2 is marked as a feasible shift)

algorithm is satisfied. This shift will be marked as feasible. The overlap section algorithm compares the number of chips in the signal on the line overlapped by the codeword and the number of non-overlapped chips (the 1 chips) in the resulting signal. It prevents the further overlapping by requiring the overlapped chips to be fewer than the non-overlapped chips [44]. In Figure 9.24, as explained in the previous paragraph, this condition is met. Therefore, Shift = 2 is marked as a feasible shift.

## 9.5.2 Network Performance Metrics

The performance of the above algorithms will be measured in terms of their normalized network throughput, packet error rate and the average number of codewords allowed into the line against the normalized offered load. These parameters are defined in the following subsection and can be determined afterwards as a queuing problem.

### 9.5.2.1 Normalized Offered Load

The offered load from a queuing system is calculated by dividing the arrival rate $\lambda$ by the transmission rate $\mu$, as:

$$\rho = \lambda/\mu \tag{9.22}$$

The transmission rate can be calculated by:

$$\mu = C/8L \ (packets/s) \tag{9.23}$$

where $C$ is the chip-rate (chips/s) and $L$ denoted average packet size in bytes (note that 1 byte = 8 bits).

Hence the normalized offered load $\rho$ can be rewritten as:

$$\rho = \frac{8L\lambda}{C} \tag{9.24}$$

For an OCDMA system in which the codeset is optical orthogonal code (OOC) and fully specified by $(N, w, k)$, the transmission rate can be rewritten as:

$$\mu = \frac{C}{8LN} \ (packets/s) \tag{9.25}$$

where $N$ is the code-length, $w$ is the code-weight and $k$ is the maximum collision value (i.e. cross-correlation). From Equations (9.24) and (9.25) we have:

$$\rho = \frac{\lambda}{\mu N} \tag{9.26}$$

### 9.5.2.2 Probability of Packet Error

When there are large numbers of codewords on the line, there is a high probability of error hence the packet error rate goes up. The key to successful transmission rests heavily on the following two variables:

- The number of zeros in the line, which provides a chance to scheduling a codeword into the line.
- The number of ones in the line, which means that the codewords can be successfully recovered by the receiver. The design of the optical receiver makes it such that overlaps usually result in erroneous detection.

In a given state, the number of 1 chips on the line is equal to the maximum number of correctly detectable codewords. Hence the probability of error can be approximated by:

$$P_E = 0 \ if \ N_{OL} \leq n(1) \tag{9.27}$$

$$P_E = \frac{N_{OL} - n(1)}{N_{OL}} \ if \ N_{OL} > n(1) \tag{9.28}$$

where $N_{OL}$ is the number of codeword on the line and $n(1)$ is the number of 1 chips on the line.

### 9.5.2.3 Normalized Throughput

Normalized throughput is the average number of error-free packets through the system. The different scheduling algorithms restrict the number of codewords that get into the line and denoted by $N_{OL}$. The normalized throughput can be calculated by:

$$Th_{NO} = (N_{OL}(1 - P_E)/(\lambda/\mu))\rho \tag{9.29}$$

Substituting for $\rho$ from (9.26) into (9.29) and simplifying, we obtain:

$$Th_{NO} = N_{OL}(1 - P_E)/N \tag{9.30}$$

where $N$ is the code-length, $P_E$ is the probability of error and $N_{OL}$ is the number of codewords on the line. It can be seen that the normalized throughput is a function of the number of codewords on the line ($N_{OL}$).

If Equation (9.30) is simplified further for the case where the number of codewords on the line is greater than the number of 1 s on the line (i.e. $N_{OL} > n(1)$) then the normalized throughput can be found as:

$$Th_{NO} = \left( N_{OL} \left( 1 - \frac{N_{OL} - n(1)}{N_{OL}} \right) \right) / N \qquad (9.31)$$

Further simplification gives:

$$Th_{NO} = \frac{n(1)}{N} \qquad (9.32)$$

From Equation (9.32), we can see that when the number of codewords on the line is greater than the number of ones, the throughput becomes dependent on the number of 1 chips in the line. Accordingly, the cooperative scheduling algorithms are based on trying to find ways to increase the number of 1 chips in the line.

Let us discuss some of the results based on the aforementioned analysis using ALOHA-based CDMA as a reference which basically means that there is no particular media access protocol. Nodes can transmit asynchronously with no media access protocol similar to unslotted ALOHA. The detailed analysis of ALOHA-CDMA can be found in [44]. The codeset may be selected to maximize throughput as we will also compare for various codes later. Note that the parameters used to construct a codeset may be varied to control the interference between codewords. As Figure 9.25 shows, when the normalized throughput is less than 0.3, all the media access methods (scheduling or random access protocols) have almost the same throughput and this is due to the fact that the probability of error is very small and also the number of codewords on the line is relatively small hence the number of 1 chips on the line is the same for all schemes. As the offered load increases, it can be seen that the ALOHA-CDMA starts to have a lower gradient when compared to other scheduling methods. When the system reaches 50 per cent of its capacity, the ALOHA algorithm begins to degrade while the other scheduling schemes do not degrade but maintain their throughput or experience slight increase in throughput.

At maximum load, the other scheduling schemes have about 50 per cent throughput but the ALOHA has crashed to 10 per cent. This means that only one in every ten packets inserted into the line by ALOHA is correctly decoded at the receiver. Whereas, with threshold scheduling algorithm, throughput is at about 52 per cent for the same amount of load.

The reason for this can be found when we look at the average number of codewords on the line. The scheduling scheme prevents throughput degradation by limiting the number of codewords that get into the line. The various schemes do this with various levels of aggression with the selfish scheduling scheme trying to insert as many packets as possible into the line by filling up all the 0 chips in the line. The cooperative scheme does the same by applying a threshold or barrier on the number of chips overlap and hence on the number of codewords on the line and also the delay time.

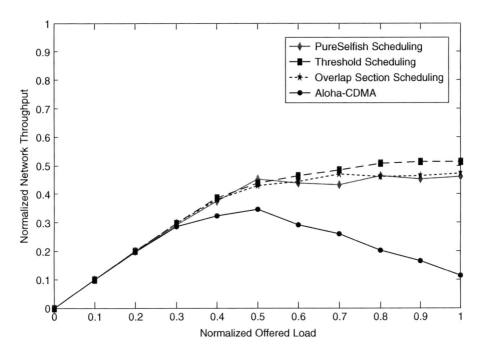

**Figure 9.25** The normalized offered load against the normalized network throughput when the codeset is OOC (10,3,1) and the threshold is 0.5 (for the threshold algorithm)

### 9.5.2.4 Average Number of Codewords on the Line

The number of codewords on the line is controlled by the scheduling algorithm in use. The threshold and the overlap section are cooperative; hence they do not allow so many overlaps. This is the main reason that their throughput is higher. By restricting the number of overlaps on the line, they also effectively reduce the average number of codewords on the line.

Unlike the cooperative scheduling schemes, as shown in Figure 9.26, ALOHA allows all the nodes to transmit whenever and however they want to. Hence, as the offered load increases, the number of codewords on the line increases and so does the number of overlaps. Obviously the probability of errors increases too.

The pure selfish algorithm determines when a node should transmit but does nothing about the scheduling effect on the line in terms of overlaps. Hence, it is able to insert more codewords on the line. This in turn affects the probability of error and ultimately the throughput.

### 9.5.2.5 Packet Error Rate

Figure 9.27 shows a clear difference between the different scheduling algorithms. The cooperative scheduling algorithms have better performance in terms of packet error rate.

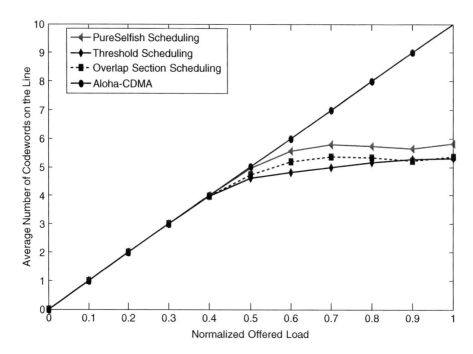

**Figure 9.26** Normalized offered load against the average number of codewords in the line when the codeset is OOC (10,3,1) and the threshold is 0.5

This can be linked directly to the fact that they try to limit the number of overlaps and thereby limit the probability of errors.

The pure selfish algorithm which does not care about overlaps in the line, therefore does not perform as well as the cooperative algorithms. However, its performance is not as bad as the ALOHA because it at least determines when nodes transmit. This act helps it to give a small reduction in the probability of errors. In ALOHA-CDMA when nodes are transmitted there is no control on the number of overlaps on the line.

Initially, the error rates are not so different from the other scheduling schemes but as the offered load increases, the errors begin to increase and, at maximum load, the error rate is as high as 90 per cent which is more than four times that of the other scheduling algorithm.

### 9.5.2.6   Sensitivity Analysis

A study was carried out on how the network responds to a wide range of changes to its following parameters:

- Code-weight
- Code-length
- Effect of varying the average packet size
- Performance under a bimodal packet distribution

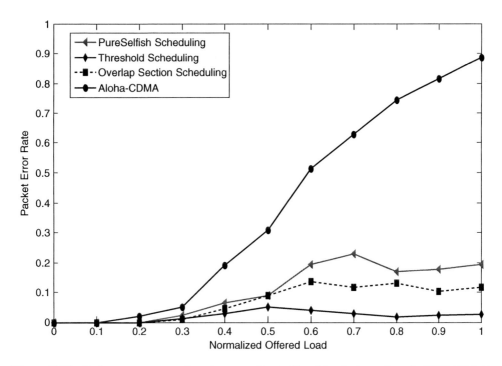

**Figure 9.27**  Packet error rate against the normalized offered load; the codeset is OOC (10,3,1) and the threshold is 0.5

The general parameters for the OOC codeset are given in Table 9.3 and are used unless otherwise stated.

The effect of varying the code-weight is shown in Figure 9.28. It can be seen that the throughput degrades by increasing the code-weight. It implies an increase in the resistance to interference from other codewords on the line, while the scheduling algorithm has more difficulty in successfully scheduling codewords onto the line. The effects of the two factors may cancel each other at low offered load but as the normalized offered load approaches one (i.e. maximum), the effect of the reduced number of codewords on the line begins to take effect and the throughput begins to drop significantly. As seen in the normalized throughput Equation (9.29), it will be simplified in that the throughput will be a function

**Table 9.3**  OOC codeset parameters

| Descriptions | Symbols | Values |
|---|---|---|
| Code-length | $N$ | 10 |
| Code-weight | $w$ | 3 |
| Cross-correlation parameter | $k$ | $\leq 2$ |
| Codeword allocation | | Uniform random |
| Code cardinality or number of nodes | $C_c$ | $\geq 10$ |

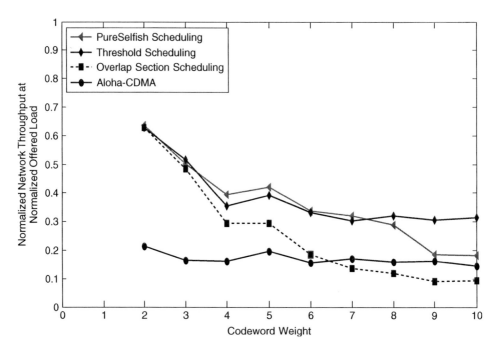

**Figure 9.28** Normalized network throughput against the code-weight when the code-length is 10, $k = 1$ and the threshold is 0.5

of the number of codewords in the line, $N_{OL}$. The problem of scheduling of the codewords comes with an increase in code-weight, and also a decrease in the number of codewords on the line leads to a decrease in the throughput.

Figure 9.29 demonstrates the network normalized throughput sensitivity against the code-length. As the code-length increases, the scheduling algorithm has more opportunities to schedule more codewords on the line, hence that improves the throughput. But with the increase in the code-length comes the reduction in the data rate. If chipping rate is $C$ chips/s, then the data rate will be $C/(8LN)$ packets/s where $L$ is the average packet size in bytes and $N$ is the code-length. When $N$ is increased and all other variables are kept constant, the data rate starts to reduce, and this reduces the overall network throughput. Another analytical proof can be found in Equation (9.30) where the normalized network throughput is given. From Equation (9.30), it can be seen that the throughput is proportional to the number of codewords on the line ($N_{OL}$) and inversely proportional to the code-length ($N$). The relatively proportional growth in both parameters $N_{OL}$ and $N$ can cancel out each other's impact on the system throughput. As observed, the increase in the code-length has no particular impact on the network throughput for all scheduling algorithms.

The effect of varying the average packet size ($L$) can be also depicted analytically to have no effect on the normalized network throughput. As Equations (9.22)–(9.30) indicate, the effect of $L$ is totally cancelled out in the final Equation (9.30) for the normalized network throughput. Consequently, any change in the average packet size has no effect on the normalized network throughput.

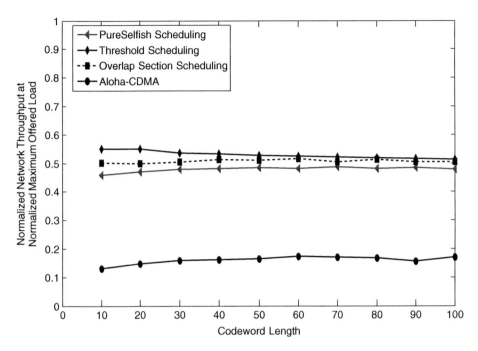

**Figure 9.29** Normalized network throughput against the code-length when the code-weight is 3, $k = 1$ and the threshold is 0.5

Let us now analyse the throughput performance under the binomial distribution of the average packet size. In reality, it is impossible that the average packet size can be found to be a particular value without a large variation. In this analysis, we try to see the performance of the network when the average packet size can be grouped largely into two. Since we now have two different packet sizes, the probability of error can be estimated by considering it as a binomial distribution. The binomial distribution gives the probability of error as:

$$P_e = \binom{n}{k} p^k (1-p)^{n-k} \tag{9.33}$$

where $\binom{n}{k} = n!/(k!(n-k)!)$ is the binomial coefficient and $p$ is the probability of selecting of the each codeword.

In this analysis, we have considered two packet sizes of $L_1$ and $L_2$ in bytes where $L_1 < L_2$. If $x$ is the fraction of packets of size $L_1$, it implies that the weighted average packet size could be:

$$L_{av} = xL_1 + (1-x)L_2 \tag{9.34}$$

The employed transmission scheduling scheme is the pure selfish algorithm. With this scheduling scheme, $N_{OL}$ codewords are added to the line and by extension $N_{OL}$ 1 chips are also added selfishly to the line. The problem is with the remaining $N_{OL}(w-1)$ 1 chips that are added randomly to the line, they are the 1s that cause overlaps and hence interfere with the codewords on the line.

$p$ is the probability that a single chip in the codeword overlaps with a 1 chip on the line when the codeword is scheduled. Thus,

$$p = \frac{w}{(N-1)} \tag{9.35}$$

Error in the packet transmission occurs when there are more than $(w-1)$ overlaps in the codeword. Hence the probability of error can be calculated as:

$$P_e = 1 - \sum_{k=0}^{w-1} \binom{N_{OL}(w-1)}{k} p^k (1-p)^{(N_{OL}(w-1)-k)} \tag{9.36}$$

Since the generic number of codewords on the line $N_{OL}$ depends on the packet size here, the number of codewords on the line with packet size $L_1$, $N_{OL}(L_1)$, is:

$$N_{OL}(L_1) = \frac{t_1}{t_{arr}} \times \frac{N_{OL}}{N} \tag{9.37}$$

where $t_1 = L_1/(C_{/N})$ and $t_{arr} = L_{av}/C$ and similarly for $N_{OL}(L_2)$. Accordingly, the normalized network throughput can be then calculated as:

$$Th_{norm} = N_{OL/N} \times (x(1 - P_e(L_1)) + (1-x)(1 - P_e(L_2))) \tag{9.38}$$

It can be seen from Figure 9.30 that when there is a higher fraction of the larger packet size, the throughput is the same as when there is only one packet size (when $x = 0$

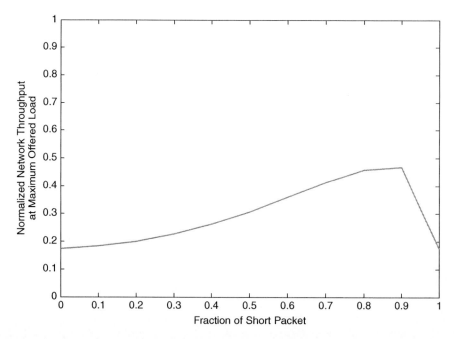

**Figure 9.30** The effect of bimodal packet distribution when the codeset is OOC (10, 3, 1) and the scheduling algorithm is the pure selfish scheduling – the average packet sizes are assumed from 50 to 1000 bytes

and $x = 1$). However, as the fraction of the smaller packets increases, it can be seen that the throughput increases and peaks accordingly when the fraction of small packet reaches 0.9 before dropping off to the state when there is only one type of packet size.

### 9.5.3 Performance of Prime Code Families in Random Access Protocols

It has been established that the different transmission scheduling algorithms perform better than the simple ALOHA-CDMA. The performance of the prime code families needs to be investigated since they are most common and well-used spreading codes. Possible prime codes are generated as introduced in Chapter 2 and then employed in the OCDMA network. Their overall performance is measured in terms of normalized throughput and the packet error rate. Table 9.4 shows the various codeset properties based on prime code families.

The most comparable codesets are chosen for evaluation and comparison purposes. The criteria are based on the codeset with closest code-length and code-weight to OOC (10, 3, 1) where the variation in code-weight is minimal.

All three previously introduced scheduling algorithms are considered except the ALOHA. This is because, as mentioned earlier, ALOHA's throughput is not as good as the others and there is no control on the number of overlaps on the line.

Figure 9.31 illustrates the normalized network throughput against the normalized offered load when different codesets have been employed and the scheduling is based on the pure selfish algorithm. It is observed that the algorithm behaviour for all codesets is pretty similar due to the fact that the codeset properties are very similar in nature. It can be seen that prime code family outperforms the others at the maximum offered load. This is due to the fact that prime code has the lowest code-weight and the lowest cross-correlation value which helps in reducing the collisions naturally.

Also it is apparent from Figure 9.31 that GPMPC provides the same code-weight and correlation properties as n-MPC but in a shorter length. On the other hand, we noted that the increase in the code-length under the same correlation constraint reduces the chance of collisions and accordingly enhances the throughput. Here we also indirectly see the effect of various codeset properties on the network throughput.

The variations of normalized network throughput against the normalized offered load are depicted in Figure 9.32 when the scheduling is the overlap section algorithm and

**Table 9.4** Prime code families codeset parameters

| $P$ | Prime Code $(P^2, P, 1)$ | n-MPC $(P^2 + P, P + 1, 2)$ | GPMPC $(P^2 + 2P, P + 2, 2)$ |
|---|---|---|---|
| 2 | – | – | (8,4,2) |
| 3 | (9,3,1) | (12,4,2) | (15,5,2) |
| 5 | (25,5,1) | (30,6,2) | (35,7,2) |
| 7 | (49,7,1) | (56,8,2) | (63,9,2) |
| 11 | (121,11,1) | (132,12,2) | (143,13,2) |

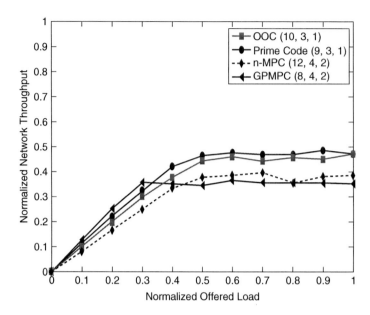

**Figure 9.31** Normalized network throughput against the normalized offered load using the different codesets for pure selfish scheduling

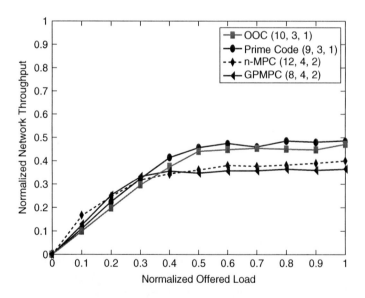

**Figure 9.32** Normalized network throughput against the normalized offered load using the different codesets for overlap section scheduling

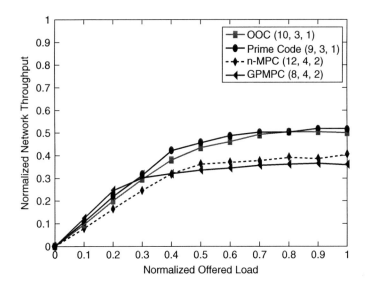

**Figure 9.33** Normalized network throughput against the normalized offered load using the different codesets for threshold scheduling when the threshold parameter is 0.5

in Figure 9.33 when the scheduling is threshold algorithm. The normalized throughput's trends in Figures 9.31–9.33 are very similar regardless of the scheduling algorithms. Prime code family codeset is comparable with OOC family since the properties are closer and the only slight difference (i.e. lower cross-correlation value) makes prime codes outperform in all scheduling algorithms as shown in Figures 9.31–9.33.

As mentioned in the sensitivity analysis, the effect of the code-length variation has no effect on the network throughput. Hence the throughputs are almost equal. For the n-MPC and GPMPC codesets, the throughput is significantly lower than that obtained from the prime codes and the OOC. This is mainly due to the difference in the code-weight. Therefore, the normalized throughput decreases by the increase in the code-weight.

Figures 9.34–9.36 illustrate again the same trends for the scheduling algorithms of pure selfish, overlap section and threshold, respectively, when the code properties are different from those in Figures 9.31–9.33. As seen, the threshold scheduling has a normalized throughput of 0.23 at the maximum offered load, while the overlap section and pure selfish scheduling have 0.205 and 0.22, respectively, for prime codes. It is also noted that the prime codes have the better throughput in the threshold and overlap section algorithms. In the pure selfish scheduling algorithm, all the codes have almost the same throughput at maximum offered load though the GPMPC codes exceeds that of the prime code by 0.02. The OOC has the lowest throughput in this case. In other scheduling schemes, the prime codes perform better and this could be attributed to the lower weight.

It can be observed that there is no significant difference between the throughput of the prime code families and that of the OOC. In order to be able to draw a definite line in the performance of these two families of optical spreading codes, their packet error rate is also studied.

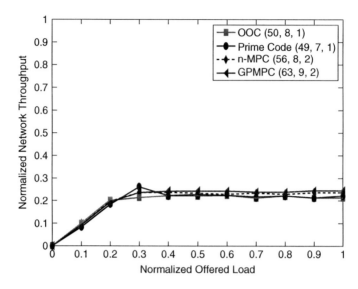

**Figure 9.34** Normalized network throughput against the normalized offered load using the different codesets for pure selfish scheduling

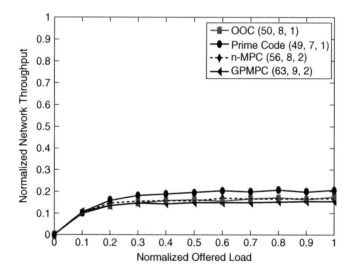

**Figure 9.35** Normalized network throughput against the normalized offered load using the different codesets for overlap section scheduling

Figure 9.37 plots the packet error rates of different codesets when the scheduling is the pure selfish scheduling. It is seen that the GPMPC has the worst packet error rate due to the shortest code-length at maximum offered load. Obviously, n-MPC has enhanced packet error rate at the maximum offered load because it has the longest code-length. It has been shown that the code-length impact on the normalized throughput is insignificant (see Figure 9.29). However, the increase in the code-length provides enhanced packet error rate due to the decrease in the chips overlap.

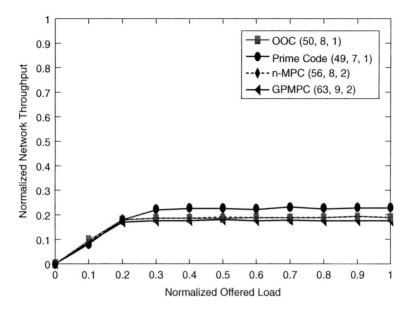

**Figure 9.36** Normalized network throughput against the normalized offered load using the different codesets for threshold scheduling when the threshold parameter is 0.5

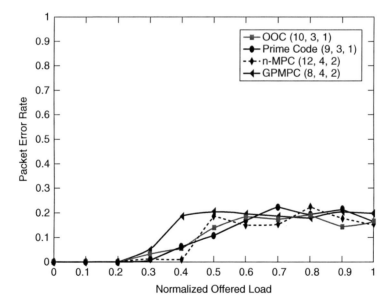

**Figure 9.37** Packet error rate against the normalized offered load using the different codesets for pure selfish scheduling

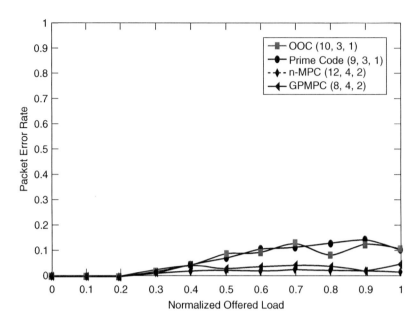

**Figure 9.38** Packet error rate against the normalized offered load using the different codesets for overlap section scheduling

The packet error rates of the overlap section algorithm employing different codesets are displayed in Figure 9.38. The packet error rates at the maximum offered load are lower than the pure selfish but still higher than the threshold scheduling scheme as seen in Figure 9.39. There is significant difference of packet error rates in OOC and prime codes with n-MPC and GPMPC codesets. As observed, in this case the OOC and prime codes have closer code-length and the same code-weight while n-MPC and GPMPC have different code-length but the same code-weight. Here we can conclude that the code-weight plays a more important role than code-length in the overlap section scheduling algorithm

In the threshold algorithm shown in Figure 9.39, the packet error rates are at a minimum compared with other algorithms at the maximum offered load and also as an average. Here also we see the impact of code-length and code-weight on the packet error rates. The prime code and OOC are above the n-MPC and GPMPC codesets in terms of packet error rates at the maximum offered loads. It is indicated that the threshold algorithm outperforms in terms of the packet error rates as well as normalized network throughput, see Figures 9.31–9.36.

Figure 9.40 magnifies the 0.0–0.1 region of packet error rate in Figure 9.39. It is apparent that the packet error rates at maximum load for OOC are $3.5 \times 10^{-2}$ and $5.8 \times 10^{-2}$ for prime codes. The OOC is about 1.5 times better than the prime code in terms of the packet error rate. The other members of the prime code family, i.e. n-MPC, has a packet error rate of $8 \times 10^{-3}$, while the packet error rate for GPMPC has extremely low error rate.

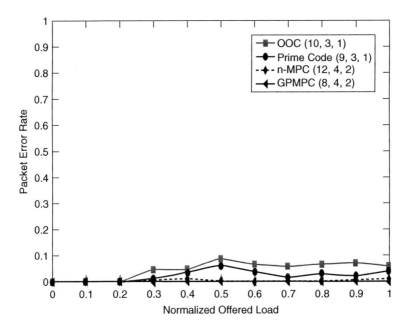

**Figure 9.39**   Packet error rate against the normalized offered load using the different codesets for threshold scheduling when the threshold parameter is 0.5

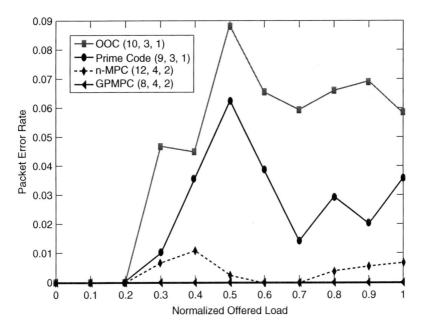

**Figure 9.40**   Packet error rate against the normalized offered load using the different codesets for threshold scheduling when the threshold parameter is 0.5

## 9.6    Multi-Protocol Label Switching

There is a simpler technique to engineer the traffic in the packet switched networks called 'multi-protocol label switching (MPLS)'. The term 'multi-protocol' refers to a technique that can be employed in various network layer protocols. However, since Internet protocol (IP) is the dominant network protocol, in this section we also regard IP as the main network layer protocol. MPLS can be seen as an extension to the existing IP architecture. It is the latest step in the evolution of routing and forwarding technology for the Internet core. It has been standardized by the Internet Engineering Task Force (IETF) and designed to enhance the speed, scalability and service provisioning capabilities in the core Internet network. As a technology in the backbone networks, MPLS can be used for IP as well as other network-layer protocols. It has been the prime candidate for IP-over-ATM backbone networks [45, 46].

This section reviews the fundamental features and architecture of MPLS as well as the generalized MPLS (GMPLS) and its compatibility with optical domain switching techniques.

### 9.6.1    MPLS Fundamentals

Within the IETF, Cisco has been working to develop a standard for invented tag switching since they first shipped it in 1998 [47]. That standard is now MPLS. Thus, tag switching is a pre-standard implementation of MPLS. A solid framework is provided by MPLS to support the operation of advanced routing services to accordingly solve a number of complex problems. MPLS concentrates on the scalability problems corresponding to the currently deployed IP-over-ATM overlay model and significantly simplifies the complexity of network operation.

MPLS is a means of forwarding data at a very high rate. It merges the speed and performance of Layer 2 (data link) with the scalability and IP intelligence of Layer 3 (networking). MPLS is based on the following key concepts [48–50]:

- A collection of distributed control protocols are used to set up paths in IP networks.
- IP routers are then responsible for forwarding and control tasks.
- A single-forwarding paradigm, called label swapping, is employed in MPLS at the data plane to support multiple-routing paradigms at the control plane.
- MPLS maps IP addresses to simple fixed-length protocol-specific identifiers called 'labels', which are distinguishing forwarding information (label) from the content of the IP headers.
- MPLS supports guaranteed quality of services (QoS) and thus is capable of integrating voice, video and data services.
- The need for interrogating the IP header of every packet at every intermediate router is eliminated.
- MPLS is based on constraint-based routing which distributes the traffic load on alternative routes when the shortest path is unavailable.
- MPLS offers a standardized solution to promote multivendor interoperability.

- Since MPLS is an integration of Layer 2 and Layer 3 technologies, it has traffic engineering capability when Layer 2 features are introduced to Layer 3.
- Traffic aggregation as the main issue of access networks scalability is facilitated by MPLS.

MPLS technology improves the scalability and performance of packet switched networks by preventing hop-by-hop routing and forwarding across networks. Its primary goal is to standardize a technology that integrates the label-switching-based forwarding method with the networking layer's routing.

Wavelength division multiplexing (WDM) was originally conceived as a means of radically increasing the transmission capacity in a single fibre but has evolved to provide other desirable features including protocol transparency, dynamic reconfigurability and improved survivability, to outperform the traditional SDH/SONET networks [51, 52]. While rising Internet traffic brought up huge bandwidth requirements, dense WDM (DWDM) has emerged with even narrower bandwidth.

To accommodate packet-based data traffic in WDM networks, they need to support intelligent optical routing and switching. By doing so, efficient routing can be achieved through the collapse of the protocol stacks [53], which leads to the design of slim protocol stacks with flexible bandwidth-on-demand and dynamic resource allocation [54]. Optical MPLS needs to be employed allowing efficient packet forwarding while decoupling the packet routing and forwarding operations to carry multiple-routing services [49]. MPLS will be a dominant unifying network management and traffic engineering protocol in the future for a variety of transport systems as it is extended to generalized MPLS (GMPLS) [55].

Figure 9.41(a) shows the traditional network designs incorporating several standard layers like IP over ATM over SONET/SDH over optical to implement optical networks [52]. Two more efficient solutions are illustrated in Figures 9.41(b) and (c) taking advantage of MPLS-over-optical and implementing the MPLS directly into the optical domain, respectively. The IP/MPLS performs interoperability and end-to-end QoS, whereas the optical layer provides high-capacity reliable transport and bandwidth management. The latest approach in Figure 9.41(c) has MPLS in the optical layer and provides an optical MPLS interface to upper layers such as IP layer. The later scheme has great potential since it simplifies data forwarding and all-optical label swapping. An overview of the optical MPLS techniques on different multiple access methods with optical labels are introduced in the following sections.

### 9.6.2 Optical MPLS Techniques

A generic photonic MPLS network block diagram is shown in Figure 9.42 [56] where IP packets, generated from the source node (client A), are routed by electronic routers to an ingress router in the photonic MPLS network. An optical edge switch, located at the ingress router, adds labels onto IP packets. Label swapping routing and forwarding are performed by optical core switches. The edge routers collect and remove the labels and forward the packets to their destination when packets leave the photonic MPLS at the egress router.

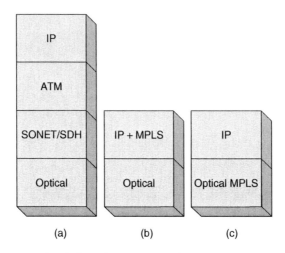

**Figure 9.41** Stack protocol solutions for transport of IP-over-optical networks: (a) traditional flow; (b) IP/MPLS-over-optical; and (c) IP-over-optical MPLS (Reprinted from: On architecture and limitation of optical multiprotocol label switching (MPLS) networks using optical-orthogonal code (OOC)/wavelength label, Y.G. Wen, Y. Zhang and L.K. Chen, *Optical Fiber Technology*, **8** (1), 2002, with permission from Elsevier.)

The physical layer network elements, as depicted in Figure 9.42(b), are connected by fibre links and the packet routing and forwarding hierarchy [35]. IP packets are generated at the electronic routing layer (i.e. IP routers) and processed in an optical MPLS adaptation layer that encapsulates IP packets with an optical label without modifying the original packet structure. The packet and labels are converted in the adaptation layer to a new assigned wavelength from local routing tables.

An optical multiplexing layer seems necessary in order to multiplex labelled packets onto a shared fibre. Several optical multiplexing approaches may be used including direct insertion onto an available WDM channel [57], packet compression through optical time multiplexing or time interleaving through optical time division multiplexing [58], and optical code division multiplexing [59].

In the optical MPLS networks, core routers can be divided into two fundamental parts: control and forwarding components in which routing and forwarding functionalities are performed, respectively. In the control components, the routing algorithm computes a new label and wavelength assignment from an internal routing table. The routing tables at egress and core routers are generated by mapping IP addresses into smaller pairs of labels and wavelengths and distributing them across the network by piggybacking them on top of routing protocols, e.g. open shortest path first (OSPF), or a separate protocol such as label distribution protocol [60]. On the other hand, the forwarding components have the responsibility of swapping the original label with the new label and physically converting the labelled packet to a new wavelength. The reverse process of optical demultiplexing, adaptation and electronic routing are performed at the egress node. As in the following sections, the labels embedded in the IP packets can be carried by (a) timeslots, (b) wavelengths, (c) subcarriers, or (d) codes and these usually denote critical

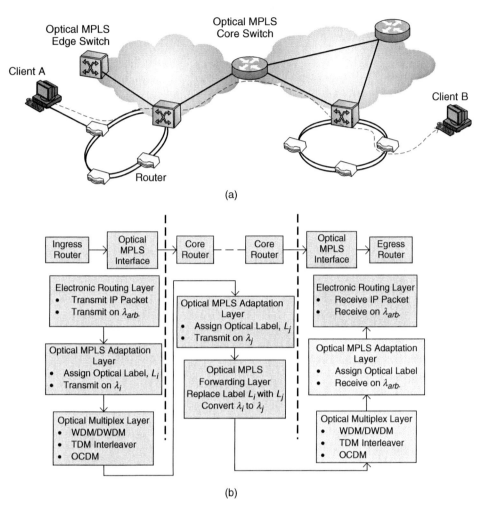

**Figure 9.42** (a) Generic optical MPLS network with label-switched path and (b) block diagram of layered routing and forwarding hierarchy and associated network element connection (Reprinted from: On architecture and limitation of optical multiprotocol label switching (MPLS) networks using optical-orthogonal code (OOC)/wavelength label, Y.G. Wen, Y. Zhang and L.K. Chen, *Optical Fiber Technology*, **8** (1), 2002, with permission from Elsevier.)

label-switched path (LSP) information such as the source and destination. Accordingly, the optimum performing labels should have the following features:

- Wavelength resources dedicated as data carriers should not be used as optical labels.
- It is preferred to implement optical label swapping in the optical, rather than the electronic, domain.
- Distortions from dispersion, timing jitter and interference from other propagating labels should be minimized in optical labels.

As introduced earlier, four basic optical MPLS including time-division multiplexed labels, wavelength-division multiplexed labels, subcarrier-division multiplexed labels and code-division multiplexed labels are discussed in the following sections.

### 9.6.2.1   Time-Division Multiplexed Optical MPLS

Figure 9.43 illustrates a schematic diagram of the TDM optical MPLS scheme. In this TDM-based labelling technique, optical labels are implemented by concatenating a routing header before the payload in the time domain. The bit rate of the routing header is usually either similar to or lower than the data rate [61]. In the similar bit rate scenario, much of the buffering time for label processing is saved whereas some challenges on the capacity of the label processor remain. Since the label rate is the same as the data rate, for the current transport networks the label processor must switch at a multi-gigabit rate to process label recovery, label swapping and packet forwarding. However, in the lower bit rate scenario, the labels are processed at a comparably lower rate, giving more accuracy while more buffering time is required for the fixed number of optical labels (i.e. bits). Since powerful electronic processors, such as FPGA and ASIC, are popular and dominant at lower rates, currently this approach is preferable. All-optical label processing, i.e. recovery, swapping and reinserting, is vital for optical MPLS networks. For the TDM-based labelling, a traditional header processor can be introduced for label processing. However, with the coherent nature of optical routing labels, label swapping can be realized by optical label conversion that can be simply done by XOR gates. The operational principle is just to use another bit stream as the swapping label to XOR with the old label's bit steam [62].

### 9.6.2.2   Wavelength-Division Multiplexed Optical MPLS

WDM-based optical labels are also called wavelength-coded labels. An example is shown in Figure 9.44. Three public wavelengths of $\lambda_2$, $\lambda_3$ and $\lambda_4$ are reserved for label coding, where the data payload is carried only by the $\lambda_1$ carrier. Different combinations of presence or absence of these three public wavelengths represent different destination labels. It is apparent from Figure 9.44 that $\lambda_2$ and $\lambda_4$ are present where $\lambda_3$ is absent,

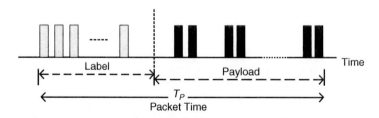

**Figure 9.43**   Time-division labelling technique (Reprinted from: On architecture and limitation of optical multiprotocol label switching (MPLS) networks using optical-orthogonal code (OOC)/wavelength label, Y.G. Wen, Y. Zhang and L.K. Chen, *Optical Fiber Technology*, **8** (1), 2002, with permission from Elsevier.)

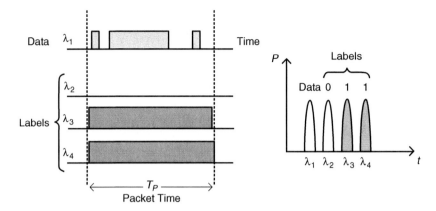

**Figure 9.44** Wavelength-division labelling technique (Reprinted from: On architecture and limitation of optical multiprotocol label switching (MPLS) networks using optical-orthogonal code (OOC)/wavelength label, Y.G. Wen, Y. Zhang and L.K. Chen, *Optical Fiber Technology*, **8** (1), 2002, with permission from Elsevier.)

thus the wavelength-coded label is '1 0 1'. For $N$-public wavelength-coded labels, the total number of wavelength labels is $2N$. Therefore, only $N$ wavelengths are reserved for label encoding; this number is much smaller than the supportable data channels of $2N$. This is a huge enhancement as compared with that of the multi-protocol lambda label switching networks [54].

For the WDM-based labelling, the all-optical label-switched router in the core router can be implemented through a fibre Bragg grating (FBG). One WDM-based optical MPLS was introduced in [63], although the architecture has very low spectral and wavelength efficiency. It is shown in Figure 9.45 that the header carries a wavelength-coded label that corresponds to a destination address. Here, multiwavelength labels are denoted by $\lambda_{1A}, \lambda_{1B}, \lambda_{1C}$ and $\lambda_{1D}$. The data payload uses another different wavelength $\lambda_{1E}$. Wavelengths ranging from $\lambda_{1A}$ through $\lambda_{1E}$ are defined as a $\lambda$-band $\lambda_{1A-E}$.

### 9.6.2.3 Subcarrier-Division Multiplexed Optical MPLS

The operational principle of the subcarrier-division multiplexed (SDM)-based optical labelling technique is depicted in Figure 9.46. Here, the optical label is carried by an out-of-band subcarrier [64, 65]. The data payload and labels are sent at different rates, and label recovery is an independent process from the data rate using microwave filtering techniques. Optical and microwave amplitude detection techniques are used in subcarrier label recovery, increasing the need for RF coherent techniques and phase synchronization across the network. In addition, the mature electronics guarantees the practicability of the optical subcarrier labelling scheme. Accordingly, test beds around the world pay a lot of attention to this labelling implementation [34, 35, 65, 66]. Setting a fixed label bit rate and modulation format independent of the packet bit rate can provide the packet transparency. The electrical switching speed of the burst-mode label recovery and the duration of the labels relative to the shortest packets at the fastest packet bit rates can help determining

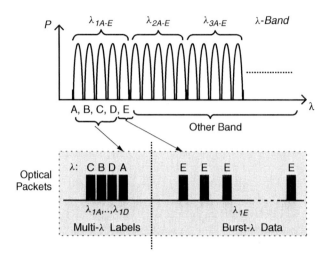

**Figure 9.45** Packet format for multiwavelength label switching (Reprinted from: On architecture and limitation of optical multiprotocol label switching (MPLS) networks using optical-orthogonal code (OOC)/wavelength label, Y.G. Wen, Y. Zhang and L.K. Chen, *Optical Fiber Technology*, **8** (1), 2002, with permission from Elsevier.)

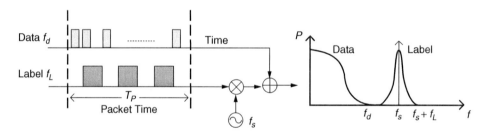

**Figure 9.46** Subcarrier-division labelling technique (Reprinted from: On architecture and limitation of optical multiprotocol label switching (MPLS) networks using optical-orthogonal code (OOC)/wavelength label, Y.G. Wen, Y. Zhang and L.K. Chen, *Optical Fiber Technology*, **8** (1), 2002, with permission from Elsevier.)

the label bit rate setting. Furthermore, running the label at a lower bit rate allows the use of less expensive electronics. The label and packet bits can be encoded using different modulations to ease the data and clock recovery. On the other hand, SDM-based labelling limits the data rate due to the interferences. In order to avoid the labels interfering with the payload, guard-bands between the data and subcarrier spectra seem necessary.

Currently, researchers prefer deploying 'remove and reinsert' mode label swapping for subcarrier label schemes [34, 35]. Basically, this is a three-stage label processing. At the first stage, the subcarrier label is extracted and removed all-optically. At the second stage, the label is recovered and label information is fed to the control plane to determine the next label. In the final stage, the new label is remodulated at the same microwave carrier frequency for the IP packets.

### 9.6.2.4 Code-Division Multiplexed Optical MPLS

The advanced approaches in high-bit-rate optical code division multiplexing (OCDM) open up a new way to use optical codes as optical labels [67]. In this approach, there is no need for logic operations following a look-up table where the toughest challenge for optical processing resides, but a simpler function of optical correlation is required instead [68]. In general, CDM-based optical encoding schemes are classified into coherent and incoherent schemes according to the degree of coherency of the light source, as introduced in Chapter 3. Thus, there are also two alternative solutions for incorporating OCDM into CDM-based optical MPLS techniques. One frontier solution was the coherent scheme introduced by Ken-Ichi Kitayama, *et al.* [29]. The operation principle of this approach is shown in Figure 9.47.

In this system the label swapping can be implemented in an all-optical domain. A schematic diagram of photonic label swapping is shown in Figure 9.48; its principle and functionality are explained and illustrated in Figures 9.49 and 9.50. With respect to the cross phase modulation (XPM) effect, the operating principle of photonic label swapping is illustrated in Figure 9.49. This shows that when the input and control pulses propagate in the fibre, the cross-phase modulation (XPM) effect causes the input pulses to experience different refractive indices. Consequently, the input signal pulse experiences a phase shift $\varphi_{max}$, which is determined by the intensity of the control light and the interaction length. By setting the wavelength and intensity of the control pulse appropriately, the signal pulse will have the desired phase shift at the output of the fibre. For example, to swap

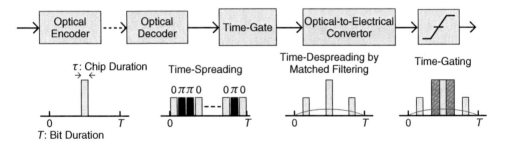

**Figure 9.47** Coherent OCDM principle (© 2000 IEEE. Reprinted with permission from Architectural considerations for photonic IP router based upon optical code correlation, K. Kitayama, N. Wada and H. Sotobayashi, *J. Lightw. Technol.*, **18** (12), 2000.)

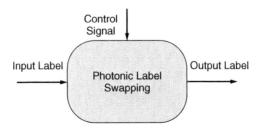

**Figure 9.48** Schematic of photonic label swapping function for CDM labels

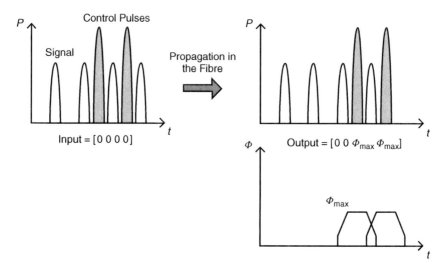

**Figure 9.49**  Principle of photonic label swapping for CDMA labels (© 2000 IEEE. Reprinted with permission from Architectural considerations for photonic IP router based upon optical code correlation, K. Kitayama, N. Wada and H. Sotobayashi, *J. Lightw. Technol.*, **18** (12), 2000.)

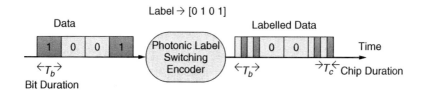

**Figure 9.50**  Code-division labelling technique

the four-chip-long BPSK optical code 0000 to a different code $00\pi\pi$, as illustrated in Figure 9.49, the intensity of the control pulses has to be set so that the total phase shift becomes $\pi$. Therefore, all-optical label swapping is realized by using control pulses to change the phase of the signal pulses.

The incoherent scheme has also been studied as an alternative to the coherent scheme [52, 69, 70]. Figure 9.50 depicts how a data bit is modulated by an optical orthogonal code (OOC) in the time domain. When the data bit is 1 an OOC sequence is sent out, and alternatively for the data bit 0 either no code sequence is sent or the complement of the previously sent OOC is transmitted. Accordingly, the OOC sequence can be seen as an optical label which can forward the IP packet to its destination. It should be noted that the sequence inversion keyed (SIK) modulation format along with direct sequence (DS)-CDMA was mainly considered for label switching implementation [52, 69]. In an optical SIK modulation, the data bits are modulated by either a unipolar signature sequence $C_i$ or its complement $\overline{C_i}$, depending on whether the transmitting bit is 0 or 1. With this in mind, code conversion which is realized by processing the incoming code label and a control code with an optical XOR gate, is able to implement all-optical label swapping.

The input signal in the router is SIK-DS-CDMA modulated by the code $C_i$, associated with the virtual optical code channel $V_{OCC}(i)$. An optical code converter converts the code $C_i$ to code $C_j$, corresponding to the routing from $V_{OCC}(i)$ to $V_{OCC}(j)$. Let us first introduce the concept of conversion code $Conv(ij)$. It acts like a transfer function to convert the information carried in the $V_{OCC}(i)$ to the $V_{OCC}(j)$. Thus the proposed code converter in [52] realizes the all-optical CDMA switching function.

To finalize the code conversion task, the conversion function $Conv(ij)$ must have the following properties:

$$C_i \oplus Conv(ij) = C_j$$

and

$$\overline{C_i} \oplus Conv(ij) = \overline{C_j}$$

where $\oplus$ is the XOR operator. It is simply found that $Conv(ij) = C_i \oplus C_j$ as we have:

$$C_i \oplus Conv(ij) = C_i \oplus (C_i \oplus C_j) = (C_i \oplus C_i) \oplus C_j = 0 \oplus C_j = C_j \qquad (9.39)$$

$$\overline{C_i} \oplus Conv(ij) = \overline{C_i} \oplus (C_i \oplus C_j) = (\overline{C_i} \oplus C_i) \oplus C_j = 1 \oplus C_j = \overline{C_i} \qquad (9.40)$$

For example, for four-chip Walsh code sequences of $C_1 = [1\ 0\ 0\ 1]$ and $C_1 = [1\ 1\ 0\ 0]$, the code conversion function from $C_1$ to $C_2$ can be verified as:

$$Conv(12) = [1\ 0\ 0\ 1] \oplus [1\ 1\ 0\ 0] = [0\ 1\ 0\ 1] \qquad (9.41)$$

$$C_1 \oplus Conv(12) = [1\ 0\ 0\ 1] \oplus [0\ 1\ 0\ 1] = [1\ 1\ 0\ 0] = C_2 \qquad (9.42)$$

$$\overline{C_1} \oplus Conv(12) = [0\ 1\ 1\ 0] \oplus [0\ 1\ 0\ 1] = [0\ 0\ 1\ 1] = \overline{C_2} \qquad (9.43)$$

The code converter configuration is illustrated in Figure 9.51. It is observed that this configuration is a kind of modified terahertz optical asymmetric demultiplexer in which

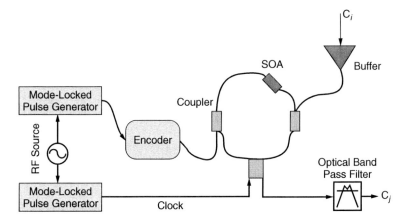

**Figure 9.51** Code converter for OCDM-based optical MPLS networks (Reprinted from: On architecture and limitation of optical multiprotocol label switching (MPLS) networks using optical-orthogonal code (OOC)/wavelength label, Y.G. Wen, Y. Zhang and L.K. Chen, *Optical Fiber Technology*, **8** (1), 2002, with permission from Elsevier.)

two arms of the control pulse input exist [52]. Conversion function $Conv(ij)$ and input code $C_i$ are used in the two branches. $Conv(ij)$ is constructed by mode-locked pulse generators (MLPG). Since the two signals are on the same wavelength, and the XOR clock is on a different wavelength, an optical band-pass filter is used at the output to separate the wavelengths. After the clock enters the loop through the main coupler, it splits into two pulses, clockwise (CW) and counter-clockwise (CCW). Each of the two pulses passes through the semiconductor optical amplifier (SOA) once, and they return to the main coupler at the same time.

When the arms carry the same signal level of either 1 or 0, both CW and CCW signals experience the same transmission properties of the SOA. The clock train is then totally reflected into the input port, and no pulse appears in the output port. On the other hand, when the arms carry different signal levels (one carries 1 and the other carries 0), both CW and CCW pulses experience different transmission properties of the SOA. Accordingly, the clock pulse train comes out from the output port while the CW and CCW have $\pi$ phase difference. Thus, the output channel code $C_j = C_i \oplus Conv(ij)$ appears at the output port of the code converter.

Alternatively, another code converter based on the same principle is shown in Figure 9.52. The device also has XOR-oriented operation, using a symmetric Mach–Zehnder (SMZ) interferometer [71]. Nevertheless, reported code converters ignored the degrading issue of multiple access interference (MAI) and the cross-correlation between the selected code channel and the other co-propagation code channels. It is possible to reduce the effect of MAI by the aid of switching and contention resolution techniques such as the switch fabric reported in [70] for an OOC MPLS core router in which the MAI was taken into account. It should be noted that, in a CDMA-based optical MPLS scheme the optical labels are actually extended into two-dimensional codes in that the combination of spreading codeset in time, for example, and a wavelength-set are promising to provide a larger label pool.

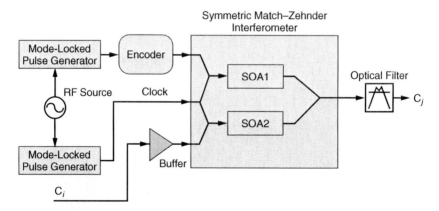

**Figure 9.52**   Code converter using SMZ for OCDM-based optical MPLS network (Reprinted from: On architecture and limitation of optical multiprotocol label switching (MPLS) networks using optical-orthogonal code (OOC)/wavelength label, Y.G. Wen, Y. Zhang and L.K. Chen, *Optical Fiber Technology*, **8** (1), 2002, with permission from Elsevier.)

## 9.6.3   *Generalized MPLS (GMPLS)*

Employing a variation of MPLS to support different data rates at the transport layer will be the next logical evolutionary step towards all optical networks. This step has already been taken as generalized multi-protocol label switching (GMPLS) which is an enhancement of multi-protocol label switching (MPLS), and is also known as multi-protocol lambda switching [50, 72, 73]. GMPLS is supposed to support multiple switching tasks such as packet switching, time-division switching, wavelength-division switching and space-division switching. The GMPLS fundamental qualities include the following:

- GMPLS consists of a set of traffic engineering and optical extensions to existing MPLS routing and signalling protocols.
- GMPLS represents an integral part of next-generation data and optical networks as an optical networking standard.
- The necessary bridge between IP and optical layers is provided by GMPLS enabling networks to be interoperable and scalable.
- More intelligent optical devices implementing GMPLS help network operators better manage restoration and bandwidth utilization.
- GMPLS allows edge networking devices such as routers and switches to directly request bandwidth from the optical layer.
- Hierarchical label-switched paths (LSPs) and optical LSPs or lightpaths are supported in GMPLS.
- Simultaneous multiple traffic types are managed by GMPLS. It creates a control plane to support multiple switching layers as [50]:
  - Packet switching for forwarding data, based on packet or cell headers
  - TDM switching for forwarding data, based on the data time-slot in a repeating cycle (e.g. SONET/SDH, PDH)
  - WDM switching for forwarding data, based on the wavelength on which it was received (e.g. wavelength converters and add/drop multiplexing)
  - Spatial switching for forwarding data, based on the position of the data in real-world physical spaces (e.g. incoming port or fibre to outgoing port or fibre).

GMPLS is focused on the control plane of various layers because each layer can use physically diverse data. The goal is to cover both the routing and signalling portions of the control plane. The term generalized extends the MPLS LSP mechanisms to generalized labels and generalized LSPs. These extensions of MPLS include [55, 67, 73, 74]:

- For signalling protocols, the resource reservation protocol (RSVP) or constraint-based routing label distribution protocol (CR-LDP) is employed. It is required in GMPLS signalling that an LSP starts and ends on similar types of devices. Unlike MPLS, where LSPs are unidirectional, lightpaths in GMPLS can be unidirectional or bidirectional. GMPLS signalling enables an upstream node to assign a label, although the assignment may be overridden by downstream nodes.
- For routing, explicit paths can be established across network layers by GMPLS. Open shortest path first (OSPF) or intermediate system to intermediate system (IS-IS) is utilized as an interior gateway routing protocol (IGP).

- GMPLS introduces link management protocol (LMP), which is a new protocol designed to resolve issues related to link management of optical networks. A router uses LMP to check the availability of a link between itself and another router. It provides four basic functions for node pairing: control channel management, link connectivity verification, link property correlation and fault isolation.

Activities such as label distribution, traffic engineering, protection and restoration are influenced by these extensions to GMPLS. They also enable rapid provisioning and management of network services. Although MPLS was designed exclusively for packet services, the primary goal of GMPLS is to provide a single suite of protocols that would be applicable to all kinds of traffic, and here again the term generalized could be used. By consolidating different traffic types, GMPLS permits simplification of networks and improves their scalability in ways unheard of until now. It offers the means by which networks can be scaled and simplified by deployment of a new class of network element (aggregation node) designed to handle multiple traffic types simultaneously.

Currently, GMPLS network architecture offers a common mechanism for routing, data forwarding and traffic engineering on transport networks with dense wavelength division multiplexing (DWDM) [55]. Next-generation all-optical networks will consist of optical elements such as routers, switches, reconfigurable optical add-drop multiplexers (ROADMs), optical cross-connects (OXCs) etc. that will employ GMPLS to dynamically provision resources to offer network survivability using protection and restoration techniques. These optical network elements which have full control of the wavelengths and codes can create self-connecting and self-regulating networks.

Since the label switching and swapping technologies were introduced alongside the optical CDMA, these two technologies have much common potential to take advantage of each other's key benefits. Optical CDMA as a physical layer multiplexing technique utilizes the code-switching technique to multiplex and transmit several distinct data rates over the same fibre on same wavelength entirely in the optical domain. The concept of using optical code as a labelling mechanism in MPLS systems has been discussed earlier. Labelling in the case of MPLS is performed using the optical code as a method of encoding addresses into the packet-header-only so-called explicit optical code labelling method [67]. GMPLS currently defines five label mapping spaces: packet switch capable, Layer 2 switch capable, time-slot switch capable, wavelength switch capable and fibre switch capable. It should be noted that only the last two mapping methods can be utilized in an all-optical switching device.

## 9.7  Summary

This chapter started with a review of existing solutions for access networks depending on different applications such as CATV and video-on-demand services. Different flavours of passive optical networks (PON) have been introduced including TDM- and WDM-based PONs such as ATM-PON, Gigabit PON, Ethernet PON and WDM-PON. The remaining part of the chapter was mostly dedicated to the optical CDMA networking perspectives by taking OCDMA as an access technique in PON, IP and MPLS networks. OCDMA-PON architecture including tuneable optical line terminals and tuneable optical network units with coherent transceivers have been studied and analysed in detail.

The scalability of OCDMA-PON architecture based on the power-budget analysis has been studied. The supportable number of nodes in the network and the accessible number of users against the fibre span (i.e. how far they can be from central office) have been analysed. The overall SNR has been analysed based on degradation of the received signal by (i) fibre link noise e.g. amplifier spontaneous emission noise, (ii) thermal noise, (iii) photodetectors' shot-noise, (iv) phase to intensity noise and (v) mainly multiple access interference. However, it should be noted that the overall promising performance in this architecture is a trade-off between driving an active optical external modulator (OEM) and complex implementation to compare with spectral amplitude coding (SAC) schemes introduced in Chapter 4. On the other hand, this architecture does not need unique transceivers for every user due to the generic time-spreading manner imposed by the electric signals, whereas every transceiver needs a specific fibre Bragg grating (FBG) or spectrum-sensitive components in SAC architectures.

Since the IP traffic is and will be the dominant traffic pattern in communications networks, the IP traffic over the OCDMA network has also been analysed in this chapter, taking advantage of both coherent modulation and incoherent demodulation. The network performance has been studied for different cases in terms of the channel utilization. Since each IP packet is buffered only twice at the edge of the network node, similar to multi-protocol label switching (MPLS), the buffer delay is significantly reduced compared with traditional routing schemes where IP packets are buffered at each hop. Also, the optical encoder is adjusted for the number of packets belonging to the same user instead of being tuned to incoming IP packet individually. Accordingly, the encoder adjusting time is significantly reduced.

The major degrading issue in the OCDMA networks is multiple access interference (MAI) that occurs when more than one user attempt to share the same channel or medium at the same time. Accordingly, there is a need for media access control, also called random access protocol or transmission scheduling, to control the collisions and support asynchronous communications.

Three recently proposed transmission scheduling algorithms – pure selfish, threshold and overlap section – have been studied and analysed. The network metrics in terms of packet error rate, offered load and network throughput, for all these three algorithms, have been evaluated. Various code families including prime codes and optical orthogonal codes were considered in the analysis. Interestingly, the sensitivity analysis has also been taken into account to show how much dependency exists between the network metrics and the parameters involved at the design stage.

Finally, to cover the recent technological aspects of optical networking, simpler traffic engineering technology in the packet switched networks, so-called multi-protocol label switching (MPLS), has also been studied. The term 'multi-protocol' refers to a technique that can be utilized in different network layer protocols. However, as mentioned earlier, IP is the dominant network protocol so in this section IP was assumed as the main network layer protocol, where MPLS can be an extension to the existing IP architecture. This chapter also covered MPLS in combination with various switching techniques such as time-division, wavelength-division and code-division switching paradigms. Code conversion techniques as the main task in the label switching/swapping technology have also been covered. In the next chapter we discuss the service-oriented optical networks based on optical CDMA.

# References

1. Diab, W. W. and Frazier, H.M. (2006) *Ethernet in the first mile: access for everyone*. Standards Information Networks. IEEE Press, New Jersey, USA.
2. Kitayama, K., Wang, X. and Wada, N. (2006) OCDMA over WDM PON – solution path to gigabit symetric FTTH. *J. Lightw. Technol.*, **24** (4), 1654–1662.
3. Reed, D. (2003) *Copper evolution*. Federal Communications Commission, Technological Advisory Council III.
4. Goralski, W. (1998) *ADSL and DSL Technologies*. McGraw-Hill, USA.
5. Kramer, G. (2005) *Ethernet passive optical network*. McGraw-Hill, New York, USA.
6. Killat, U. (1996) *Access to B-ISDN via PON – ATM communication in practice*. Wiley Teubner Communications, Chichester, England.
7. Lam, C. F. (2007) *Passive optical network: principles and practice*. Academic Press, Elsevier, USA.
8. Beck, M. (2005) *Ethernet in the first mile: the IEEE 802.3ah standard*. McGraw-Hill, USA.
9. Dutta, R. and Rouskas, G.N. (2000) A survey of virtual topology design algorithms for wavelength routed optical networks. *Opt. Networks Mag.*, **1** (1), 73–89.
10. Ramaswami, R. and Sivarajan, K.N. (1998) *Optical Networks: a practical perspective*. Morgan Kaufmann, Burlington, MA, USA.
11. Yamamoto, F. and Sugie, T. (2000) Reduction of optical beat interference in passive optical networks using CDMA technique. *IEEE Photonics Tech. Letters*, **12** (12), 1710–1712.
12. Mukherjee, B. (1997) *Optical communication networks*. McGraw-Hill, New York, USA.
13. Sivalingam, K. M. and Subramanian, S. (2005) *Emerging optical network technologies*. Springer Science+Business Media Inc., USA.
14. Perros, H. G. (2005) *Connection-oriented networks: SONET/SDH, ATM, MPLS, and optical networks*. John Wiley & Sons, Chichester, England.
15. Hernandez-Valencia, E., Scholten, M. and Zhu, Z. (2002) The generic frame procedure (GFP): an overview. *IEEE Comm. Mag.*, **40** (5), 63–71.
16. Ueda, H. *et al.* (2001) Deployment status and common technical specifications for a B-PON system. *IEEE Comm. Mag.*, **39** (12), 134–141.
17. Ohara, K. (2003) Traffic analysis of Ethernet-PON in FTTH trial service. In: *OFC*, CA, USA.
18. Assi, C., Ye, Y. and Dixit, S. (2003) Dynamic bandwidth allocation for quality of service over Ethernet PON. *IEEE J. on Selected Areas in Comm.*, **21** (11), 1467–1477.
19. Sala, D. and Gummalla, A. (2001) PON function requirements: services and performance. In: *IEEE 802. Meeting*, USA.
20. Iwatsuki, K., Kani, J.I. and Suzuki, H. (2004) Access and metro metworks based on WDM technologies. *J. Lightw. Technol.*, **22** (11), 2623–2630.
21. Karbassian, M. M. and Ghafouri-Shiraz, H. (2009) Analysis of scalability and performance in passive optical CDMA network. *J. Lightw. Technol.*, **27** (17), 3896–3903.
22. Ahn, B. and Park, Y. (2002) A symmetric-structure CDMA-PON system and its implementation. *IEEE Photonics Tech. Letters*, **14** (9), 1381–1383.
23. Guennec, Y. L., Maury, G. and Cabon, B. (2003) BER Performance Comparison Between an Active Mach–Zehnder Modulator and Passive Mach–Zehnder Interferometer for Conversion of Microwave Sub-carrier of BPSK Signals. *J. Microw. & Opt. Tech. Let.*, **36** (6), 496–498.
24. Gnauck, A. H. (2003) 40-Gb/s RZ-differential phase shift keyed transmission. In: *OFC*, CA, USA.
25. Zhang, C., Qui, K. and Xu, B. (2006) Investigation on performance of passive optical network based on OCDMA. In: *IEEE ICC, Circuits and Systems*, Istanbul, Turkey.
26. Zhang, C., Qui, K. and Xu, B. (2007) Passive optical networks based on optical CDMA: design and system analysis. *Chinese Science Bulletin*, **52** (1), 118–126.
27. Benedetto, S. and Olmo, G. (1991) Performance evaluation of coherent code division multiple access. *Electronics Letters*, **27** (22), 2000–2002.
28. Wang, X. and Kitayama, K. (2004) Analysis of beat noise in coherent and incoherent time-spreading OCDMA. *J. Lightw. Technol.*, **22** (10), 2226–2235.
29. Kitayama, K., Wada, N. and Sotobayashi, H. (2000) Architectural considerations for photonic IP router based on optical code correlation. *J. Lightw. Technol.*, **18** (12), 1834–1844.
30. Meenakshi, M. and Andonovic, I. (2006) Code-based all optical routing using two-level coding. *J. Lightw. Technol.*, **24** (4), 1627–1637.

31. Teixeira, A. L. J. *et al.* (2003) All-optical routing based on OCDMA header. In: *LEOS, the 16th Annual Meeting of the IEEE*, Tuscon, Arizona, Oct. 26–28.

32. Xu, R., Gong, Q. and Ya, P. (2001) A novel IP with MPLS over WDM-based broadband wavelength switched IP network. *J. Lightw. Technol.*, **19** (5), 596–602.

33. Yao, S., Yoo, S.J.B. and Mukherjee, B. (2001) All-optical packet switching for metropolitan area networks: opportunities and challenges. *IEEE Comm. Mag.*, **39** (3), 142–148.

34. Seddighian, P. *et al.* (2008) Time-stacked optical labels: an alternative to label-swapping. In: *OFC*, San Diego, USA, Feb. 26–28.

35. Seddighian, P. *et al.* (2007) All-Optical Swapping of Spectral Amplitude Code Labels for Packet Switching. In: *Photonics in Switching*, San Diego, USA.

36. Gumaste, A. *et al.* (2008) Light-mesh: A pragmatic optical access network architecture for IP-centric service oriented communication. *Opt. Switching and Networking*, **5** (2,3), 63–74.

37. Karbassian, M.M. *et al.* (2011) Experimental demonstration of transparent QoT-aware cross-layer lightpath protection switching. In: *Proc. WCECS (ICCST)*, San Francisco, USA, pp. 808–813.

38. Wei, Z. and Ghafouri-Shiraz, H. (2002) IP routing by an optical spectral-amplitude-coding CDMA network. *IEE Proc. Communications*, **149** (5), 265–269.

39. Lemieux, J. F. *et al.* (1999) Step-tunable (100 GHz) hybrid laser based on Vernier effect between Fabry–Perot cavity and sampled fibre Bragg grating. *Electronics Letters*, **35** (11), 904–906.

40. Schröder, J. *et al.* (2006) Passively mode-locked Raman fiber laser with 100 GHz repetition rate. *Optics Letters*, **31** (23), 3489–3491.

41. Karbassian, M. M. and Ghafouri-Shiraz, H. (2009) IP routing and transmission analysis over optical CDMA networks: coherent modulation with incoherent demodulation. *J. Lightw. Technol.*, **27** (17), 3845–3852.

42. Wei, Z. and Ghafouri-Shiraz, H. (2002) IP transmission over spectral-amplitude-coding CDMA links. *J. Microw. & Opt. Tech. Let.*, **33** (2), 140–142.

43. Kamath, P., Touch, J.D. and Bannister, J.A. (2004) The need for medium access control in optical CDMA networks. In: *IEEE InfoCom*, Hong Kong.

44. Kamath, P., Touch, J.D. and Bannister, J.A. (2004) Algorithms for interference sensing in optical CDMA networks. In: *Proc. ICC*, Paris, France.

45. Armitage, G. (2000) MPLS: the magic behind the myths. *IEEE Comm. Mag.*, Jan. 124–131.

46. Li, T. (1999) MPLS and the evolving Internet architecture. *IEEE Comm. Mag.*, Dec. 38–41.

47. Rekhter, Y. *et al.* (1997) *Cisco systems' tag switching architecture overview*. IETF – Network Working Group.

48. Lawrence, J. (2001) Designing multi-protocol label switching networks. *IEEE Comm. Mag.*, July. 134–142.

49. Viswanathan, A. *et al.* (1998) Evolution of multi-protocol label switching. *IEEE Comm. Mag.*, May. 165–173.

50. Ilyas, M. and Moftah, H.T. (2003) *Handbook of optical communication networks*. CRC Press, Florida, USA.

51. Kitayama, K. (1998) Code division multiplexing lightwave networks based upon optical code converter. *IEEE J. on Selected Areas in Comm.*, **16** (9), 1309.

52. Wen, Y. G., Zhang, Y. and Chen, L.K. (2002) On architecture and limitation of optical multi-protocol label switching (MPLS) networks using optical-orthogonal-code (OOC)/wavelength label. *Optical Fiber Technology*, **8** (1), 43–70.

53. Pattavina, A. (2005) Architectures and performance of optical packet switching nodes for IP networks. *J. Lightw. Technol.*, **23** (3), 1023–1032.

54. Zheng, Q. and Gurusamy, M. (2009) LSP partial spatial-protection in MPLS over WDM optical networks. *IEEE Trans on Comm.*, **57** (4), 1109–1118.

55. Yoshikane, N. *et al.* (2007) GMPLS-based multiring metro WDM networks employing OTN-based client interfaces for 10 GbE services. In: *OFC*, Anaheim, CA, USA.

56. Kitayama, K. and Wada, N. (1999) Photonic IP routing. *IEEE Photonics Tech. Letters*, **11** (12), 1689–1691.

57. Brackett, C. A. (1990) Dense wavelength division multiplexing networks: principle and applications. *IEEE J. on Selected Areas in Comm.*, **8** (8), 948–964.

58. Toliver, P. *et al.* (1999) Simultaneous optical compression and decompression of 100-Gb/s OTDM packets using a single bidirectional optical delay line lattice. *IEEE Photonics Tech. Letters*, **11** (9), 1183–1185.

59. Menendez, R. *et al.* (2005) Network applications of cascaded passive code translation for WDM-compatible spectrally phase-encoded optical CDMA. *J. Lightw. Technol.*, **23** (10), 3219–3230.

60. Mohamad, A. S. and Asano, S. (2008) Generalized multi-protocol label switching-based all-optical network for optical quality control. In: *Proceedings of the Fifth IASTED*, Anaheim, USA.
61. Singh, R. K. and Singh, Y.N. (2006) An overview of photonic packet switching architectures. *IETE Technical Review*, **23** (1), 15–34.
62. Yan, N., Monroy, I.T. and Koonen, T. (2005) All-optical label swapping node architectures and contention resolution. In: *Proc. of Optical Network Design and Modelling*, Milan, Italy.
63. Wada, N. *et al.* (2000) Photonic packet switching based on multi-wavelength label switching using fiber Bragg gratings. In: *ECOC*, Munich, Germany.
64. Jourdan, A. *et al.* (2001) The perspective of optical packet switching in IP-dominant backbone and metropolitan networks. *IEEE Comm. Mag.*, **39** (3), 136–141.
65. Capmany, J. *et al.* (2003) Subcarrier multiplexed optical label swapping based on subcarrier multiplexing: a network paradigm for the implementation of optical Internet. In: *ICTON*, Warsaw, Poland.
66. Carena, A. *et al.* (1998) OPERA: An optical packet experimental routing architecture with label swapping capability. *J. Lightw. Technol.*, **16** (12), 2135–2145.
67. Khattab, T. and Alnuweiri, H. (2007) Optical CDMA for all-optical sub-wavelength switching in core GMPLS networks. *J. on Selected Areas in Comm.*, **25** (5), 905–921.
68. Wada, N. and Kitayama, K. (1999) 10 Gb/s optical code division multiplexing using 8-chip optical bipolar code and coherent detection. *J. Lightw. Technol.*, **17** (10), 1758–1765.
69. Wen, Y. G. *et al.* (2000) An all-optical code converter scheme for OCDM routing networks. In: *ECOC*, Munich, Germany.
70. Wen, Y. G., Chen, L.K. and Tong, F. (2001) Fundamental limitation and optimization on optical code conversion for WDM packet switching networks. In: *OFC*, Anaheim, CA, USA.
71. Nakamura, S., Ueno, Y. and Tajima, K. (2001) 168-Gb/s all-optical wavelength conversion with a symmetric Mach–Zehnder-type switch. *IEEE Photonics Tech. Letters*, **13** (10), 1091–1093.
72. Banjeree, A. *et al.* (2001) Generalized multi-protocol label switching: an overview of routing and management enhancements. *IEEE Comm. Mag.*, Jan. 144–150.
73. Banjeree, A. *et al.* (2001) Generalized multi-protocol label switching: an overview of signaling enhancements and recovery techniques. *IEEE Comm. Mag.*, July. 144–151.
74. Perelló, J. *et al.* (2007) Resource discovery in ASON/GMPLS transport networks. *IEEE Comm. Mag.*, **45**(10), 86–92.

# 10

# Services Differentiation and Quality of Services in Optical CDMA Networks

## 10.1 Introduction

The Internet of the future requires higher bit rate and ultrafast services such as streaming over the Internet protocol (IP) like video-on-demand and IPTV. Due to its tremendous resources of bandwidth and extremely low loss, fibre-optic can be the best physical transmission medium for telecommunications and computer networks. Among optical access technologies, we need to pick the most suitable one that can make full use of the large bandwidth in the fibre-optic. Meanwhile, it has to have the potential to support random access protocols, different services with different data rates (i.e. differentiated services – DiffServ) with bursty traffic behaviours such as IP traffic.

Label switched paths could be used via the routing to establish end-to-end paths. Multi-protocol label switching (MPLS) is a switching protocol between Layers 2 and 3, adding labels in packet headers and forwarding labelled packets in corresponding paths using switching instead of routing; see Chapter 9 for more details. This transmission is an identical performance to that of the optical CDMA (OCDMA) concept, providing that it is utilized as a network access protocol since it has the potential to support label switching. Major applications of MPLS are network traffic engineering. Generalized MPLS (GMPLS) extends MPLS to add a signalling and routing control plane for devices in the packet domain, time domain, wavelength domain and fibre domain, all providing end-to-end provisioning of connections, resources and quality of services (QoS). There has been accelerating interests in optical transport networks (OTN) due to recent standardized Ethernet in the first-mile (IEEE 802.3ah) and also the establishment of Ethernet passive optical network (EPON) as a pragmatic solution for the fibre-to-the-home technology.

*Optical CDMA Networks: Principles, Analysis and Applications*, First Edition. Hooshang Ghafouri-Shiraz and M. Massoud Karbassian.
© 2012 John Wiley & Sons, Ltd. Published 2012 by John Wiley & Sons, Ltd.

Differentiated services (DiffServ) refer to the architecture carrying data with different rates, different qualities and different traffic patterns, for example voice, web browsing, email, video and file transfer. The DiffServ architecture for OCDMA networks focus on introducing the service plane, along with data and control planes, which is an intermediate functional plane which contains the intelligence for service provisioning.

The control plane is the part of the router architecture in which the network map exists, like the information in a routing table that defines where to route the incoming packets. In most cases, the routing table contains a list of destination addresses and the corresponding outgoing interfaces. Depending on the control plane protocol, it can also define certain packets to be discarded as well as privileged treatment of certain packets for which a high QoS is defined by such mechanisms as DiffServ [1].

On the other hand, the data plane sits in the part of the router architecture that decides what to do with packets arriving on an inbound interface. The decision could come from a decision layer or a service plane to, for example, establish a DiffServ in the network. The data plane commonly refers to a table where the destination address of the incoming packet exists and the router retrieves the information required to determine the path from the receiving element through the internal forwarding fabric of the router to the appropriate outgoing interface. The outgoing interface encapsulates the packet in the correct data link protocol. Depending on the routing protocol and router's configuration, functions usually implemented at the outgoing interface may set various packet fields such as the differentiated services code point (DSCP) field used by DiffServ architecture.

Having this concept in mind – that users (or applications) should be able to adapt the network to their requirements – users or applications can be offered management and/or control of the network resources, e.g. bandwidth allocation at the wavelength or sub-wavelength levels in WDM and Ethernet passive optical networks [2]. An important step towards this concept is the dynamic resource management capability in terms of data rate, bandwidth and QoS immediately when needed or in advance for a period of time in the future. These features can be implemented in the OCDMA networks much more easily than other architectures due to processing in the code domain rather than in the frequency or time domain. Optical switching has the ability of bandwidth manipulation at the wavelength levels, e.g. in optical packet/burst switching, and also the capability to accommodate a wide range of traffic distributions. It is observed that the state of the art of each individual technology of OCDMA [3], service-enabled network [4] and DiffServ [5] have been extensively studied; however, bringing them all under one umbrella has great potential in optical networking.

A conceptual diagram of the network resource management (NRM) architecture is illustrated in Figure 10.1. Network service components are displayed as a service to the service-oriented architecture (SOA)-bus, and then the network service can be adapted to the business processes. Network service clients, such as provisioning systems or other operation support systems (OSS), invoke network services following the application's requirements. The virtualization of a network resource is a key technology of this architecture to abstract and represent the service typically using Web services description language (WSDL) [6]. This means that network service clients can coordinate different network resources independent of the underlying network architecture in the same manner, through the SOA-bus. Moreover, virtualized network resources can be aggregated to a single network service, shown as network service-2 in Figure 10.1. The roles, functions

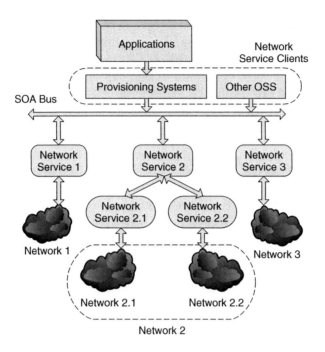

**Figure 10.1**  Diagram of service-oriented bus network resource management

or procedures of a particular network service can be also manipulated by going through additional service components as shown in network service-3 in Figure 10.1. NRMs are supposed to correspond approximately to each network service component in Figure 10.1. NRMs are also expected to play the role of virtualizing each network resource and exposing the network service to the SOA-bus [7]. In this manner, a reconfigurable system can be established with good maintainability and reusability.

A network that consists of nodes connected to each other with minimal connection to the central office, as shown in Figure 10.2, would result in significant fibre savings [4].

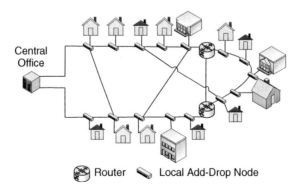

**Figure 10.2**  Diagram of SOA-bus enabled optical network nodes in ring/mesh architecture

The nodes are naturally able to provide bypass to traffic in addition to being able to add/drop traffic. The condition for a node implementation in an access network to result in lower fibre requirement than a star-based PON was earlier investigated in [8]. A node in an access network supporting the SOA-bus has to have the following networking properties:

- *Sub-wavelength/packet-mode behaviour*: In the access area, a single-/two-wavelength solution (like OCDMA-PON) is preferred, due to cost considerations and simplicity. Furthermore, multiple nodes share the same wavelength bandwidth efficiently, implying that each node will receive sub-wavelength granularity. To enable efficient statistical multiplexing we desire packet-centric communication, so a network solution in the access has to support packet-switched communication, leading to: (i) sub-wavelength granularity, and (ii) good statistical multiplexing.
- *Largely passive standalone equipment*: At the access network end, the electrical power seems unpredictable as the end-user behaviour is uncertain. On the other hand, the network operators are, in general, reluctant to be dependent on end-users' behaviour, thus considering the end-users' equipment as part of provider's equipment makes power consumption better under control. The traffic can hop across multiple nodes (using bypass properties) while travelling from a source to a destination, thus it is imperative that the node should be able to forward traffic even while it is switched OFF. Simple bypass using optical drop-and-continue techniques for optical circuits will not work because it does not support the packet-switched mode in which it is an essential technology for the node in the access area to enable SOA-bus activity.
- *Simple routing in a node with a high throughput*: For the same reason, the node power behaviour can be uncertain, resulting in uncertainty of the routing capability of a node. This implies the need for a simplistic routing mechanism with minimal dependence on the behaviour of each node. The best behaviour in such a situation is to have a virtual $K^2$ connectivity (assuming $K$ nodes), whereby a node that is passive and gets switched off does not affect the overall routing mechanism. However, issues such as flooding and learning preferred routes (to avoid duplication) are critical and need to be solved in a network.
- *Fault tolerance*: An SOA-bus comprising the end-user nodes can also provide a protection path in case of a fibre/node snap. This is due to the ability to provide alternate routes resulting from the $K^2$ connectivity.
- *Simplified protocol and ability to support services*: A $K^2$ connected graph which is largely passive implies a shared medium. Another constraint on the graph is the need for efficient statistical multiplexing. Passive $K^2$ connectivity combined with good statistical multiplexing results in a paradox that implies the requirement of an efficient protocol that can support new services. The protocol needs to be simple; for example, it has to say something that uses carrier-sense type access without unnecessary overheads, while maintaining low delay at higher loads. The service-enabled network consists of all-optical type of nodes. The central idea behind the choice of transparent network is that all-optical nodes are much lower in cost than the opaque networks due to less processing and can function for bypass traffic in the network even when they are powered off [4].

In order to make the optical domain compatible with IP networks, mapping universal IP addresses to generic frame protocol (GFP) can be a solution [9] or direct mapping

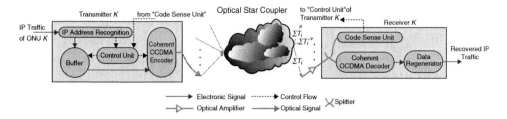

**Figure 10.3**   Example of OCDMA network node architecture for IP traffic

to address code sequences [10]. A basic structure for routing IP traffic over an optical network is illustrated in Figure 10.3 similar to the one discussed in Chapter 9. At each transmitter, the destination of each incoming IP packet is recognized and the packet is saved into a buffer which is divided into $N$ subparts [11].

IP packets that are destined for different receivers are stored in different subparts. If IP packets should be routed to the same receiver, they are saved in the same subpart sequentially. The control unit monitors the total amount of accumulated data in each sub-part. When this exceeds a predetermined threshold, the control unit adjusts the tuneable optical encoder to the desired address and sends the accumulated data like a burst. Con-sequently, the optical encoders do not needed to be tuned for each IP packet; however, they are required to be tuned for a number of packets [10, 12]. Thus, the requirement for tuning-speed of the encoder is significantly alleviated. Although the buffer delay increases in this scheme, especially when the tuning threshold is large, setting a proper threshold will make this delay acceptable even for real-time services. It should be noted that when two (or more) transmitters were to send signals to the same receiver at the same time, a collision would occur. To avoid such a collision, a corresponding MAC or scheduling protocol should also be employed carefully, as introduced and analysed in Chapter 9.

In the following, we study the OCDMA coding properties that can be interpreted for service differentiation and for connectivity that enables the network control plane to manage user-code allocation in higher layers than physical.

## 10.2   Differentiated Services in Optical CDMA

As mentioned in previous chapters, OCDMA networks take advantage of optical spread-ing codes with low cross-correlation as destination addresses to allow multiple users to simultaneously access a common channel. Due to the intensity modulation techniques in incoherent OCDMA systems, the unipolar codes comprising (0,1) are used. Various families of unipolar codes with low (optimal) correlation value have been introduced in Chapter 2. Accordingly, optical orthogonal codes (OOC) are considered with respect to identical code-length and code-weight (i.e. the number of 1s in the sequence) to satisfy correlation properties and to accommodate a certain number of simultaneous users with equal data rate and equal performance [13–15]. To utilize the spreading codes for distin-guishing various services with different data rates and qualities, the correlation values as a distinguishing parameter can be assigned to a specific service.

In doing so, variable-weight OOCs have been studied. Variable-weight OOC sequences have the same code-length but different code-weights. Code-weight represents the signal power and, accordingly, variable-weight sequences refer to variations in the signal power, which has a direct effect on the overall performance in the OCDMA network. A code sequence with a greater code-weight is less sensitive to interference than that with a lower code-weight. Variable-weight OOCs are based on various combinatorial techniques [16, 17] where the codes are available only for two different weights, which are able to support only two different services. However, multiweight OOCs relax the number of weights in the codeset for synchronous and asynchronous applications [18], the cross-correlation $C_c$ of the codeset being 'two', which can degrade system performance. Another multiweight single-length OOC with strict correlation property (i.e. $C_a = C_c = C = 1$) was studied in [19] where they are only strict for code sequences with similar code-weight, where $C_a$ and $C$ are the auto-correlation and the maximum correlation value, respectively. On the other hand, when several different code-weights in the codeset coexist to support multiservice in the network, their correlation values are violated, and this results in high multiple access interference (MAI) degrading the quality of service [20]. Therefore, strict and flexible variable-weight OOCs constructions are required for differentiated QoS in OCDMA networks.

A typical OCDMA network architecture is illustrated in Figure 10.4 [21]. There are $M$ users in the network where any pair of transceivers wants to send data through a star coupler to share a channel using a multiple access technique. Every user is assigned with unipolar (0,1) code sequences from a strict variable-weight OOC, for example.

The laser source at the transmitter produces very short pulses which are encoded by using an optical encoder. For example, in the time-spreading scheme studied in Chapter 5, the encoders include optical tapped-delay lines where the number of encoder taps determines the address code and the code-weight. The data at active transmitters,

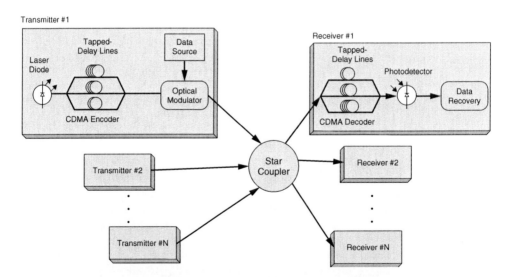

**Figure 10.4** Typical OCDMA network topology (Reprinted from: Design of strict variable-weight optical orthogonal codes for differentiated quality of service in optical CDMA networks, Nasaruddin and T. Tsujioka, *J. Computer Networks*, **52** (10), 2008, with permission from Elsevier.)

using an on-off keying (OOK) modulation, are encoded with the intended address code. Thereafter, the active transmitters superimpose their transmissions over a (common) channel and are then distributed to each receiver. The decoder should match the desired address code at the receiver end with optical tapped-delays which are the complement of the delay coefficients. If any pair of transceiver address codes matches exactly, the output of the correlator will be an auto-correlation peak, which recognizes that the transmitter has transmitted a 1 data bit. Otherwise, the correlator output will be the cross-correlation between other user signals. The output of the photodetector is fed to a threshold detector for data recovery. This network should be able to support multiple services recognized by different performance levels and varied QoS when employing VW-OOCs.

On the other hand, all the spreading codes we have studied have a single-rate feature. This means that all the users are communicating at the same data rate. We have learned that the code-weight variation has something to do with signal power or signal-to-noise ratio (i.e. the correlation ratio) that provides different qualities of service, but in multimedia communications we need to establish variable data rate links for several applications such as voice, data, video, email etc. The question is how to establish a variable data rate in OCDMA networks. In doing so, we need to understand the coding properties and their interpretations. By recalling code-length, data-rate and chip-rate relationships, as discussed throughout this book, we learned that the code-length represents the OCDMA system data rate; accordingly, a variable code-length coding scheme can be promising to implement this concept in OCDMA networks. Now in variable code-length code construction, similar to variable weight OOCs, the code-weight between any different code sequences are the same, while the code-length will vary.

The growing applications of multimedia transmission result in a requirement for multirate fibre-optic networks. For example, a system may need to provide some 64 kbps channels for voice-over-IP applications and some multi-gigabit-per-second channels for data services. Historically, to overcome this issue in radio CDMA networks, multicode and variable-length spreading techniques were studied [22, 23]. Therefore, the same techniques have also been tested in the optical domain for fibre-optic CDMA networks [24–26]. Petrovic et al. [25] proposed the idea of variable-length codes for optical fibre local area networks (LANs) in 1990. Then another multilength code construction based on different code-weights was proposed by Maric et al. [26]. However, the system using the coding technique proposed by Maric et al. had a high error probability for high-data-rate users. In 2002, Kwong et al. [24] proposed another family of multilength OOCs where the inter-cross-correlation constraint increases and accordingly increases the MAI between variable-rate users. On the other hand, as part of OOC design introduced in Chapter 2 and as will be seen in the following sections, code-length selection is relatively limited with respect to the code-weight and correlation values, thus the transmission rates will be limited to particular rates depending on the code-length constraints [27].

Two-dimensional (2D) wavelength–time codes, as introduced in Chapter 2, with good correlation properties were also considered to increase the number of subscribers and simultaneous users [28–30], for example concatenating the well-known prime codes [28] and OOCs [31].

The 2D 'extended' carrier-hopping prime codes (ECHPCs) [30, 32] constructed by expanding the carrier-hopping prime codes, have asymptotically optimal cardinality and possess zero auto-correlation side lobes as well as cross-correlation value of one. The

high-bit-rate OCDMA systems with broadband lasers take advantage of ECHPC in which
the number of time-slots is limited, and the number of available wavelengths exceeds the
code-length; also, note that the codeset cardinality depends on the number of wavelengths.
However, most current studies on 2D wavelength–time codes are based on the assumption
of a single-rate communication link. It is expected that future OCDMA networks will
simply support a large variety of services, e.g. Internet, voice-over-IP, video-on-demand,
email, IPTV etc., in a symmetric and scalable way. It is also expected that users with very
different bit rate and qualities-of-service will be accommodated simultaneously [24, 33].

By employing multiple-length 2D wavelength–time-spreading codes, the rate
can be dynamically matched to the users' requirements through the code matrices
assignment of different lengths. One possible approach to designing multiple-length 2D
wavelength–time codes is to merge several wavelength–time code sequences of different
lengths. However, the resulting cross-correlation function will be destructive and cause
strong interference if the code matrices are picked carelessly. Another possible approach
is to remove elements of some of the matrices (i.e. to puncture them) in single-length
wavelength–time codes. To maintain the cross-correlation property the weights of these
punctured matrices will be different from the original ones resulting in non-identical
auto-correlation peaks. The power control and correlation process will be complicated
in this scheme [5, 34]. Multiple-length codes can be alternatively generated by padding
zeros to the end of some codeset matrix sequences. Yet, to maintain the correlation
properties many zeros are required to break a 'periodic' correlation function into an
'aperiodic' one, resulting in very long code-length. Therefore, the codeset is forced to
start with short matrices which have small cardinality in the first place. As we see, there
is always a trade-off between correlation values, code-length and codeset cardinality.
And every code parameter has a practical meaning in the real world. For example, the
correlation values refer to the signal similarities in that they can be better distinguished;
code-weight is interpreted as signal power and signal-to-noise ratio to enhance the bit
error rate; and codeset cardinality denotes the number of subscribers in the network and
obviously the more the better.

Consequently, the coding construction techniques and algorithms play significant
roles in the OCDMA system performance and scalability where this field of research is
extremely active and thriving, as seen in the literature. In the following sections, we will
review and study a few recent and up-to-date variable-weight and/or variable-length optical spreading codes supporting service differentiation in the optical CDMA networking.

## 10.3   Variable-Weight Optical Spreading Codes

Here in this section, we introduce one of the recent, comprehensive variable-weight optical
orthogonal code (VW-OOC) constructions and algorithms. The concept here is based on
Nasaruddin *et al.*'s [21] design of strict VW-OOC to differentiate quality of service in
OCDMA networks.

To start with, we first need to come to the field and get familiar with the terms. The
VW-OOC codeset $C$ includes sequences comprising (0,1) with same lengths but different
weights denoted by $(n, W, \lambda, Q)$ where $n$ is the code-length, $W = \{w_1, w_2, \ldots, w_L\}$
is a set of code-weights, $\lambda$ is the maximum value of auto- and cross-correlation
$Q = \{q_1, q_2, \ldots, q_L\}$ denotes the fraction of code sequences $L$, is the number of different

weights in the codeset and $w_l$ is the $l^{th}$ code-weight, where $l = \{1, 2, \ldots, L\}$. $C$ can also be viewed as a family of $w_l$-sets of integer modulo $n$ in which each $w_l$-set corresponds to a codeword, and the integers within each $w_l$-set specify the positions of the 1s in the code sequence. The cardinality of a VW-OOC codeset, denoted by $|C|$, is the number of sequences in the codeset. Here, the strict VW-OOC is considered as meaning that $\lambda_a = \lambda_c = \lambda = 1$. Now a strict VW-OOC must fulfil the following conditions:

- Every $n$-tuple in $C$ has a code-weight $w_l$ contained in the set $W$.
- For any code sequence $x \in C$ and any integer $\tau \in \{1, 2, \ldots, n-1\}$, the auto-correlation is bounded by:

$$R_{xx} = \sum_{t=0}^{n-1} x_t x_{(t+\tau) \bmod n} \leq \lambda_a \qquad (10.1)$$

where $x = \{x_0, x_1, \ldots, x_{n-1}\}$ and $x_t \in \{0, 1\}$ for $t = \{0, 1, 2, \ldots, n-1\}$.

- For any two code sequences $x, y \in C$ and $x \neq y$ for any integer $\tau \in \{0, 1, 2, \ldots, n-1\}$ the cross-correlation is bounded by:

$$R_{xy} = \sum_{t=o}^{n-1} x_t y_{(t+\tau) \bmod n} \leq \lambda_c \qquad (10.2)$$

where $y = \{y_0, y_1, \ldots, y_{n-1}\}$ and $y_t \in \{0, 1\}$ for $t = \{0, 1, 2, \ldots, n-1\}$.

It is noted that the non-shifted auto-correlation value (i.e. $\tau = 0$) is equal to the code-weight which should be maximized to ensure that the received signal is stronger than background and interfering signals, while the shifted version of auto-correlation values (i.e. the side lobes) should be minimized as set to 1 in this construction method. This implies that the system operates fully asynchronously. The cross-correlation value in Equation (10.2) is also bounded to 1, thus all code sequences are distinguishable from each other leading to quite insignificant multiple access interference (MAI). To make strict VW-OOC work, sequences should have no repeated interval between two 1s within a sequence and among other sequences. If $x$ is a sequence of $C$ of weight $w_l$ where $x$ consists of $x_{j_0} = x_{j_1} = \cdots = x_{j_{w_l-1}} = 1$, then a set of intervals between the positions of two 1s associated with $x$ is denoted by $D_{x,w_l} = \{d_0, d_1, \ldots, d_{w_l-1}\}$ where:

$$d_i = \begin{cases} j_{i+1} - j_i & \text{for } i = 0, 1, \ldots, w_l - 2 \\ n + j_0 - j_{w_l-1} & \text{for } i = w_l - 1 \end{cases} \qquad (10.3)$$

where $w_l$ is the $l^{th}$ code-weight. Now, $R_{x,w_l} = [r_x(i,j)]$ denotes the $(w_l - 1) \times w_l$ array of integers of which the $(i, j)^{th}$ interval is given by:

$$r_x(i, j) = \sum_{k=0}^{i} d_{(j+k) \bmod w_l} \qquad (10.4)$$

where $i = \{0, 1, \ldots, w_l - 2\}$ and $j = \{0, 1, \ldots, w_l - 1\}$.

To ensure that the code sequences are strict VW-OOC, the used intervals should obey the following theorem; let $C = \{C_{w_1}, \ldots, C_{w_l}, \ldots, C_{w_L}\}$ be a codeset of a strict VW-OOC of size $|C|$, where $C_{w_l} = \{c_{1,w_l}, \ldots, c_{m,w_l}, \ldots, c_{|C_{w_l}|,w_l}\}$ is a sub codeset with code-weight $w_l$ and $c_{m,w_l}$ is $m^{th}$ sequence for $l = \{1, 2, \ldots, L\}$. $D$ is the collection of used intervals by a codeset in $C$. The following conditions lead to the code sequence being a strict VW-OOC:

- The intervals used by a sub-codeset $C_{w_l}, D_{w_l}$ with elements of $\{1, 2, \ldots, n-1\}$ must be distinct and formally $D_{1,w_l} \cap \ldots \cap D_{m,w_l} \cap \ldots \cap D_{|c_{w_l}|,w_l} = \emptyset$, where $D_{m,w_l}$ intervals are used by the $m^{th}$ code sequence in the $C_{w_l}$

- All intervals used by the codeset $C, D = \{D_{w_1}, \ldots, D_{w_l}, \ldots, D_{w_L}\}$ and $D$ with elements of $\{1, 2, \ldots, n-1\}$ must be distinct and formally $D_{w_1} \cap \ldots \cap D_{w_l} \cap \ldots \cap D_{w_L} = \emptyset$ such that $w_l \neq w_{l'}$.

Equations (10.3) and (10.4) can analyse all the used intervals in the codeset of a strict VW-OOC. The first condition means that there is no repeated used interval among code sequences with the same code-weight of $w_l$. The second condition will be met when the used intervals among code sequences with different code-weights $w_l \neq w_{l'}$ are not replicated. Accordingly, the strict VW-OOC codeset will be obtained.

The upper-bound for the codeset cardinality for strict $(n, W, \lambda, Q)$ VW-OOC is modified as [16]:

$$|C| \leq \frac{(n-1)(n-2)\ldots(n-\lambda)}{\sum_{l=1}^{L} q_l w_l (w_l - 1)(w_l - 2)\ldots(w_l - \lambda)} \tag{10.5}$$

By considering the strict condition of $\lambda = 1$, the upper-bound will be simplified as:

$$|C| \leq \frac{(n-1)}{\sum_{l=1}^{L} q_l w_l (w_l - 1)} \tag{10.6}$$

where $q_l$ denotes the fraction of code sequence with weight $w_l$. A codeset is called optimal strict VW-OOC when it satisfies Equation (10.6), with equality being the maximum system capacity. After simple manipulation with Equation (10.6), the bound for code-length will be obtained as:

$$n \geq \sum_{l=1}^{L} w_l \times (w_l - 1) \times q_l \times |C| + 1 \tag{10.7}$$

We are now familiar with the terms and conditions of a strict VW-OOC codeset. Among many methods, we study two approaches to construct this codeset as discussed in the following sections.

## 10.3.1  Distinct Set Approach

A distinct set consists of integer intervals where each interval denotes the distance between a specific pair of ones in that all intervals in the set must be distinct. For example, with $w$ 1s there will be $w(w-1)/2$ entries in a distinct set. First order differences are the

| First order............... | $m_1$ | $m_2$ | $m_3$ |
| Second order............. | $m_1 + m_2$ | $m_2 + m_3$ | |
| Third order............. | $m_1 + m_2 + m_3$ | | |

**Figure 10.5**  Order differences of a distinct set (Reprinted from: Design of strict variable-weight optical orthogonal codes for differentiated quality of service in optical CDMA networks, Nasaruddin and T. Tsujioka, *J. Computer Networks*, **52** (10), 2008, with permission from Elsevier.)

intervals for every pair of adjacent 1s in the set. Second order differences are the intervals between 1s placed two apart on the set. A set with $w$ 1s will have $(w - 1)$ first order differences, $(w - 2)$ second order differences and so on, up to a single $(w - 1)^{th}$ order difference [21]. If there are $w = 4$ 1s, and the terms $m_1, m_2$ and $m_3$ are the intervals for the first order differences, then the second and third order differences form a distinct set as shown in Figure 10.5.

For example, suppose a distinct set has $w = 4$ 1s and the intervals for order differences are $\{m_1, m_2, m_3\} = \{1, 3, 2\}$. As shown in Figure 10.5, the second and third orders differences are $\{4, 5\}$ and $\{6\}$ respectively. All intervals (entries) in the set will be then $I = \{1, 3, 2, 4, 5, 6\}$ which are distinct. The maximum interval in this set is denoted by $I_{max} = 6$. The $I_{max}$ also denotes the length of the set, given by:

$$I_{max} \geq \frac{w(w - 1)}{2} \tag{10.8}$$

As observed, this distinct set is an optimal set since the $I_{max} = 6$ satisfies Equation (10.8) equality.

Let $S$ be a family of distinct sets and $s_l$ is a distinct with $w_l$ ones. Each $s_l$ has $\binom{w_l}{2}$ possible intervals. This includes intervals for first order differences $I^{(1)}_{w_l, s_l} = \{m_{1, s_l}, m_{2, s_l}, \ldots, m_{w_l - 1, s_l}\}$ or in other word, $I^{(1)}_{w_l, s_l} = \{m_{r, s_l} | r = 1, 2, \ldots, w_l - 1\}$ for each $s_l$ in that:

- All the differences, $m_{r, s_l}$ and $m_{r, s'_l}$ when $s_l \neq s'_l$, are distinct, and
- Other order differences, $I^{(2)}_{w_l, s_l} = \{m_{r, s_l} + m_{r+1, s_l}\}, \ldots, I^{(w_l - 1)}_{w_l, s_l} = \{m_{r, s_l} + m_{r+1, s_l} + \ldots + m_{w_l - 1, s_l}\}$ when $s_l \neq s'_l$ are also distinct.

After all these definitions as well as terms and conditions, we are ready to describe the Nasaruddin's construction algorithm [21] of strict VW-OOC codeset in the following steps:

1. At the beginning, we need to define the number of services, $L$ and the number of subscribers, $Q = \{q_1, q_2, \ldots, q_L\}$ at each service which are to be realized in the OCDMA network. Accordingly, $L$ different code-weights $W = \{w_1, w_2, \ldots, w_L\}$ and the number of codesets $C = \{c_{w_1}, c_{w_2}, \ldots, c_{w_L}\}$ with $Q$ number of users in each codeset are defined. Then, the first order differences of the distinct sets with specific adjacent intervals for the $s_l^{th}$ code sequence are chosen as given by:

$$I^{(1)}_{w_l, s_l} = \{m_{1, s_l}, m_{2, s_l}, \ldots, m_{w_l - 1, s_l}\} \tag{10.9}$$

where $l = (1, 2, \ldots, L)$, $s_l = \left(1, 2, \ldots, |C_{w_l}|\right)$ and $|C_{w_l}|$ is the number of code sequences with code-weight $w_l$. The remaining order differences are specified with no repeated interval within a set and among other sets.

2. The code sequences for $C_{w_l}$ that contain $w_l$-set of 1s are constructed whose 1 positions are given by:

$$\{0, b_{1,s_l}, b_{2,s_l}, \ldots, b_{w_l-1,s_l}\} \tag{10.10}$$

where $b_{1,s_l} = m_{1,s_l}$, $b_{2,s_l} = m_{1,s_l} + m_{2,s_l}$, $\ldots, b_{w_l-1,s_l} = m_{1,s_l} + m_{2,s_l} + \ldots + m_{w_l-1,s_l}$. It should be noted that the first pulse of all code sequences is always located in the first chip of the code sequence.

3. The maximum interval in the distinct set is given by $I_{\max} \geq w(w-1)/2$. Thus, the largest chip number of 1 positions in $C$ is denoted by $b_{\max}$ which should satisfy:

$$b_{\max} \geq \sum_{l=1}^{L} \frac{w_l(w_l-1)}{2} \times q_l \times |C| \tag{10.11}$$

4. Thus, the code-length for the strict VW-OOC must fulfil:

$$n \geq 2b_{\max} + 1 \tag{10.12}$$

which is the same result given in Equation (10.7) by substituting Equation (10.11) into Equation (10.12).

5. Finally, a strict VW-OOC codeset $\{n, \{w_1, w_2, \ldots, w_L\}, 1, \{q_1, q_2, \ldots, q_L\}\}$ is constructed.

Here we have an example to clearly explain the algorithm:

1. Let us select $L = 3$, $W = \{w_1, w_2, w_3\} = \{4, 3, 2\}$ and $|C| = \left\{|C_{w_1}|, |C_{w_2}|, |C_{w_3}|\right\} = \{1, 1, 1\}$ for $Q = \{q_1, q_2, q_3\} = \{1/3, 1/3, 1/3\}$. There are three code sequences in this codeset, $|C| = 3$ and then $s_1 = s_2 = s_3 = 1$. Next, we choose the first order of three distinct sets with different number of ones according to $W$, e.g. $\{I_{w_1,s_1}^{(1)}, I_{w_2,s_2}^{(1)}, I_{w_3,s_3}^{(1)}\} = \{I_{4,1}^{(1)}, I_{3,1}^{(1)}, I_{2,1}^{(1)}\} = \{\{1, 4, 7\}, \{2, 6\}, \{3\}\}$. Other order differences for the set is shown in Table 10.1 where $I_{w_l,s_l}$ is a collection of all order differences in $s_l$.

2. A code sequence, $C_{w_1}$ of code-weight $w_1 = 4$ is constructed and by using Equation (10.10) and the $I_{w_1,s_1}^{(1)}$ where a $w_1$-set ones positions is $C_{w_1} = \{(0, 1, 5, 12)\}$; and similarly $C_{w_2}$ and $C_{w_3}$ are obtained with the sequences of $C_{w_2} = \{(0, 2, 8)\}$ and $C_{w_3} = \{(0, 3)\}$ respectively. A codeset for VW-OOC is produced as $C = \{C_{w_1}, C_{w_2}, C_{w_3}\} = \{(0, 1, 5, 12), (0, 2, 8), (0, 3)\}$. The signalling format of the three code sequences are illustrated in Figure 10.6.

3. It is found from $C$ that $b_{\max} = 12$.

4. The code-length will then be $n = 25$ according to Equation (10.12).

5. Finally, a strict VW-OOC codeset of $\{25, \{4, 3, 2\}, 1, \{1/3, 1/3, 1/3\}\}$ is constructed.

It is noted that the intervals of a distinct set are related to the used intervals of a code sequences of strict VW-OOC. Equations (10.3) and (10.4) calculate all the used intervals in the code sequence of a strict VW-OOC. Thus, the intervals used

**Table 10.1**  Order differences of the distinct set
(Reprinted from: Design of strict variable-weight optical
orthogonal codes for differentiated quality of service in
optical CDMA networks, Nasaruddin and T. Tsujioka,
*J. Computer Networks*, **52** (10), 2008, with permission
from Elsevier.)

| $I_{w_1,s_1} = I_{4,1}$ | | $I_{w_2,s_2} = I_{3,1}$ | | $I_{w_3,s_3} = I_{2,1}$ | |
|---|---|---|---|---|---|
| 1 | 4 | 7 | 2 | 6 | 3 |
| 5 | | 11 | | 8 | |
| | 12 | | | | |

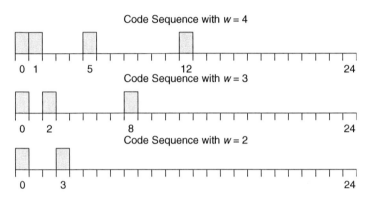

**Figure 10.6**  The code sequences of (25,{4,3,2},1,{1/3,1/3,1/3}) VW-OOC (Reprinted from: Design of strict variable-weight optical orthogonal codes for differentiated quality of service in optical CDMA networks, Nasaruddin and T. Tsujioka, *J. Computer Networks*, **52** (10), 2008, with permission from Elsevier.)

by a $\{25, \{4, 3, 2\}, 1, \{1/3, 1/3, 1/3\}\}$ strict VW-OOC according to Equations (10.3) and (10.4) are $D_{w_1} = \{1, 4, 7, 13, 5, 11, 20, 14, 12, 24, 21, 18\}$, $D_{w_2} = \{2, 6, 17, 8, 23, 19\}$ and $D_{w_3} = \{3, 22\}$. As seen, there is no repetition of intervals in the codeset and $D_{w_1} \cap D_{w_2} \cap D_{w_3} = \emptyset$. Therefore, the cross-correlation value is now bounded by one and the theorem is satisfied. As expected, by using this triplet-weight OOC, three different services can be supported simultaneously in the OCDMA network.

Another example of a strict VW-OOC constructed with four different code-weights, $L = 4$ in the codeset is illustrated in Table 10.2 [21]. The strict VW-OOC codeset is defined by $\{427, \{5, 4, 3, 2\}, 1, \{6/50, 8/50, 12/50, 24/50\}\}$. The cardinality of the codeset is 50 (users) supporting four different services; six users on code-weight five, eight users on code-weight four, 12 users on code-weight three, and 24 users on code-weight two. As seen in Table 10.2, more code sequences are produced with the smallest code-weight to optimize the codeset without increasing the code-length. Yet, the codeset is still poor when comparing its code-length with the upper-bound of the code-length in Equations (10.7) and (10.12) with respect to the upper-bound of $b_{max}$ in Equation (10.11). An explanation for

**Table 10.2** A distinct set of $\{427, \{5, 4, 3, 2\}, 1, \{6/50, 8/50, 12/50, 24/50\}\}$ (Reprinted from: Design of strict variable-weight optical orthogonal codes for differentiated quality of service in optical CDMA networks, Nasaruddin and T. Tsujioka, *J. Computer Networks*, **52** (10), 2008, with permission from Elsevier.)

| $w_1 = 5$ | $w_2 = 4$ | $w_3 = 3$ | $w_4 = 2$ |
|---|---|---|---|
| | | (0,85,185) | (0,116)(0,118) |
| | | (0,86,187) | (0,119)(0,120) |
| | (0,39,98,172) | (0,87,186) | (0,121)(0,123) |
| (0,18,25,53) | (0,44,105,180) | (0,88,191) | (0,124)(0,125) |
| (0,2,11,29,60) | (0,46,108,184) | (0,89,193) | (0,126)(0,127) |
| (0,3,13,33,68) | (0,48,112,190) | (0,91,197) | (0,129)(0,131) |
| (0,4,16,38,70) | (0,50,117,196) | (0,92,199) | (0,132)(0,134) |
| (0,5,19,42,82) | (0,51,122,202) | (0,93,203) | (0,135)(0,137) |
| (0,6,21,47,90) | (0,56,128,209) | (0,94,205) | (0,139)(0,140) |
| | (0,57,130,213) | (0,95,208) | (0,141)(0,143) |
| | | (0,96,210) | (0,144)(0,145) |
| | | (0,97,212) | (0,147)(0,148) |

this is that the strict VW-OOC requires a relaxing code-length to maintain the correlation value between the code sequences with the same code-weight and between the code sequences with the different code-weights.

## 10.3.2 Random Approach

The random method is, in a minimax sense, more powerful than the deterministic method. It typically involves an iterative process in which every iteration searches the progress of a new solution in the neighbouring existing solution. The local optimization procedure starts from an initial value that is selected in the first step. The 'stop' step will be used when a global optimum is found because the computer time is an exhausted resource. Moreover, the random approach can be altered and adapted to specific requirements so that the required strict correlation values for VW-OOC can be satisfied. To generate VW-OOCs, it starts with an empty codeset, and a new code sequence is attempted to include the codeset repeatedly.

The algorithm for generating of strict VW-OOCs based on the random method can be described in the following steps:

1. The positive integers of $W = \{w_1, w_2, \ldots, w_L\}$ for $L$ different code-weights and the number of code sequences of $C = \{C_{w_1}, C_{w_2}, \ldots, C_{w_L}\}$ with $Q = \{q_1, q_2, \ldots, q_{L_{w_1}}\}$ based on the network specifications are selected as inputs.
2. According to Equation (10.7), the code-length $n$ can be obtained from the parameters given in step (1); while, $n$ refers to the number of possible assignments of positive integers in the search process.

3. The first chip position is set to 0 for each code sequence, then a random integer of $(1, 2, \ldots, n - 1)$ is assigned to $(b_{1,t_l}, b_{2,t_l}, \ldots, b_{w_l-1,t_l})$ by computer search, where $t_l = (1, \ldots, |C_{w_l}|)$ and $l = (1, \ldots, L)$; $|C_{w_l}|$ is the number of code sequences with code-weight $w_l$. The $w_l$-set of 1 positions for the $t_l^{th}$ code sequence is represented by:

$$(0, b_{1,t_l}, b_{2,t_l}, \ldots, b_{w_l-1,t_l}) \bmod n \tag{10.13}$$

4. The used intervals in the code sequences are checked as:

$$d_{hk,t_l} = b_{k,t_l} - b_{h,t_l} \text{ and } d_{hk,t_l} \neq d'_h k', t'_l \tag{10.14}$$

where $t_l = (1, \ldots, |C_{w_l}|)$ and $t'_l = (1, \ldots, |C_{w_{l'}}|)$; $d_{hk,t_l} \in D_{w_l}$ and $d_{h'k',t_{l'}} \in D_{w_{l'}}$ where $h = (0, 1, \ldots, k - 1)$ and $k = (1, 2, \ldots, w_l - 1)$; $h' = (0, 1, \ldots, k' - 1)$ and $k' = (1, 2, \ldots, w_{l'} - 1)$. The intervals in each code sequence are computed and then they are compared with their own used intervals and intervals in other sequences. If the correlation constraints for strict VW-OOC are not satisfied, it returns to the previous step and continues.
5. When all correlation constraints are met, the search stops, otherwise it returns to step (3) and continues.

Here we can also look into an example of generating VW-OOC through a random approach. The algorithm should be implemented in a computer program to illustrate the steps in generating strict VW-OOC codeset as follows:

1. $W = \{5, 3, 2\}, C = \{1, 1, 1\}$ with $Q = \{1/3, 1/3, 1/3\}$ are selected as input parameters.
2. The code-length will then be $n = 29$.
3. A codeset is generated with a different number of integers such as $C = \{\{0, 12, 14, 20, 25\}, \{0, 7, 26\}, \{0, 28\}\}$ and stored in a look-up table.
4. The used intervals in the codeset are checked; all used intervals in the codeset are $D_{w_1} = \{12, 2, 6, 5, 4, 14, 8, 11, 9, 16, 20, 13, 15, 21, 18, 25, 17, 27, 23, 24\}$, $D_{w_2} = \{7, 19, 3, 26, 22, 10\}$ and $D_{w_3} = \{28, 1\}$, which fulfil the correlation constraints of strict VW-OOC.
5. Once all the parameters are computed and conditions are met, searching stops. The codeset is now strict VW-OOC $\{29, \{5, 3, 2\}, 1, \{1/3, 1/3, 1/3\}\}$.

Since the code sequences are generated randomly, the algorithm can also be flexibly altered to find many different codesets under arbitrary constraints. The stopping conditions in step (5) can be changed as well as the random seeds. Therefore, different codesets may be used for code reconfiguration. A reconfiguration of a strict VW-OOC $\{29, \{5, 3, 2\}, 1, \{1/3, 1/3, 1/3\}\}$ is $\{\{0, 1, 3, 12, 25\}, \{0, 15, 23\}, \{0, 19\}\}$.

Tables 10.3 and 10.4 [21] show another example of strict VW-OOC based on the random method and its reconfiguration, respectively. As seen, the 1 positions in the codesets are different from each other. To further increase system security and eliminate any vulnerability to interception, a strategy of code reconfiguration seems necessary [35]. Frequent code allocation enhances the network security markedably. Thus for this purpose, a large

**Table 10.3**  Strict VW-OOC $\{101, \{4, 3, 2\}, 1, \{5/15, 5/15, 5/15\}\}$ based on the random approach (Reprinted from: Design of strict variable-weight optical orthogonal codes for differentiated quality of service in optical CDMA networks, Nasaruddin and T. Tsujioka, *J. Computer Networks*, **52** (10), 2008, with permission from Elsevier.)

| $w_1 = 4$ | $w_2 = 3$ | $w_3 = 2$ |
|-----------|-----------|-----------|
| (0,21,70,99) | (0,1,91) | (0,95) |
| (0,14,79,82) | (0,45,61) | (0,8) |
| (0,7,34,64) | (0,4,9) | (0,51) |
| (0,46,63,89) | (0,18,66) | (0,76) |
| (0,32,73,86) | (0,20,59) | (0,24) |

**Table 10.4**  A reconfiguration of strict VW-OOC $\{101, \{4, 3, 2\}, 1, \{5/15, 5/15, 5/15\}\}$ based on random approach (Reprinted from: Design of strict variable-weight optical orthogonal codes for differentiated quality of service in optical CDMA networks, Nasaruddin and T. Tsujioka, *J. Computer Networks*, **52** (10), 2008, with permission from Elsevier.)

| $w_1 = 4$ | $w_2 = 3$ | $w_3 = 2$ |
|-----------|-----------|-----------|
| (0,36,74,82) | (0,42,94) | (0,13) |
| (0,30,67,81) | (0,44,54) | (0,45) |
| (0,1,17,41) | (0,2,5) | (0,80) |
| (0,26,35,58) | (0,29,62) | (0,12) |
| (0,48,73,79) | (0,4,15) | (0,83) |

codeset cardinality is important. It should also be noted that the random approach has a code reconfiguration feature that cannot be found in deterministic algorithms producing fixed codesets [16, 18, 19]. On the other hand, searching all the possible codesets in the algorithm would be very time-consuming. Obviously, the search time increases when $|C|$ or $W$ increases and accordingly to reduce the search time a relaxing code-length is required, nevertheless the codeset would be less optimal.

## 10.3.3  Performance Analysis

The performance of an OCDMA network is inherently influenced by the multiple access interference (MAI) from the other users. Here, we analyse the performance of a system deploying a VW-OOC codeset in terms of bit error rate (BER) considering the MAI as the major degrading factor to the overall performance. Other performance impacts such as the effects of shot noise, thermal noise and dispersions are not taken into account since we only evaluate the coding schemes and their properties in terms of 'hit' occurrence (i.e. code overlap) in the system. Removing the other noise sources, an error takes place only when the cumulative effect of MAI, which means the received 0 data bit reaches above the decision threshold. There is no frame synchronization between the transmitters, whereas the chip is assumed to be synchronous between users. This implies that each chip of a codeword must be perfectly aligned with a chip of another codeword. When $M$ users are transmitting code sequences simultaneously, $M - 1$ users contribute to MAI. If one user transmits a pulse or a 1 data bit, an encoder generates a code sequence but no code sequence for a 0 data bit.

In a strict VW-OOC $(n, W, \lambda, Q)$, each interfering user may contribute only to one chip overlap (hit) with the intended receiver (i.e. it is strict). Then, the probability that one of the 1 positions of a user's code sequence with code-weight $w_l$ will overlap with a 1 position of the intended user's code sequence with code-weight $w_{l'}$ is given by:

$$P_{l,l'} = \frac{w_l w_{l'}}{2n} \tag{10.15}$$

where $l, l' \in \{1, \ldots, L\}$, $w_l, w_{l'} \in W = \{w_1, \ldots, w_L\}$ and $w_l \neq w_{l'}$.

Let us assume $L$ different code-weights of code sequences that coexist in the OCDMA network. Then there are $M = \{M_{w_1}, M_{w_2}, \ldots, M_{w_L}\}$ users on $L$ different services in the OCDMA network. Thus, a user with code-weight $w_1$, for example, could be interfered with by other users of code-weights $\{w_1, w_2, \ldots, w_L\}$ at one chip position of each code-weight, due to the strict feature of the codesets. The bit-error probability of a user with code-weight $w_l$ of the strict VW-OOC can be then evaluated as [21, 36]:

$$
\begin{aligned}
P_{E,w_l} &= \frac{1}{2} \sum_{g_1+g_2+\cdots+g_L=w_l}^{M-1} \left\{ \binom{M_{w_l} - 1}{g_l} \cdot (P_{l,l})^{g_l} \cdot (1 - P_{l,l})^{M_{w_l}-1-g_l} \right. \\
&\quad \left. \cdot \prod_{\substack{l'=1 \\ l' \neq l}}^{L} \binom{M_{w_{l'}}}{g_{l'}} \cdot (P_{l,l'})^{g_{l'}} \cdot (1 - P_{l,l'})^{M_{w_{l'}}-g_{l'}} \right\} \\
&= \frac{1}{2} - \frac{1}{2} \sum_{g_1+g_2+\cdots+g_L=0}^{w_l-1} \left\{ \frac{(M_{w_l} - 1)!}{(M_{w_l} - 1 - g_l)! \cdot g_l!} \cdot (P_{l,l})^{g_l} \cdot (1 - P_{l,l})^{M_{w_l}-1-g_l} \right. \\
&\quad \left. \cdot \prod_{\substack{l'=1 \\ l' \neq l}}^{L} \frac{M_{w_{l'}}!}{(M_{w_{l'}} - g_{l'})! \cdot g_{l'}!} \cdot (P_{l,l'})^{g_{l'}} \cdot (1 - P_{l,l'})^{M_{w_{l'}}-g_{l'}} \right\} \tag{10.16}
\end{aligned}
$$

where $g_l$ is the number of interferers with code sequences of code-weight $w_l$. For optimal operation, the threshold decision $(Th)$ is set to the code-weight $w_l$ [13]. To verify Equation (10.16) and its validity for $\lambda = 1$, the parameters of $M_{w_{l'}} = 0, g_{l'} = 0, P_{l,l'} = 0$ can be inserted into it. The result will be then an error probability of a single-weight constant-length OOC as:

$$P_{E,w_l} = \frac{1}{2} - \frac{1}{2} \sum_{g_l=0}^{w_l-1} \left\{ \frac{(M_{w_l} - 1)!}{(M_{w_l} - 1 - g_l)! \cdot g_l!} \cdot (P_{l,l})^{g_l} \cdot (1 - P_{l,l})^{M_{w_l}-1-g_l} \right\} \tag{10.17}$$

By removing the irrelevant indices to avoid confusion in Equation (10.17) since $P_{E,w_l}$ has only a single-weight, $P_E$ for single-weight is given by:

$$P_E = \frac{1}{2} \sum_{g=Th}^{M-1} \left\{ \binom{M-1}{g} \cdot (P_s)^g \cdot (1 - P_s)^{M-1-g} \right\} \qquad (10.18)$$

where $P_s = w^2/(2n)$ for single-weight constant-length strict OOC [13, 21, 36]. For further numerical analysis on the VW-OOC, readers are encouraged to see [16–18, 21, 36–39].

## 10.4   Variable-Length Optical Spreading Codes

As discussed in Section 10.2, there is a need for multirate multimedia communications in OCDMA networks, but how to establish a multirate channel over a single channel is challenging, since the OCDMA signal processing is in the code domain rather than the frequency or time domains. Accordingly, most of the OCDMA network specifications, like signal power, data rate and bit error rate, can be interpreted in the coding properties. The link data rate is directly related to the code-length since the number of chips in the code structure implies the spreading factor, meaning of data rate. Implementing multirate OCDMA leads us to construct the optical variable-length spreading coding scheme which is introduced in this section.

Let us review some fundamentals relevant to variable-length spreading code construction. The OCDMA technique is basically settled and is based on the assignment of orthogonal codes to each user. It has been realized that the different codesets with their various code properties perform differently. The correlation functions play an important role in system design. There are usually two correlation functions: one is the auto-correlation function, determining how well a code sequence can be detected when interference and noise exist. The auto-correlation function is usually used for the precise data acquisition and synchronization in synchronous schemes. The other correlation function is cross-correlation, which indicated mutual interference between two different code sequences.

Consider any two one-dimensional (1D) code sequences $X = (x_0, x_1, \ldots, x_{L-1})$ and $Y = (y_0, y_1, \ldots, y_{L-1})$ in codeset $C$, where $x_i, y_i \in \{0, 1\}$ and $L$ is the code-length. For any integer $0 \leq \tau \leq L - 1$, the 1D cyclic discrete correlation function between $X$ and $Y$ is defined by [14]:

$$R_{XY}(\tau) = \sum_{i=0}^{L-1} x_i y_{i \oplus \tau} \qquad (10.19)$$

where $\oplus$ denotes modulo-$L$ addition. When $X \neq Y$, $R_{XY}(\tau)$ represents the cross-correlation function, whereas for $X = Y$, $R_{XY}(\tau)$ represents the auto-correlation function denoted by $R_X(\tau)$.

On the other hand, the correlation function of two-dimensional (2D) code sequences becomes a summation over all the two dimensions such as time and wavelength. Consider any two code sequences $X = \left( x_{0,0}, \ldots, x_{0,L_2-1}, x_{1,0}, \ldots, x_{1,L_2-1}, \ldots, x_{L_1-1,0}, \ldots, x_{L_1-1,L_2-1} \right)$ and $Y = \left( y_{0,0}, \ldots, y_{0,L_2-1}, y_{1,0}, \ldots, y_{1,L_2-1}, \ldots, y_{L_1-1,0}, \ldots, \right.$

$y_{L_1-1,L_2-1})$ in a 2D codeset where $x_{i,j}, y_{i,j} \in \{0,1\}$, $L_1$ and $L_2$ are 2D code-lengths. For any integers $0 \le \tau_1 \le L_1 - 1$ and $0 \le \tau_2 \le L_2 - 1$, the 2D cyclic discrete correlation function between $X$ and $Y$ is defined by:

$$R_{XY}(\tau_1, \tau_2) = \sum_{i=0}^{L_1-1} \sum_{j=0}^{L_2-1} x_{i,j} y_{i \oplus \tau_1, j \widehat{\oplus} \tau_2} \tag{10.20}$$

where $\oplus$ and $\widehat{\oplus}$ denote a modulo-$L_1$ and modulo-$L_2$ additions respectively. Similarly when $X \ne Y$, $R_{XY}(\tau_1, \tau_2)$ represents the cross-correlation function, whereas for $X = Y$, $R_{XY}(\tau_1, \tau_2)$ indicates the auto-correlation function as $R_X(\tau_1, \tau_2)$.

As we have seen earlier, in Chapter 2, a spreading codeset is characterized by a quadruple $(L, w, \lambda_a, \lambda_c)$ denoting code-length $L$, code-weight $w$, maximum out-of-phase auto-correlation value $\lambda_a$ and maximum cross-correlation value $\lambda_c$. Accordingly, the correlation values of a 1D codeset must satisfy the following properties:

$$R_X(\tau) = \begin{cases} = w & \text{for } \tau = 0 \\ \le \lambda_a & \text{for } 1 \le \tau \le L - 1 \end{cases} \tag{10.21}$$

and,

$$R_{XY}(\tau) \le \lambda_c \text{ for } 0 \le \tau \le L - 1 \tag{10.22}$$

On the other hand, 2D unipolar codes use two features and therefore their presentation will be different. One feature considers the product of the two dimensions code-lengths as an actual code-length [40] whereas another feature uses the fifth parameter in the codeset characterization [41]. Since we are looking at using variable-length code sequences to implement multirate services, one dimension should be time, as a platform for data rate variation [27]. Therefore, the forms of the correlation functions will be slightly changed. The two code sequences are changed into $X = (x_0, x_1, \ldots, x_{L-1})$ and $Y = (y_0, y_1, \ldots, y_{L-1})$, where $x_i, y_i \in \{0, 1, \ldots, r\}$ and $r$ denotes the length of another dimension, for example space or wavelength.

Accordingly, a 2D unipolar codeset will be characterized by $(L, r, w, \lambda_a, \lambda_c)$ where $L$ is the code-length in the time domain and $r$ is the length of another dimension and the rest remain the same. The correlation constraints $\lambda_a$ and $\lambda_c$ must satisfy the following properties:

$$R_X(\tau) = \sum_{i=0}^{L-1} \delta(x_i, x_{i \oplus \tau}) \begin{cases} = w & \text{for } \tau = 0 \\ \le \lambda_a & \text{for } 1 \le \tau \le L - 1 \end{cases} \tag{10.23}$$

$$R_{X,Y}(\tau) = \sum_{i=0}^{L-1} \delta(x_i, y_{i \oplus \tau}) \le \lambda_c \text{ for any } \tau \tag{10.24}$$

where $\delta(z, t) = \begin{cases} 1 & \text{for } z = t \ne 0 \\ 0 & \text{otherwise} \end{cases}$ and $\oplus$ denotes a modulo-$L$ addition.

As mentioned, the auto-correlation function is utilized for the initial data acquisition and synchronization, and the cross-correlation function is referred to the mutual interference

between two users; accordingly, two essential criteria for designing multilength unipolar codes must be considered [42]. The first is to maintain the identical auto-correlation
function peaks of different code-lengths, and the other is to keep the cross-correlation
functions among different code-lengths low in a way that does not substantially reduce
the codeset cardinality. The code construction algorithm explained here is based on
J.-Y. Lin *et al.* [27] and will follow these two criteria in the following steps:

**Step 1.** A conventional unipolar constant-length codeset $C$ with code-length $L_1$, code-
weight $w$ and codeset cardinality of $|C|$ is constructed. The code construction can
follow either of the two coding techniques introduced in Chapter 2 for 1D or 2D
spreading codes. Let $X$ denote a code sequence of codeset $C$ and be represented
by a notation of $\left\{ x_i^{(P)} \mid 1 \le i \le w \right\}$ where $x_i^{(P)}$ is the time position of the $i^{th}$
non-zero value of the code sequence in the codeset.

**Step 2.** A codeset $C'$ with code-length $L_m = m \cdot L_1$ and $|C|$ code sequences is constructed
by concatenating method [31]. The concatenating method refers to constructing
a code sequence $Z \in C'$ by concatenating $m$ copies of $X \in C$. It should be noted
that only the length of the time domain is extended in a 2D coding scheme.
Therefore, the notation for $Z$ becomes $\left\{ z_{i,j}^{(P)} \mid z_{i,j}^{(P)} = x_i^{(P)} + j \cdot L_1 \right\}$ where $z_{i,j}^{(P)}$
is the time position of the $(i + jw)^{th}$ non-zero value in the new code sequence
when $1 \le i \le w, 0 \le j \le m - 1$.

**Step 3.** Another codeset $C''$ with code-length $L_m = m \cdot L_1$ and $|C|$ code sequences is
constructed based on $C'$. The code sequences in $C''$ can be obtained from one-
to-one mapping of sequences in $C'$. For example, let $Y$ be a code sequence in
$C''$, a correspondent code sequence of $Z$. A set notation is used to explain the
$Y$ construction. To begin with, it is required to divide the time positions of the
non-zero value of $Z$ into $w$ groups, so that the set notation for the $i^{th}$ group
is $\left\{ z_{i,j}^{(P)} \mid z_{i,j}^{(P)} = x_i^{(P)} + j \cdot L_1, 0 \le j \le m - 1 \right\}$. Each group has $m$ elements in
which only one of the elements can be preserved while the others are removed.
Accordingly, each group will contain only one element while the total number
of elements in $w$ groups reduces from $mw$ to $w$. Finally, the $w$ elements which
are from the $w$ groups form a new codeset representing the time positions of the
non-zero value of code sequence $Y$. Since only the number of elements (i.e. the
time positions of non-zero values in a code sequence) is changed from $mw$ (i.e.
in $Z$) to $w$ (i.e. in $Y$), the time length for a code sequence will not be changed.
Thus, $C''$ is a codeset with time-length (code-length) $L_m$ and code-weight $w$.
In addition, since the code sequences of $C''$ are obtained from those of $C'$ by
one-to-one mapping, the codeset cardinality of $C''$ must also equal $|C|$ [27].

As observed so far, the overall OCDMA performance depends on the spreading codes
designed for the network; moreover the performance of the codes is defined by the
codeset correlation functions. The cross-correlation function between any two constant-
weight code sequences should be identical. However, in a variable-length codeset, the
cross-correlation function between two code sequences, representing the multiple access
interference, depends on the lengths of the desired code sequence (intended user) and the
interfering code sequence(s) (interfering user[s]) and may not be the same. To distinguish

the differences, the cross-correlation functions are classified into two types according to the variable code-lengths as the intra-cross-correlation function and the inter-cross-correlation function. The intra-cross-correlation function is defined as the cross-correlation between any two same-length code sequences while the inter-cross-correlation function is the cross-correlation between any two different-length code sequences. It is noted that the number of summation terms in the inter-cross-correlation function equals the time-length (code-length) of the desired code sequence. Here are a few theories relating to the correlation function constraints.

**Theory I.** In an optical unipolar codeset $C$ with code-length $L_1$, code-weight $w$, auto-correlation constraint $\lambda_a$ and cross-correlation constraint $\lambda_c$, each code sequence $X_i \in C$ can generate a code sequence $Z_i \in C'$ by concatenating $m$ copies of $X_i$. Thus, the 1D codeset $C'$ is $\{mL_1, mw, mw, m\lambda_c\}$ and 2D codeset will be $\{mL_1, r, mw, mw, m\lambda_c\}$.

The proof of this theory is based on the correlation functions as introduced by:

$$R_X(\tau) = \sum_{i=0}^{L_1-1} x_i \cdot x_{i \oplus \tau} \begin{cases} = w & for\ \tau = 0 \\ \leq \lambda_a & for\ 1 \leq \tau \leq L_1 - 1 \end{cases} \tag{10.25}$$

and

$$R_{X,T}(\tau) = \sum_{i=0}^{L_1-1} x_i \cdot t_{i \oplus \tau} \leq \lambda_c\ for\ any\ \tau \tag{10.26}$$

It is obvious that since the code sequence $Z$ is constructed by $m$ copies of $X$, the $Z$ code-length and code-weight equals $mL_1$ and $mw$ respectively. Now, the auto-correlation of code sequence $Z \in C'$ is [27]:

$$\sum_{i=0}^{mL_1-1} z_i \cdot z_{i \oplus \tau} = \sum_{i=0}^{L_1-1} z_i \cdot z_{i \oplus \tau} + \sum_{i=L_1}^{2L_1-1} z_i \cdot z_{i \oplus \tau} + \cdots + \sum_{i=(m-1)L_1}^{mL_1-1} z_i \cdot z_{i \oplus \tau}$$

$$= \sum_{i=0}^{L_1-1} z_i \cdot z_{i \oplus \tau} + \sum_{i=0}^{L_1-1} z_{L_1+i} \cdot z_{(L_1+i) \oplus \tau} + \cdots$$

$$+ \sum_{i=0}^{L_1-1} z_{(m-1)L_1+i} \cdot z_{((m-1)L_1+i) \oplus \tau} \tag{10.27}$$

On the other hand, $z_{jL_1+i}$ equals $x_i$ for all $i$ and $0 \leq j \leq m-1$ due to the concatenating process. Therefore, Equation (10.27) becomes:

$$\sum_{i=0}^{mL_1-1} z_i \cdot z_{i \oplus \tau} = \sum_{i=0}^{L_1-1} z_i \cdot z_{i \oplus \tau} + \sum_{i=0}^{L_1-1} z_{L_1+i} \cdot z_{(L_1+i) \oplus \tau} + \cdots$$

$$+ \sum_{i=0}^{L_1-1} z_{(m-1)L_1+i} \cdot z_{((m-1)L_1+i) \oplus \tau}$$

$$= \sum_{i=0}^{L_1-1} x_i \cdot x_{i\oplus\tau} + \sum_{i=0}^{L_1-1} x_i \cdot x_{i\oplus\tau} + \ldots + \sum_{i=0}^{L_1-1} x_i \cdot x_{i\oplus\tau}$$

$$= m \sum_{i=0}^{L_1-1} x_i \cdot x_{i\oplus\tau} \tag{10.28}$$

For any integer $0 \le \tau \le mL_1$, if $\tau = jL_1$ and $0 \le j \le m$ is an integer, Equation (10.28) becomes:

$$\sum_{i=0}^{mL_1-1} z_i \cdot z_{i\oplus\tau} = m \sum_{i=0}^{L_1-1} x_i \cdot x_{i\oplus\tau} = mw \tag{10.29}$$

However, if $\tau \ne jL_1$, Equation (10.28) becomes:

$$\sum_{i=0}^{mL_1-1} z_i \cdot z_{i\oplus\tau} = m \sum_{i=0}^{L_1-1} x_i \cdot x_{i\oplus\tau} \le m\lambda_a \tag{10.30}$$

Since $\lambda_a \le w$ the codeset $C'$ auto-correlation constraint will be $mw$. Similarly, the cross-correlation function of two $Z, Z' \in C'$ for any integer $\tau$ is:

$$\sum_{i=0}^{mL_1-1} z_i \cdot z'_{i\oplus\tau} = \sum_{i=0}^{L_1-1} z_i \cdot z'_{i\oplus\tau} + \sum_{i=L_1}^{2L_1-1} z_i \cdot z'_{i\oplus\tau} + \ldots + \sum_{i=(m-1)L_1}^{mL_1-1} z_i \cdot z'_{i\oplus\tau}$$

$$= \sum_{i=0}^{L_1-1} z_i \cdot z'_{i\oplus\tau} + \sum_{i=0}^{L_1-1} z_{L_1+i} \cdot z'_{(L_1+i)\oplus\tau} + \cdots$$

$$+ \sum_{i=0}^{L_1-1} z_{(m-1)L_1+i} \cdot z'_{((m-1)L_1+i)\oplus\tau}$$

$$= \sum_{i=0}^{L_1-1} x_i \cdot x'_{i\oplus\tau} + \sum_{i=0}^{L_1-1} x_i \cdot x'_{i\oplus\tau} + \cdots + \sum_{i=0}^{L_1-1} x_i \cdot x'_{i\oplus\tau}$$

$$= m \sum_{i=0}^{L_1-1} x_i \cdot x'_{i\oplus\tau} \le m\lambda_c \tag{10.31}$$

Therefore, the codeset $C'$ is $\{mL_1, mw, mw, m\lambda_c\}$. It can be similarly proven for 2D spreading codes.

**Theory II.** A codeset $C'$ was constructed from $C$ in *Step 2* of the coding algorithm with time length $L_1$, code-weight $w$, auto-correlation constraint $\lambda_a$ and cross-correlation constraint $\lambda_c$. Furthermore, codeset $C''$ is constructed by *Step 3* of the algorithm. The codeset $C''$ is $\{mL_1, w, \lambda_a, \lambda_c\}$ and $\{mL_1, r, w, \lambda_a, \lambda_c\}$ for 1D and 2D codes, respectively.

As with the proof for previous theory, here the analysis is for 1D coding. As indicated, code-weight $w$ denotes the number of 1s in the sequences.

Suppose $x_i^{(1)}$ as the time position of the $i^{th}$ 1 in the code sequence $X \in C$ and $1 \le i \le w$. By reducing the product of 0 by 0 in the code sequence rotation in the correlation functions, the correlation functions will become:

$$R_X(\tau) = \sum_{i=1}^{w} \left( x_{x_i^{(1)}} \cdot x_{x_i^{(1)} \oplus \tau} \right) \begin{cases} = w & \text{for } \tau = 0 \\ \le \lambda_a & \text{for } 1 \le \tau \le L_1 - 1 \end{cases} \quad (10.32)$$

and

$$R_{X,T}(\tau) = \sum_{i=1}^{w} x_{x_i^{(1)}} \cdot t_{t_i^{(1)} \oplus \tau} \le \lambda_c \text{ for } 0 \le \tau \le L_1 - 1 \quad (10.33)$$

Since the code sequences in $C''$ and $C'$ are based on the one-to-one mapping, let $Y \in C''$ be a code sequence and be the correspondent sequence of $Z \in C'$. Since $Z$ is constructed by a concatenating method from $X \in C$ and $C''$ is constructed from $C'$, the code sequences $X, Y$ and $Z$ can be represented by $X \rightarrow x_0 x_1 \ldots x_{L_1-2} x_{L_1-1}$, $Z \rightarrow z_0 z_1 \ldots z_{L_1-1} z_{L_1} \ldots z_{2L_1-1} \ldots z_{(m-1)L_1-1} \ldots z_{mL_1-1}$ and finally $Y \rightarrow y_0 y_1 \ldots y_{L_1-1} y_{L\_1} \ldots y_{2L_1-1} \ldots y_{(m-1)L_1-1} \ldots y_{mL_1-1}$ where $z_i = x_{i \bmod L_1}$. If $x_i = 0$ then $y_{i+jL_1} = 0$ for $0 \le j \le m - 1$ and $0 \le i \le L_1 - 1$.

If $x_i = 1$, then one element of the codeset $\{y_{i+jL_1} | 0 \le j \le m - 1\}$ equals 1 and the others will be set to 0. Accordingly, there are $wx_i$'s and $wy_i$'s equal to 1. Therefore, $C''$ is a codeset with code-length $mL_1$, code-weight $w$. The auto-correlation of $Y$ for any integer $0 \le \tau \le mL_1$ is written then as:

$$\sum_{i=0}^{mL_1-1} y_i \cdot y_{i \oplus \tau} = \sum_{i=0}^{L_1-1} y_i \cdot y_{i \oplus \tau} + \sum_{i=L_1}^{2L_1-1} y_i \cdot y_{i \oplus \tau} + \cdots + \sum_{i=(m-1)L_1}^{mL_1-1} y_i \cdot y_{i \oplus \tau}$$
$$(10.34)$$

Suppose $z_i^{(1)}$ the time position of the $i^{th}$ 1 in the code sequence $Z$ and $0 \le i \le mw$. For $0 \le j \le m - 1$ and $i \le w$, it is apparent that $z_i^{(1)} = x_i^{(1)}$ and $z_{i+jL_1}^{(1)} = x_i^{(1)} + jL_1$.

By reducing the products of 0 by 0 in the code sequence rotation in the correlation function, and also because $Y$ is constructed from $Z$, the auto-correlation of $Y$ will be the sum of the $mw$ possible non-zero values. Thus, Equation (10.34) is rewritten as:

$$\sum_{i=0}^{mL_1-1} y_i \cdot y_{i \oplus \tau} = \sum_{i=0}^{L_1-1} y_i \cdot y_{i \oplus \tau} + \sum_{i=L_1}^{2L_1-1} y_i \cdot y_{i \oplus \tau} + \cdots + \sum_{i=(m-1)L_1}^{mL_1-1} y_i \cdot y_{i \oplus \tau}$$

$$= \sum_{i=1}^{w} y_{z_i^{(1)}} \cdot y_{z_i^{(1)} \oplus \tau} + \sum_{i=w+1}^{2w} y_{z_i^{(1)}} \cdot y_{z_i^{(1)} \oplus \tau} + \cdots$$

$$+ \sum_{i=(m-1)w+1}^{mw} y_{z_i^{(1)}} \cdot y_{z_i^{(1)} \oplus \tau}$$

$$= \sum_{i=1}^{w} y_{z_i^{(1)}} \cdot y_{z_i^{(1)} \oplus \tau} + \sum_{i=1}^{w} y_{z_{i+w}^{(1)}} \cdot y_{z_{i+w}^{(1)} \oplus \tau} + \cdots$$

$$+ \sum_{i=1}^{w} y_{z_{i+(m-1)w}^{(1)}} \cdot y_{z_{i+(m-1)w}^{(1)} \oplus \tau}$$

$$= \sum_{j=0}^{m-1} y_{z_{1+jw}^{(1)}} \cdot y_{z_{1+jw}^{(1)} \oplus \tau} + \sum_{j=0}^{m-1} y_{z_{2+jw}^{(1)}} \cdot y_{z_{2+jw}^{(1)} \oplus \tau} + \cdots$$

$$+ \sum_{j=0}^{m-1} y_{z_{w+jw}^{(1)}} \cdot y_{z_{w+jw}^{(1)} \oplus \tau}$$

$$= \sum_{i=1}^{w} \sum_{j=0}^{m-1} y_{z_{i+jw}^{(1)}} \cdot y_{z_{i+jw}^{(1)} \oplus \tau} \qquad (10.35)$$

The time positions of 1s in $Z$ are divided to $w$ groups, referring to *Step 3*, and the set notation for the $i^{th}$ group is $\{z_{i,j}^{(1)} | z_{i,j}^{(1)} = x_i^{(1)} + jL_1, 0 \le j \le m - 1\}$. Then, only one of the elements can be preserved and the others will be removed. Consequently, among $m y_{z_i^{(1)} + jw}$ where $0 \le j \le m - 1$ only one $y_{z_i^{(1)}}$ is set to 1 and the others are set to 0. Let $J(i)$ be the $j$ value making $y_{z_i^{(1)} + jw}$ equal one in the $i^{th}$ group. Thus, $\sum_{j=0}^{m-1} y_{z_i^{(1)} + jw} \cdot y_{(z_i^{(1)} + jw) \oplus \tau} = y_{z_i^{(1)} + J(1)w} \cdot y_{(z_i^{(1)} + J(1)w) \oplus \tau}$. Moreover, if $x_i = 0$, then $y_{i+jL_1}$ must be 0 for $0 \le j \le m - 1$, $0 \le i \le L_1 - 1$. However, if $x_i \ne 0$, then $y_{i+jL_1}$ may be equal to 0. Accordingly, $y_{z_i^{(1)} + J(i)w} \cdot y_{(z_i^{(1)} + J(i)w) \oplus \tau} \le x_{x_i^{(1)}} \cdot x_{x_i^{(1)} \oplus \tau}$ and Equation (10.35) for any integer $0 \le \tau \le mL_1 - 1$ is rewritten as:

$$\sum_{i=0}^{mL_1-1} y_i \cdot y_{i \oplus \tau} = \sum_{i=1}^{w} \sum_{j=0}^{m-1} y_{z_i^{(1)} + jw} \cdot y_{(z_i^{(1)} + jw) \oplus \tau}$$

$$= \sum_{i=1}^{w} y_{z_i^{(1)} + J(i)w} \cdot y_{(z_i^{(1)} + J(i)w) \oplus \tau}$$

$$\le \sum_{i=1}^{w} x_{x_i^{(1)}} \cdot x_{x_i^{(1)} \oplus \tau} \le \lambda_a \qquad (10.36)$$

With similar procedure for the cross-correlation function, any two code sequences $Y, Y' \in C''$ which are constructed from different code sequences $X, X' \in C$ are cross-correlated for any integer $0 \le \tau \le mL_1 - 1$ as:

$$\sum_{i=0}^{mL_1-1} y_i \cdot y'_{i \oplus \tau} = \sum_{i=0}^{L_1-1} y_i \cdot y'_{i \oplus \tau} + \sum_{i=L_1}^{2L_1-1} y_i \cdot y'_{i \oplus \tau} + \cdots + \sum_{i=(m-1)L_1}^{mL_1-1} y_i \cdot y'_{i \oplus \tau}$$

$$= \sum_{i=1}^{w} y_{z_i^{(1)}} \cdot y'_{z_i^{(1)} \oplus \tau} + \sum_{i=w+1}^{2w} y_{z_i^{(1)}} \cdot y'_{z_i^{(1)} \oplus \tau} + \cdots$$

$$+ \sum_{i=(m-1)w+1}^{mw} y_{z_i^{(1)}} \cdot y'_{z_i^{(1)} \oplus \tau}$$

$$= \sum_{i=1}^{w} y_{z_i^{(1)}} \cdot y'_{z_i^{(1)} \oplus \tau} + \sum_{i=1}^{w} y_{z_{i+w}^{(1)}} \cdot y'_{z_{i+w}^{(1)} \oplus \tau} + \cdots$$

$$+ \sum_{i=1}^{w} y_{z_{i+(m-1)w}^{(1)}} \cdot y'_{z_{i+(m-1)w}^{(1)} \oplus \tau}$$

$$= \sum_{j=0}^{m-1} y_{z_{1+jw}^{(1)}} \cdot y'_{z_{1+jw}^{(1)} \oplus \tau} + \sum_{j=0}^{m-1} y_{z_{2+jw}^{(1)}} \cdot y'_{z_{2+jw}^{(1)} \oplus \tau} + \cdots$$

$$+ \sum_{j=0}^{m-1} y_{z_{w+jw}^{(1)}} \cdot y'_{z_{w+jw}^{(1)} \oplus \tau}$$

$$= \sum_{i=1}^{w} \sum_{j=0}^{m-1} y_{z_{i+jw}^{(1)}} \cdot y'_{z_{i+jw}^{(1)} \oplus \tau} = \sum_{i=1}^{w} y_{z_{i+J(i)w}^{(1)}} \cdot y'_{z_{i+J(i)w}^{(1)} \oplus \tau}$$

$$\leq \sum_{i=1}^{w} x_{x_i^{(1)}} \cdot x'_{x_i^{(1)} \widehat{\oplus} \tau} \leq \lambda_c \qquad (10.37)$$

where $\oplus$ and $\widehat{\oplus}$ denote modulo-$mL_1$ and modulo-$L_1$ addition, respectively. As a result, $C''$ is $\{mL_1, w, \lambda_a, \lambda_c\}$ codeset.

**Theory III.** A codeset $C$ with time length $L_1$, code-weight $w$, auto- and cross-correlation constraints $\lambda_a, \lambda_c$ respectively is constructed. Now, let $C''_m$ with code-length $mL_1$ and $C''_n$ with code-length $nL_1$, where $m, n$ are positive integers, be two codesets constructed by the steps introduced previously. Thus, the inter-cross-correlation value between any two code sequences constructed from different code sequences in $C$ is less than or equal to $\lambda_c$.

To prove this theory, similar to previous ones, suppose that $Y \in C''_n$ and $Y' \in C''_m$ are code sequences from codesets $C''_m$ and $C''_n$ constructed from different code sequences of $C$. Assuming $Y$ as a intended user, the inter-cross-correlation between two $Y$ and $Y'$ in a period $nL_1$ for any integer $0 \leq \tau \leq mL_1 - 1$ becomes:

$$\sum_{i=0}^{nL_1-1} y_i \cdot y'_{i \oplus \tau} = \sum_{i=0}^{L_1-1} y_i \cdot y'_{i \oplus \tau} + \sum_{i=L_1}^{2L_1-1} y_i \cdot y'_{i \oplus \tau} + \cdots + \sum_{i=(n-1)L_1}^{nL_1-1} y_i \cdot y'_{i \oplus \tau}$$

$$= \sum_{i=1}^{w} y_{z_i^{(1)}} \cdot y'_{z_i^{(1)} \oplus \tau} + \sum_{i=w+1}^{2w} y_{z_i^{(1)}} \cdot y'_{z_i^{(1)} \oplus \tau} + \cdots$$

$$+ \sum_{i=(n-1)w+1}^{nw} y_{z_i^{(1)}} \cdot y'_{z_i^{(1)} \oplus \tau}$$

$$= \sum_{i=1}^{w} y_{z_i^{(1)}} \cdot y'_{z_i^{(1)} \oplus \tau} + \sum_{i=1}^{w} y_{z_{i+w}^{(1)}} \cdot y'_{z_{i+w}^{(1)} \oplus \tau} + \cdots$$

$$+ \sum_{i=1}^{w} y_{z_{i+(n-1)w}^{(1)}} \cdot y'_{z_{i+(n-1)w}^{(1)} \oplus \tau}$$

$$= \sum_{j=0}^{n-1} y_{z_{1+jw}^{(1)}} \cdot y'_{z_{1+jw}^{(1)} \oplus \tau} + \sum_{j=0}^{n-1} y_{z_{2+jw}^{(1)}} \cdot y'_{z_{2+jw}^{(1)} \oplus \tau} + \cdots$$

$$+ \sum_{j=0}^{n-1} y_{z_{w+jw}^{(1)}} \cdot y'_{z_{w+jw}^{(1)} \oplus \tau}$$

$$= \sum_{i=1}^{w} \sum_{j=0}^{n-1} y_{z_{i+jw}^{(1)}} \cdot y'_{z_{i+jw}^{(1)} \oplus \tau} = \sum_{i=1}^{w} y_{z_{i+J(i)w}^{(1)}} \cdot y'_{z_{i+J(i)w}^{(1)} \oplus \tau}$$

$$\leq \sum_{i=1}^{w} x_{x_i^{(1)}} \cdot x'_{x_i^{(1)} \widehat{\oplus} \tau} \leq \lambda_c \qquad (10.38)$$

where $\oplus$ and $\widehat{\oplus}$ denote modulo-$mL_1$ and modulo-$L_1$ addition respectively.

## 10.4.1 Performance Analysis

In this section, the performance of the multirate system based on the error probabilities as a function of the number of different rate users is analysed. Here we report only the performances of one-dimensional codes, but the analysis procedures can be adapted to the two-dimensional codes as well [27]. The first multirate system is referred to a multirate code system that uses the OOCs based on the construction introduced in the algorithmic steps above. To simplify this, two different lengths of OOCs are considered for the analysis. Again, the construction can be generalized to more different lengths. A codeword $X \in C$ of $(n, w, \lambda_a, \lambda_c)$ is assigned to a new user allocated to the multirate system. Then, users are categorized into two classes according to their bit rates (i.e. two lengths). Here, we name *class I* for users transmitting at the higher bit rate and *class II* for users transmitting at the lower bit rate. It can be assumed that the bit rate of *class I* users is twice as that of *class II* users, meaning that the symbol length of *class II* users is twice that of *class I* users. When a *class I* user transmits data, a codeword $X \in C$ is assigned, mapping its data bits. On the other hand, a *class II* user maps its data bits by using the constructed codeword $Y \in C''$ where $Y$ is constructed from the assigned codeword $X$.

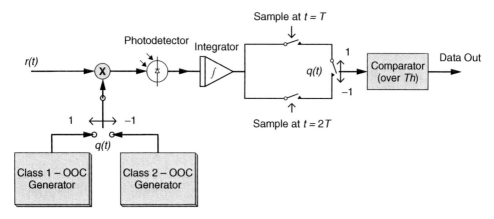

**Figure 10.7** Multirate (two-rate) receiver (Reprinted from: Variable-length code construction for incoherent optical CDMA systems, J.Y. Lin, J.S. Jhou and J.H. Wen, *Optical Fiber Technology*, **13** (2), 2007, with permission from Elsevier.)

Assume the chip-synchronized system and $\lambda_a = \lambda_c = 1$, and that the numbers of *class I* and *class II* users in the system are $N_1$ and $N_2$ respectively, where the total number of users is then $N = N_1 + N_2$ upper-bounded by the codeset cardinality $|C|$. It should also be noted that the data sequences are independent and data bits 1 and 0 are equiprobable. For the purpose of comparison, a second system which implements the multirate function by transmitting each symbol twice for *class II* users is evaluated. This system is referred to as a repeat code system [27] where the performance of the repeat code system with automatic request (ARQ) scheme is also analysed.

Figure 10.7 shows the simplified configuration of the multirate receiver in the multirate code system.

Since there is multirate data at the receiver, thus a rate control unit or function is required which is represented by function $q(t)$ where $q(t) = -1$ is for lower bit rate and $q(t) = 1$ for higher bit rate. The bit duration of *class I* users is denoted by $T$. In the following, the bit error rate (BER) of *class I* and *class II* users with and without the ARQ scheme are studied.

### 10.4.1.1  BER of High-Bit-Rate Users

Let $I$ be the total interference from all users in *class I* and *II*, and let $I_i^{(1)}$ and $I_j^{(2)}$ be the amount of interference from the $i^{th}$ user in *class I* and $j^{th}$ user in *class II* respectively. Without loss of generality, the intended user is assumed to be among *class I* users. Since there are $N_1$ *class I* users and $N_2$ *class II* users, the interference $I$ come from $(N_1 - 1)$ *class I* and $N_2$ *class II* users. The probability density function (PDF) of $I_i^{(1)}$ is introduced as [15]:

$$P\left(I_i^{(1)} = k\right) = \begin{cases} \dfrac{w^2}{2n} & k = 1 \\ 1 - \dfrac{w^2}{2n} & k = 0 \\ 0 & \text{otherwise} \end{cases} \tag{10.39}$$

where $2 \leq i \leq N_1$, and similarly the PDF of $I_j^{(2)}$ is:

$$
P\left(I_j^{(2)} = k\right) = \begin{cases} \dfrac{w^2}{4n} & k = 1 \\ 1 - \dfrac{w^2}{4n} & k = 0 \\ 0 & \text{otherwise} \end{cases} \tag{10.40}
$$

where $1 \leq j \leq N_2$. Since $I_i^{(1)}$ and $I_j^{(2)}$ are independent, the PDF of $I$, $P_I^{(Class\ I)}(I)$ becomes:

$$
P_I^{(Class\ I)}(I) = P\left(I_2^{(1)}\right) * P\left(I_3^{(1)}\right) * \ldots * P\left(I_{N_1}^{(1)}\right) * P\left(I_1^{(2)}\right) * \ldots * P\left(I_{N_2}^{(2)}\right) \tag{10.41}
$$

where $*$ is the convolution operator. The BER $P_{E_1}$ of *class I* users becomes:

$$
P_{E_1} = \Pr(R \geq Th | b = 0)\Pr(b = 0) = \frac{1}{2}\sum_{i=Th}^{N-1} P_I^{(Class\ I)}(i) \tag{10.42}
$$

where $Th, R$ and $b$ denote the threshold, the output of the desired user's integrator at time $T$ and the data transmitted by the intended user respectively.

### 10.4.1.2 BER of Low-Bit-Rate Users

The BER of *class II* users is derived by a similar method. Now, it is assumed that the intended user is among *class II* users and the probability density functions of $I_i^{(1)}$ and $I_j^{(2)}$ are the same as Equations (10.39) and (10.40), although, the interference this time comes from $N_1$ *class I* and $(N_2 - 1)$ *class II* users. Therefore, the PDF for the interference $P_I^{(Class\ II)}$ will be:

$$
P_I^{(Class\ II)}(I) = P\left(I_1^{(1)}\right) * P\left(I_2^{(1)}\right) * \ldots * P\left(I_{N_1}^{(1)}\right) * P\left(I_2^{(2)}\right) * \ldots * P\left(I_{N_2}^{(2)}\right) \tag{10.43}
$$

where $*$ is the convolution operator. The BER $P_{E_2}$ of *class I* users becomes:

$$
P_{E_2} = \frac{1}{2}\sum_{i=Th}^{N-1} P_I^{(Class\ II)}(i) \tag{10.44}
$$

The average bit error probability of multirate code system $P_E$ is [27]:

$$
P_E = \frac{N_1 R_1 P_{E_1} + N_2 R_2 P_{E_2}}{N_1 R_1 + N_2 R_2} \tag{10.45}
$$

where $R_i$ is the data rate of *class i* user, and since $R_1 = 2R_2$, Equation (10.45) is modified to:

$$
P_E = \frac{2N_1 P_{E_1} + N_2 P_{E_2}}{2N_1 + N_2} \tag{10.46}
$$

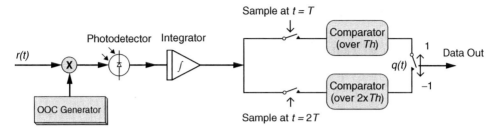

**Figure 10.8** Multirate (two-rate) receiver with ARQ scheme (Reprinted from: Variable-length code construction for incoherent optical CDMA systems, J.Y. Lin, J.S. Jhou and J.H. Wen, *Optical Fiber Technology*, **13** (2), 2007, with permission from Elsevier.)

### 10.4.1.3 BER of Repeat Code System with ARQ Scheme

Figure 10.8 illustrates the configuration of a multirate receiver with an ARQ scheme. Here also $q(t)$ is the rate control signal and it has the same functionality and value as in the previous scheme. With similar approach and notation, the error probabilities of two-class users and probability density functions of the interference can be calculated accordingly.

Since the *class I* users interference $I_i^{(1)}$ is similar to the multirate code system, the $I_i^{(1)}$ PDF is the same as Equation (10.39). Furthermore, *class II* users transmit data at twice the data rate of *class I* users, so the $I_j^{(2)}$ PDF is also similar to Equation (10.40) [27]. Thus, the PDF for $I$, $P_I^{(Class\ I)}$ can be derived by substituting the following PDF into Equation (10.41):

$$P\left(I_i^{(1)} = k\right) = P\left(I_j^{(2)} = k\right) = \begin{cases} \dfrac{w^2}{2n} & k = 1 \\ 1 - \dfrac{w^2}{2n} & k = 0 \\ 0 & \text{otherwise} \end{cases} \tag{10.47}$$

where $2 \leq i \leq N_1$ and $1 \leq j \leq N_2$. Then, the BER of *class I* users $P_{E_1}$ is derived. As the *class I* users in the repeat code system with an ARQ scheme do not use the repetition technique, the BER of *class I* users $P_{A_1}$ is the same as $P_{E_1}$ in Equation (10.42).

However, *class II* users transmit each bit twice and so the code-length and code-weight are $2n$ and $2w$, repectively, referring to the code sequence $Z$ in *Step 2*. Consequently, the distribution of $I_i^{(1)}$ will be of the form [27]:

$$P\left(I_i^{(1)} = k\right) = \begin{cases} \dfrac{w^2}{4n} & k = 2 \\ \dfrac{w^2}{2n} & k = 1 \\ 1 - \dfrac{3w^2}{4n} & k = 0 \\ 0 & \text{otherwise} \end{cases} \tag{10.48}$$

And similarly the distribution of $I_j^{(2)}$ will be of the form:

$$
P\left(I_j^{(2)} = k\right) = 
\begin{cases}
\dfrac{3w^2}{8n} & k = 2 \\[2mm]
\dfrac{w^2}{4n} & k = 1 \\[2mm]
1 - \dfrac{5w^2}{8n} & k = 0 \\[2mm]
0 & \text{otherwise}
\end{cases}
\tag{10.49}
$$

Then, the PDF of the total interference $I$, $P_I^{(Class\ 2)}(I)$ and the BER of *class II* users, $P_{E_2}$ can be derived. The average BER of the repeat code system, similar to Equation (10.46), becomes:

$$
P_E = \frac{2N_1 P_{E_1} + N_2 P_{E_2}}{2N_1 + N_2}
\tag{10.50}
$$

In an ARQ scheme where the *class II* users employ repetition as an error-detection code, the multirate repeat code system receiver checks whether the two received bits, which are coded from the same data bit, are the same or not. If the two received bits are identical, the receiver will easily make a decision, but if the two received bits are different, the receiver will request the transmitter to retransmit the data bit. To simplify this, the overhead is not considered in the ARQ scheme and we assume that each data bit will be retransmitted until it can be easily recognized. Thus, a lower-bound is calculated for the bit error probability of the system using the ARQ scheme. In order to calculate the bit error probability of *class II* users with the ARQ scheme, the error probability for each received bit $P_{B_E}$, must be calculated and is the same as $P_{E_1}$. Since each data bit is considered as an error bit after two received bit errors take place, the bit error probability of *class II* users with ARQ scheme is:

$$
\begin{aligned}
P_{A_2} &= P_{B_E} \cdot P_{B_E} + 2P_{B_E}(1 - P_{B_E}) \cdot P_{B_E} \cdot P_{B_E} + 2P_{B_E}(1 - P_{B_E}) \\
&\quad \cdot 2P_{B_E}(1 - P_{B_E}) \cdot P_{B_E} P_{B_E} + \cdots = P_{B_E}[1 - 2P_{B_E}(1 - P_{B_E})]
\end{aligned}
\tag{10.51}
$$

and the average bit error probability of repeat code system with ARQ scheme $P_A$ will be of the form:

$$
P_E = \frac{2N_1 P_{A_1} + N_2 P_{A_2}}{2N_1 + N_2}
\tag{10.52}
$$

## 10.5 Multirate Differentiated Services in OCDMA Networks

Key optical technologies employing advanced photonic devices such as optical logic gates in OCDMA systems have the advantages of all-optical processing. Optical logic gates are the fundamental elements in designing advanced OCDMA transceivers proved to be of enormous applications for all-optical multiservice networks [43].

Recalling a conventional incoherent OCDMA system using optical orthogonal codes (OOCs), the code sequences have constant code-length and code-weight, so all users have the same transmission rate and quality of services (QoS). On the other hand, new applications of multimedia such as high-definition television (HDTV), video conferencing, e-learning, interactive gaming, etc. emerged due to the growth of the Internet and diversified data traffic. Consequently, supporting multirate and differentiated-QoS transmission in a network are essential challenges in the (future) optical networking. In order to support multirate and differentiated-QoS transmission in OCDMA systems, multilength OOC (ML-OOC) and variable-weight OOC (VW-OOC) have been introduced in previous sections and extensively in [16, 17, 24, 27]. Additionally, multilength variable-weight spreading codes (MLVW-OOC) have been proposed to support multirate and differentiated services (DiffServ) simultaneously in a network [19, 21, 36]. Also, carrier hopping prime codes (CHPM) and multiwavelength OOC designed for 2D OCDMA systems [32, 44] have been revised to provide multiservice transmission in 2D OCDMA networks [37]. In incoherent OCDMA, all subscribers communicate at the same power level and, at the receiver end, AND logic gate structure can be employed [45]. Multilevel signalling techniques [43] allow subscribers to transmit at different power levels to be better distinguished and/or to differentiate services. By this technique at the receiver structure using advanced optical logic gates (e.g. AND, OR, XNOR), multilevel signals help to reduce the interference from users with different power levels. Evidently, the elimination of the interference with different power levels results in performance improvement [1].

Multilevel signalling techniques are already employed in a two-class variable-weight OCDMA system where all users have the same energy level in one bit duration [1]. It is reported that users with high code-weight transmit their corresponding optical pulses at a lower power while users with low code-weight transmit their corresponding optical pulses at a higher power level. Using the above multilevel signalling alongside the optical AND logic gates, great improvement in the performance of users with low code-weight (high-power) is obtained, while the performance of users with high code-weight (low-power) is not affected compared to a typical one-level OCDMA system. It is also indicated that by deploying a multistage receiver structure, constructed of advanced optical logic gates, multiple access interference at different power levels are distinguishable, and so the overall performance of both high-power and low-power subscribers is enhanced. This section concentrates on OCDMA using MLVW-OOC, and the performance of the system is evaluated in terms of the probability of error (i.e. $P_E$ meaning bit error rate) where the derived equation is generalized to also express the $P_E$ of systems using MLVW-OOC, OOC, ML-OOC, and VW-OOC.

In typical incoherent OCDMA systems, the optical signal intensity is encoded using spreading codes such as OOC. To recall the properties of the OOC family, here we review OOC as a family of (0,1) sequences with good auto- and cross-correlation properties, as discussed in detail in Chapter 2. An OOC codeset is characterized by $(L, w, \lambda)$ where $L$, $w$ and $\lambda$ denote the code-length, code-weight and maximum value of shifted auto- and cross-correlation value respectively. It has been shown that in a system using OOC as spreading codes, the overall performance relates to the codeset characterization and the number of interfering users so that the greater the number of interfering users the lower the performance obtained. As mentioned, the codeset parameters such as code-length, code-weight and cardinality are interpreted as data rate, signal power and system capacity respectively in a real world setting. Although the increase in code-weight will improve the overall performance (i.e. higher signal power or signal-to-noise ratio), this decreases the number of available code sequences, $N_c$ as expressed in:

$$N_c \leq \left\lfloor \frac{(L-1)(L-2)\ldots.(L-\lambda)}{w(w-1)(w-2)\ldots(w-\lambda)} \right\rfloor \tag{10.53}$$

where $\lfloor x \rfloor$ is the integer value of $x$. As observed, increasing the code-weight $w$ decreases the number of available sequences. It should also be noted that the code-length has an opposite relation with the transmission rate. On the other hand, the increase in transmission rate leads to a decrease in $N_c$ for a specific QoS, due to the interference. Additionally, as discussed, ML-OOC codesets have been introduced [24, 27] to implement multirate transmission where code sequences have constant-weight but variable lengths as discussed in Section 10.4. Furthermore, to implement the differentiated-QoS VW-OOC, codesets have been developed [17], where the code sequences have constant length but variable weights, as discussed in Section 10.3.

To simultaneously support differentiated-QoS and multirate transmissions, there is a need for multilength variable-weight OOC (MLVW-OOC) codeset where the code sequences can choose different weights and lengths at the same time. To provide the requested services in a network employing MLVW-OOC as spreading codes, subscribers requiring high QoS and high bit rate are allocated to code sequences with high code-weight and short code-length respectively, for example, or any combination of high and low code-weights for desired QoS and/or long or short code-lengths for desired transmission rate.

Assuming the MLVW-OOC codeset is characterized with $(L = \{L_1, L_2, \ldots L_Q\}, w = \{w_1, w_2, \ldots w_Q\}, N_c = \{N_{c_1}, N_{c_2}, \ldots, N_{c_Q}\}, Q, \Gamma)$ where $L_i$, $w_i$ and $N_{c_i}$ denote the code-lengths, code-weights and number of available codes in class $i$, respectively. $Q$ is also referred to as the number of specified services in the network. $\Gamma$ indicates the cross-correlation matrix defined as:

$$\Gamma = \begin{bmatrix} \lambda_{1,1} & \cdots & \lambda_{1,Q} \\ \vdots & \ddots & \vdots \\ \lambda_{Q,1} & \cdots & \lambda_{Q,Q} \end{bmatrix} \tag{10.54}$$

Now, the $k^{th}$ code sequence of class $n$ is denoted by $C_{k,n}$ and thus $\lambda_{n,m}$ is defined by:

$$\sum_{t=0}^{L_n-1} C_{k,n}(t) \cdot C_{f,m}(t \oplus_m \tau) \leq \lambda_{n,m}$$

$$\sum_{t=0}^{L_m-1} C_{f,m}(t) \cdot C_{k,n}(t \oplus_n \tau) \leq \lambda_{m,n} \tag{10.55}$$

where $0 \leq \tau \leq \min(L_m, L_n)$ is an integer, and $\oplus_n$ and $\oplus_m$ denote module-$L_n$ and module-$L_m$ additions respectively. In case $n = m$, then $\lambda_{n,m}$ will be referred to as intra-cross-correlation showing the maximum cross-correlation between the same class code sequences, while when $n \neq m$, then $\lambda_{n,m}$ will be referred to as inter-cross-correlation showing the maximum cross-correlation between two code sequences from different classes.

To construct the MLVW-OOC, it should be mentioned that the combination of two algorithms introduced in Sections 10.3 and 10.4 can generate MLVW-OOC codesets and/or sequences. For example, a constant-weight variable-length code can be generated, and every codeset of a particular length can be treated for generating variable-weight code sequences and vice versa. However, there is a specific algorithm for generating MLVW-OOC in the literature [1, 19, 32, 36, 43, 44]. If the strict MLVW-OOC codeset ($\lambda_{i,j} = 1$; $i,j = \{1, 2, \ldots Q\}$) is constructed accordingly, the number of available code sequences should satisfy the inequality:

$$\sum_{i=1}^{Q} N_{c_i} \times w_i \times (w_i - 1) \leq L_{\max} - 1 \tag{10.56}$$

where $L_{\max} = \max(L_i, i = 1, 2, \ldots, Q)$. There will be more available code sequences when the correlation constraints are $\lambda_{i,j} \geq 1$. Thus, the codeset cardinality will be Johnson-bounded to [1]:

$$\sum_{i=1}^{Q} N_{c_i} \leq \left\lfloor \frac{(L_{\min} - 1)(L_{\min} - 2) \ldots (L_{\min} - \lambda)}{w_{\max}(w_{\max} - 1)(w_{\max} - 2) \ldots (w_{\max} - \lambda)} \right\rfloor \tag{10.57}$$

where $L_{\min} = \min(L_i, i = 1, 2, \ldots, Q)$, $w_{\max} = \max(w_i, i = 1, 2, \ldots, Q)$ and $\lambda = \lambda_{i,j}$ for $i,j = \{1, 2, \ldots Q\}$. As seen, the increase in the correlation constraint leads to a greater number of available code sequences.

The well-known optical AND logic gate is employed at the receiver end in multiservice OCDMA networks [1, 43, 45]. The optical AND logic gate structure comprises an optical hard-limiter followed by a tapped-delay, line decoder and an AND logic gate element as seen in Figure 10.9 [1]. It can be observed from Figure 10.9 that the optical AND gate can be modelled by a combiner and followed by a hard-limiter.

It should be noted that the threshold of the hard-limiter at the input of the optical AND logic gate should be equal to the power level $P_{th}$, while the threshold of the hard-limiter of the AND logic gate is $w \times P_o$, where $w$ is the number of delay lines and $P_o$ is the signal power level at each delay line. Therefore, when all inputs at the AND logic gate have a pulse with a power $P_o$ then the output is 'on' or pulsed signal, otherwise 'off'. The delay lines in optical AND logic gates are revised according to the assigned code sequences.

### 10.5.1 Performance Analysis

In an incoherent modulation format such as on-off keying (OOK) modulation, a data bit 1 is encoded by the assigned code sequence whereas the data bit 0 is not encoded and not transmitted. When the Poisson shot noise at the corresponding receiver is not taken into account, the data bit 1 will be always detected correctly due to the additive and positive properties of the optical channel. However, an error can happen when the transmitter sends a data bit 0 along with the intended user's marked chips filled by interfering users causing multiple user interference.

Considering MAI as the only degrading factor, the probability of error $P_E$ of a multiservice OCDMA system's typical optical AND gate receiver is analysed. Since the emphasis of this study is to investigate the multiservice properties of the OCDMA system, $P_E$ is evaluated when the fibre impairments, shot noise and thermal noise are not taken into

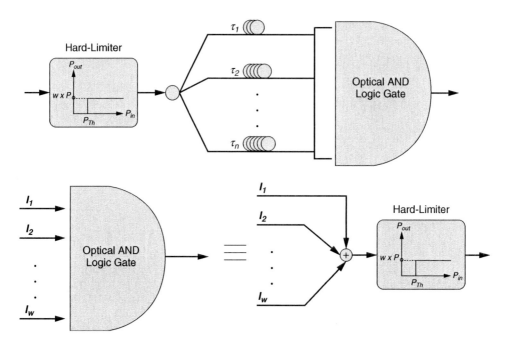

**Figure 10.9** Optical logic AND gate configuration and model (Reprinted from: Multirate, differentiated-QoS, and multilevel fiber-optic CDMA system via optical logic gate elements, H. Beyranvand, B.M. Ghaffari and J.A. Salehi, *J. Lightw. Technol.*, **27** (19), 2007, with permission from Elsevier.)

account. However, the accuracy of the performance analysis will be enhanced by considering the most relevant degrading factor, but here we only focus on the MLVW-OOC codeset performance based on the 'hit' incidents in the code sequences.

The intended user sends a data bit 0 and all its marked chips filled by interfering users' marks. Clearly, since the optical channel is additive with no loss, the data bit 1 will be decoded and detected correctly. Therefore, $P_E$ of class $k$ user, $P_E(k)$ will be evaluated as [1]:

$$P_E(k) = \frac{1}{2} \Pr(\text{error}|0) = \Pr(\alpha_1 \geq 1, \alpha_2 \geq 1, \ldots, \alpha_{w_k} \geq 1) \qquad (10.58)$$

where $\alpha_i$ denotes the number of interferences in the $i^{th}$ marked chip of the intended user and $w_k$ denotes the assigned code-weight to the intended user. Thus, Equation (10.58) can be extended to:

$$\Pr(\alpha_1 \geq 1, \alpha_2 \geq 1, \ldots, \alpha_{w_k} \geq 1) = 1 - \Pr(\alpha_1 = 0 \ or \ \alpha_2 = 0 \ or \ \ldots \ or \ \alpha_{w_k} = 0)$$

$$= \sum_{i=0}^{w_k} (-1)^i \times \binom{w_k}{i} \times \Pr(\alpha_1 = \alpha_2 = \ldots = \alpha_i = 0)$$

$$(10.59)$$

Since the users from different classes are independent, the probability of interference is then obtained from the probability of each class as:

$$\Pr(\alpha_1 = \alpha_2 = \ldots = \alpha_i = 0) = \prod_{q=1}^{Q} \Phi^{(q)}(\alpha_1 = \alpha_2 = \ldots = \alpha_i = 0) \qquad (10.60)$$

where $\Phi^{(q)}$ denotes the probability of interference of users in class $q$. Equation (10.60) can be rewritten considering the independence of users in each class as:

$$\Phi^{(q)}(\alpha_1 = \alpha_2 = \ldots = \alpha_i = 0) = [\Phi^{(q)}(\alpha_1 = \alpha_2 = \ldots = \alpha_i = 0 | one\ user)]^{N_q} \qquad (10.61)$$

where $N_q$ is the number of class $q$ interfering users. It is noted that when the intended user is in class $k$ then $\max\{N_q\} = N_{c_q}$ for $q \neq k$ and $q = 1, 2, \ldots, Q$ where $N_{c_q}$ is the number of available class $q$ codes and $\max(N_k) = N_{c_k} - 1$. Then $\Phi^{(q)}$ of one user is now rewritten as:

$$\Phi^{(q)}(\alpha_1 = \alpha_2 = \ldots = \alpha_i = 0 | one\ user) = 1 - \Phi^{(q)}$$

$$(\alpha_1 = 0\ or\ \alpha_2 = 0\ or\ \ldots\ or\ \alpha_{w_k} = 0 | one\ user) \qquad (10.62)$$

It is noted that the maximum interference made by a class $q$ user is $\lambda_{k,q}$, thus:

$$\Phi^{(q)}(\alpha_1 = 1\ or\ \alpha_2 = 1\ or\ \ldots\ or\ \alpha_i = 1 | one\ user)$$

$$= \sum_{j=1}^{\lambda_{k,q}} (-1)^{j+1} \times \binom{i}{j} \times \Phi^{(q)}(\alpha_1 = \ldots = \alpha_j = 1 | one\ user) \qquad (10.63)$$

Also, the total interference from a class $q$ user on a class $k$ user becomes $w_q w_k / 2L_q$ [1]. Note that $w_q$ marked chips (i.e. 1s) in a code sequence of class $q$ out of $L_q$ chips have pulsed signal interfering with the $w_k$ marked chips of the class $k$ users. Furthermore, the data bits are independent and, due to the OOK modulation, they are equiprobably either on or off, hence the term $1/2$ that has appeared. $P_m^{(k,q)}$ is defined as the probability that expresses a code sequence of class $q$ interfering with $m$ marked chips of code sequence of class $k$, by:

$$\sum_{m=0}^{\lambda_{k,q}} m \cdot \binom{w_k}{m} \cdot P_m^{(k,q)} = \frac{w_k w_q}{2L_q} \qquad (10.64)$$

Now, by this definition we have:

$$\Phi^{(q)}(\alpha_1 = \ldots = \alpha_j = 1 | one\ user) = P_j^{(k,q)} + \binom{w_k - j}{1} \cdot P_{j+1}^{(k,q)} + \ldots + \binom{w_k - \lambda_{k,q}}{\lambda_{k,q} - j}$$

$$\cdot P_{\lambda_{k,q}}^{(k,q)} = \sum_{m=j}^{\lambda_{k,q}} \binom{w_k - m}{m - j} \cdot P_m^{(k,q)} \qquad (10.65)$$

According to the calculations from Equation (10.58) to (10.65), the probability of error, $P_E$ of class $k$ user $P_E(k)$ is obtained by:

$$P_E(k) = \frac{1}{2} \sum_{i=0}^{w_k} \left\{ (-1)^i \binom{w_k}{i} \prod_{q=1}^{Q} \left[ 1 + \sum_{j=1}^{\lambda_{k,q}} \sum_{m=j}^{\lambda_{k,q}} \left( (-1)^j \binom{i}{j} \binom{w_k - m}{m - j} P_m^{(k,q)} \right) \right]^{N_q} \right\}$$

(10.66)

where $N_q$ denotes the number of interfering users in class $q$.

It is observed that Equation (10.66) not only states the probability of error in MLVW-OOC, but also other schemes such as OOC, ML-OOC, VW-OOC codesets can be derived from it. To obtain the $P_E$ for different schemes the codeset parameters should be revised accordingly. For example, $P_E$ for an OOC-based OCDMA system requires the codeset modification of $(L = \{L\}, w = \{w\}, N_c = \{N\}, Q = 1, \Gamma = [\lambda])$. Equation (10.66) will then be modified to:

$$P_E = \frac{1}{2} \sum_{i=0}^{w} \left\{ (-1)^i \binom{w}{i} \left[ 1 + \sum_{j=1}^{\lambda_{1,1}} \sum_{m=j}^{\lambda_{1,1}} \left( (-1)^j \binom{i}{j} \binom{w - m}{m - j} P_m^{(1,1)} \right) \right]^{N} \right\}$$

(10.67)

where in this case, $P_m^{(1,1)} = w^2/2L$. It can be observed that the Equation (10.67) is similar to the OCDMA system employing OOC [1, 43, 45].

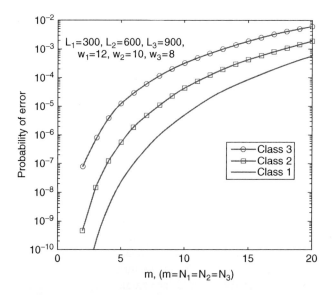

**Figure 10.10** Probability of error of one-level OCDMA system using MLVW-OOC under given conditions (Reprinted from: Multirate, differentiated-QoS, and multilevel fiber-optic CDMA system via optical logic gate elements, H. Beyranvand, B.M. Ghaffari and J.A. Salehi, *J. Lightw. Technol.*, **27** (19), 2007, with permission from Elsevier.)

Figure 10.10 [1] illustrates the performance of an OCDMA system utilizing the MLVW-OOC codeset characterized by $(L = \{300, 600, 900\}, w = \{12, 10, 8\}, N_c = \{20, 20, 20\}, Q = 3, \Gamma = [\lambda_{i,j} = 2, i, j = 1, \ldots, 3])$ against the number of interfering users in different classes denoted by $m = N_1 = N_2 = N_3$. It is apparent that the higher code-weight enhances the performance, and also the increase in the number of interfering users degrades the overall system performance.

## 10.6  Summary

This chapter is dedicated to service differentiation and quality of service in optical networks based on the OCDMA. Since the signal processing in the OCDMA is performed in the code domain rather than wavelength or time domain, coding parameters play significant roles in implementing the concept of multirate and multiservices. The spreading code properties are thoroughly interpreted with actual meaning in a real network environment. We have seen that the code-length is directly related to data rate; the code-weight denotes the signal power and the codeset cardinality implies the network capacity in terms of number of accommodated users; and the correlation properties refer to the similarity or difference of any two code sequences (i.e. users) in a codeset.

To support multimedia services with different bit rates and qualities over an OCDMA network, multiple-length multiple-weight spreading codesets opened their path as a very hot research topic in the field of OCDMA networking. A variable-length code construction providing a multirate OCDMA network has been reviewed. The overall performances of the multirate code system and repeat code systems were studied, and subscribers are categorized into two different classes with different rates. However, it can easily be generalized to more services or rates. The repeat code system with automatic request (ARQ) scheme can outperform the multirate code system when only users of *class II* are present; otherwise its performance is weaker in other conditions.

Two approaches (distinct set and random) are introduced and studied in detail for designing strict VW-OOCs with ideal in-phase cross-correlation value, considering differentiated quality of service in OCDMA networks. The desired codeset parameters (code-weights as services) can be chosen in both design approaches. Furthermore, the random approach due to its random nature is able to provide several independent codesets for given code parameters that can be deployed in code reconfiguration enhancing the security. It is noted that users with larger code-weight always perform better than those with lower code-weights, obviously due to higher signal power leading to higher signal-to-noise ratio. Then, strict VW-OOCs have shown that they can provide the differentiated QoS required for multiservice implementation in OCDMA networks.

It is indicated that the performance of a variable-weight, equal-energy OCDMA system using a two-level signalling technique is improved, but there are a few multilevel signalling techniques for distinguishing the power levels as a parameter. On the other hand, for simultaneously supporting multirate and quality of services, multilength variable-weight spreading codes (i.e. MLVW-OOC) are studied in this chapter.

It is worth remembering that although the system with MLVW-OOC is investigated, the analysis evaluated a general form of probability of error, in that the corresponding probability of error for OOC, VW-OOC and ML-OOC can also be obtained by only modifying the characterizing code parameters as shown here in this chapter.

# References

1. Beyranvand, H., Ghaffari, B.M. and Salehi, J.A. (2009) Multirate, differentiated-QoS, and multilevel fiber-optic CDMA system via optical logic gate elements. *J. Lightw. Technol.*, **27** (9), 4348–4359.
2. Assi, C., Ye, Y. and Dixit, S. (2003) Dynamic bandwidth allocation for quality of service over Ethernet PON. *IEEE J. on Selected Areas in Comm.*, **21** (11), 1467–1477.
3. Karbassian, M. M. and Ghafouri-Shiraz, H. (2009) Analysis of scalability and performance in passive optical CDMA network. *J. Lightw. Technol.*, **27** (17), 3896–3903.
4. Gumaste, A. *et al.* (2008) Light-mesh: A pragmatic optical access network architecture for IP-centric service oriented communication. *Opt. Switching and Networking*, **5** (2,3), 63–74.
5. Yang, C. C., Huang, J.F. and Hsu, T.C. (2008) Differentiated service provision in optical CDMA network using power control. *IEEE Photonics Tech. Letters*, **20** (20), 1664–1666.
6. Hayashi, M., Tanaka, H. and Suzuki, M. (2008) Advanced reservation-based network resource manager for optical network. In: *OFC*, San Diego, California, USA.
7. Draft recommendation Y.RACF (Y.2111) release 2, 'Resource and admission control functions in NGN,' ITU-T (2007).
8. Gumaste, A. and Zheng, S. (2006) Light-frames: A pragmatic solution to optical packet transport – extending the Ethernet from LAN to optical networks. *J. Lightw. Technol.*, **24** (10), 3598–3615.
9. Meenakshi, M. and Andonovic, I. (2006) Code-based all optical routing using two-level coding. *J. Lightw. Technol.*, **24** (4), 1627–1637.
10. Karbassian, M. M. and Ghafouri-Shiraz, H. (2009) IP routing and traffic analysis in coherent optical CDMA networks. *J. Lightw. Technol.*, **27** (10), 1262–1268.
11. Wei, Z. and Ghafouri-Shiraz, H. (2002) IP routing by an optical spectral-amplitude-coding CDMA network. *IEE Proc. Communications*, **149** (5), 265–269.
12. Karbassian, M. M. and Ghafouri-Shiraz, H. (2009) IP routing and transmission analysis over optical CDMA networks: coherent modulation with incoherent demodulation. *J. Lightw. Technol.*, **27** (17), 3845–3852.
13. Azizoghlu, M., Salehi, J.A. and Li, Y. (1992) Optical CDMA via temporal codes. *IEEE Trans on Comm.*, **40** (8), 1162–1170.
14. Salehi, J. A. (1989) Code division multiple-access techniques in optical fiber networks – part I: fundamental principles. *IEEE Trans. on Comm.*, **37** (8), 824–833.
15. Salehi, J. A. and Brackett, C.A. (1989) Code division multiple-access technique in optical fiber networks – part II: system performance analysis. *IEEE Trans. on Comm.*, **37** (8), 834–842.
16. Yang, G. C. (1996) Variable-weight optical orthogonal codes for CDMA network with multiple performance requirements. *IEEE Trans on Comm.*, **44** (1), 47–55.
17. Gu, F. R. and Wu, J. (2005) Construction and performance analysis of variable-weight optical orthogonal codes for asynchronous optical CDMA systems. *J. Lightw. Technol.*, **23** (2), 740–748.
18. Djordjevic, I. B., Vasic, B. and Rorison, J. (2003) Design of multiweight unipolar codes for multimedia optical CDMA applications based on pairwise balanced designs. *J. Lightw. Technol.*, **21** (9), 1850–1856.
19. Tarhuni, N. *et al.* (2005) Multiclass optical orthogonal codes for multiservice optical CDMA networks. *J. Lightw. Technol.*, **24** (2), 694–704.
20. Khaleghi, S. and Pakravan, M.R. (2010) Quality of service provisioning in optical CDMA packet networks. *J. Optical Communications and Networking*, **2** (5), 283–292.
21. Nasaruddin and Tsujioka, T. (2008) Design of strict variable-weight optical orthogonal codes for differentiated quality of service in optical CDMA networks. *J. Computer Networks*, **52** (10), 2077–2086.
22. Lee, S. J., Lee, H.W. and Sung, D.K. (1999) Capacities of single-code and multicode DS-CDMA systems accommodating multiaccess services. *IEEE Trans. on Vehic. Technol.*, **48** (no 5), 376–384.
23. Adachi, F., Sawahashi, M. and Okawa, K. (1997) Tree-structured generation of orthogonal spreading codes with different lengths for forward link of DS-CDMA mobile radio. *Electronics Letters*, **33** (1), 27–28.
24. Kwong, W. C. and Yang, G.C. (2002) Design of multilength optical orthogonal codes for optical CDMA multimedia networks. *IEEE Trans on Comm.*, **50** (8), 1258–1265.
25. Petrovic, R. and Holmes, S. (1990) Orthogonal codes for CDMA optical fiber LANs with variable bit interval. *Electronics Letters*, **26** (10), 662–664.
26. Maric, S. V., Moreno, O. and Corrada, C.J. (1996) Multimedia transmission in fiberoptic LANs using optical CDMA. *J. Lightw. Technol.*, **14** (10), 2149–2153.
27. Lin, J. Y., Jhou, J.S. and Wen, J.H. (2007) Variable-length code construction for incoherent optical CDMA systems. *Optical Fiber Techonlogy*, **12** (2), 180–190.

28. Yang, G. C. and Kwong, W.C. (2002) *Prime codes with applications to CDMA: optical and wireless networks*. Artech House, Massachusetts, USA.

29. Wei, Z., Shalaby, H.M.H. and Ghafouri-Shiraz, H. (2001) Modified quadratic congruence codes for fiber Bragg-grating-based spectral-amplitudecoding optical CDMA systems. *J. Lightw. Technol.*, **19** (9), 1274–1281.

30. Yang, G. C. and Kwong, W.C. (2004) A new class of carrier-hopping codes for code-division multiple-access optical and wireless systems. *IEEE Comm. Letters*, **8** (1), 51–53.

31. Chung, F. R. K., Salehi, J.A. and Wei, V.K. (1989) Optical orthogonal codes: design, analysis and application. *IEEE Trans. on Info. Theory*, **35** (3), 595–605.

32. Kwong, W. C. and Yang, G.C. (2005) Multiple-length extended carrier-hopping prime codes for optical CDMA systems supporting multirate multimedia services. *J. Lightw. Technol.*, **23** (11), 3653–3662.

33. Yang, G. C., Lin, S.Y. and Kwong, W.C. (2002) MFSK/FH-SSMA wireless systems with double-media services over fading channels. *IEEE Trans. on Vehic. Technol.*, **49** (3), 900–910.

34. Inaty, E. *et al.* (2002) Multirate optical fast frequency hopping CDMA systems using power control. *J. Lightw. Technol.*, **20** (2), 166–177.

35. Shake, T. H. (2005) Security performance of optical CDMA against eavesdropping. *J. Lightw. Technol.*, **23** (2), 655–670.

36. Nasaruddin and Tsujioka, T. (2007) Multiple-length variable-weight optical orthogonal codes for supporting multirate multimedia services in optical CDMA Networks. *IEICE Trans. on Commun.*, **E90-B** (8), 1968–1978.

37. Liang, W. *et al.* (2008) A new family of 2D variable-weight optical orthogonal codes for OCDMA systems supporting multiple QoS and analysis of its performance. *Photonic Network Communications*, **16** (1), 53–60.

38. Murugesan, K. (2004) Performance analysis of low-weight modified prime sequence codes for synchronous optical CDMA networks. *J. Optical Communications*, **25** (2), 68–74.

39. Ohtsuki, T. (1999) Performance analysis of direct-detection optical CDMA systems with optical hard-limiter using equal-weight orthogonal signaling. *IEICE Trans. on Comm.*, **E82-B** (3), 512–520.

40. Park, E., Mendez, A.J. and Gasmeiere, E.M. (1992) Temporal/spatial optical CDMA networks. *IEEE Photonics Tech. Letters*, **4** (10), 1160–1162.

41. Fathallah, H., Rusch, L.A. and LaRochelle, S. (1999) Passive optical fast frequency-hop CDMA communication system. *J. Lightw. Technol.*, **17** (3), 397–405.

42. Maric, S. V. and Lau, V.K.N. (1998) Multirate fiber-optic CDMA: system design and performance analysis. *J. Lightw. Technol.*, **16** (1), 9–17.

43. Ghaffari, B. M. and Salehi, J.A. (2009) Multiclass, multistage, and multilevel fiber-optic CDMA signaling techniques based on advanced binary optical logic gate elements. *IEEE Trans on Comm.*, **57** (5), 1424–1432.

44. Kwong, W. C. and Yang, G.C. (2004) Multiple-length multiple-wavelength optical orthogonal codes for optical CDMA systems supporting multirate multimedia services. *J. on Selected Areas in Comm.*, **22** (9), 1640–1647.

45. Mashhadi, S. and Salehi, J.A. (2006) Code division multiple-access techniques in optical fiber networks – Part III: Optical AND gate receiver structure with generalized optical orthogonal codes. *IEEE Trans on Comm.*, **45** (8), 1457–1468.

# Index

*Optical CDMA Networks: Principles, Analysis and Applications*, First Edition. Hooshang Ghafouri-Shiraz and
M. Massoud Karbassian.
© 2012 John Wiley & Sons, Ltd. Published 2012 by John Wiley & Sons, Ltd.